Fundamentals of Construction Project Delivery Systems

Fundamentals of Construction Project Delivery Systems advocates f transformative approach that places emphasis on integrated methodolo such as design-build and integrated project delivery. In addition, this bo embraces new approaches in the industry: framework and program-bas procurement.

Unlike conventional perspectives that view project delivery as a mer administrative process and a set of rules/features, this book challenges thi notion. It proposes a shift in perspective, positioning the project delivery system as a design endeavor to define an organizational structure. Here, the owner is envisioned as actively adjusting the degree of organizational integration and interest alignment responding to the given internal and external context. Divided into four parts, this book begins by introducing project delivery systems before covering selection methods and contract types in Part 2. The third section goes into detail on four delivery project methods: Design-Bid-Build, CM-at-Risk, Design-Build, and Integrated Project Delivery and follows a similar chapter structure with a case study to demonstrate the method. The final part looks to the future covering new procurement trends, innovation, culture, and efficiency.

This book offers practical guidance on how owners can effectively leverage integrated approaches to enhance project outcomes. For educational purposes, this book offers guidance for undergraduate and graduate students in project delivery method courses within construction management programs.

Yong-Woo Kim is a full Professor and P. D. Koon Endowed Professorship holder at the University of Washington, and a Visiting Professor at Ewha Womans University. Dr. Kim's main research interests include project-based production management, lean construction, supply chain management, and integrated project delivery systems. Over the last 25 years, he has offered teaching and training on the project delivery system, project management, and lean construction – not only in the classroom but also to industry professionals.

Fundamentals of Construction Project Delivery Systems

Yong-Woo Kim

Routledge
Taylor & Francis Group
NEW YORK AND LONDON

Designed cover image: Shutterstock

First published 2026
by Routledge
605 Third Avenue, New York, NY 10158

and by Routledge
4 Park Square, Milton Park, Abingdon, Oxon, OX14 4RN

Routledge is an imprint of the Taylor & Francis Group, an informa business

ISBN: 978-1-041-04184-9 (hbk)
ISBN: 978-1-041-04561-8 (pbk)
ISBN: 978-1-003-62884-2 (ebk)

DOI: 10.1201/9781003628842

Typeset in Sabon
by codeMantra

Access the Support Material: www.routledge.com/9781041045618

Contents

About the Author

Yong-Woo Kim is a full Professor and P. D. Koon Endowed Professorship holder at the University of Washington, and a Visiting Professor at Ewha Womans University. Dr. Kim's main research interests include project-based production management, lean construction, supply chain management, and integrated project delivery systems. Over the last 25 years, he has offered teaching and training on the project delivery system, project management, and lean construction – not only in the classroom but also to industry professionals.

His research has been supported by the Construction Industry Institute, National Electrical Contractors Association, U.S. Department of Transportation, General Association of America, National Research Foundation of Korea, State University of New York Research Foundation, University of Washington Royalty Research Fund, and various construction contractors. He has more than 100 publications in international journals and conferences.

Preface

In the construction industry, project delivery systems are the backbone for organizing project structures and defining stakeholders' roles and responsibilities. Traditionally, these systems have been perceived as mere selections made by owners to transfer/minimize risks, operating under the assumption that project delivery risks remain constant. However, the contemporary construction landscape demands a paradigm shift.

Today, project delivery systems must be understood as design processes, wherein the industry (especially owners) recognizes the importance of organizational integration and alignment of interests. As projects get complex and uncertain while demanding speed, the significance of these elements grows exponentially as they impact the way design and construction are managed, as well as how people behave.

Unlike conventional perspectives that view project delivery as a mere administrative process and a set of rules and features (as evidenced in current textbooks), this proposed book challenges this notion. It suggests a shift in perspective, positioning the project delivery system as a design endeavor to define an organizational structure. Here, the owner is envisioned as actively adjusting the degree of organizational integration and interest alignment, responding to the given internal and external context.

In essence, this book advocates for the integrated approaches such as Design-Build and Integrated Project Delivery. In addition, this book embraces new approaches in our industry: framework and program-based procurement. Framework promotes a reliable supply chain and integration while owners reduce the time and effort spent on their procurement. Program-based procurement is a new approach, especially in modular or off-site construction, where significant initial investment is needed.

By adopting this innovative approach, this book aims to explore the attributes and mechanics of project delivery systems. It delves into the dynamics of organizational integration and interest alignment. Through insightful case studies, this book offers practical guidance on how owners can effectively leverage integrated approaches to enhance project outcomes.

For educational purposes, this book guides future project managers with a solid grasp of their project's organizational structure.

The key features of this book include:

- While traditional textbooks focus on rule-based contracts, terms, and conditions, this book features the mechanics of each procurement system and relevant risks.
- This book includes case studies on each type of method.
- The book provides a new type of procurement system, which includes (1) a framework and (2) program-based procurement. They are not typically addressed in most construction textbooks.

Finally, whether you are a project management student or a professional engaged in project delivery systems, I hope this book will broaden your perspective and enhance your understanding of the subject. Wish that the insights presented here will support your academic and professional development.

Acknowledgments

Over the last 25 years, I have been privileged to engage with many industry professionals through training sessions and consulting opportunities. I am deeply grateful to them for sharing stories and perspectives I would not have encountered otherwise. Their insights have inspired me and broadened how I view project delivery in practice. Such experiences helped shape this book.

I am equally indebted to many students in my Project Delivery System classes. From them, I have learned how the concepts and principles are best received and how I should deliver and interact so that they can be more clearly understood. My graduate students, in particular, often shared their experiences as junior engineers, providing valuable perspectives that directly impacted this book. In truth, they have not only been my students but also among my teachers.

Finally, I wish to thank my family, whose love and encouragement sustained me through the demands of an academic career. My late mother, **Jeongok Lee**, who passed away in 2023, always reminded me to approach students with gratitude and compassion. My father, **Jwasang**, has been a constant source of strength and stability. They have taught me, through their lives, the meaning of respect and love for others.

To my beloved wife, **Eunsu**, my daughter, **Ashley**, and my son, **Aidan**—you are my refuge and foundation. In moments when I felt discouraged, it was your love and support that lifted me and gave me the courage to continue. This book exists because of your unwavering presence.

Part 1

Introduction

Chapter 1

Introduction to the Project Delivery System

A construction project delivery system (PDS) refers to how a project owner acquires the services to design and construct a facility to meet specific needs. The PDS outlines how these services are procured and managed, ensuring that the project goals are met effectively.

Then, why is this delivery system important? As a project owner, you may not be an expert in design and construction. That's nothing wrong! This is where external service providers come in. You'll rely on design professionals to take your needs and turn them into detailed plans. You also need to bring construction professionals to turn those plans into a physical facility. The delivery system sets up a framework for how you'll hire these experts and how they will work together to meet your specific needs.

In this chapter, I will introduce you to the intriguing world of PDSs. First, we'll start with a definition of a PDS. Next, we'll explore the major components of a PDS. These components are like the building blocks of the entire process, each playing a critical role in the PDS. Understanding these components will give you a solid foundation for grasping how different delivery systems work.

To help explain the mechanics of each project delivery method, we'll focus on two main criteria in designing the PDS: organizational integration and interest alignment. Organizational integration refers to how different parties involved in the project work together and how responsibilities are shared. Interest alignment, on the other hand, addresses how the goals and incentives of these parties are aligned to ensure everyone is working toward the same goals. These criteria will be consistently used throughout the book to analyze and compare various PDSs.

This chapter is designed to serve as a guideline for you as you read and study this book. It aims to provide you with the background knowledge you need before we dive deeper into the specifics of each PDS. By the end of this chapter, you'll have a strong understanding of the basic concepts and be well-prepared to embark on this journey with me.

DOI: 10.1201/9781003628842-2

Definition

Can you guess what the PDS is in the construction industry? Given that the construction delivery system involves procuring services (i.e., design and construction) from external providers, it must be more than just a collection of game rules and procedures. Although many efforts have been made to define the term "PDS," there is no generally accepted definition. Various definitions have been proposed by industry experts and researchers, each highlighting different aspects of the PDSs.

In this section, I want to show you various definitions used in our industry. By exploring these definitions, you can better understand the nuances that different professionals emphasize. Here, I will introduce my definition of a PDS, which this book will use consistently. This definition will help us stay aligned as we explore the various PDSs in the construction industry.

Current Definitions

Let's dive into some current definitions of PDSs. These definitions will help you understand a PDS's various perspectives and functions (Table 1.1).

First, Miller et al. defined a PDS as "a system for organizing and financing design, construction, operations, and maintenance activities that facilitate the delivery of a good or service."[1] This definition underscores the function of organizing and financing at a macro level, addressing all phases from design to maintenance.

Next, the Construction Industry Institute (CII) focused on the roles and responsibilities of project stakeholders. According to CII, the PDS is defined as a project delivery and contracting strategy that assigns roles and responsibilities to achieve the project goals.[2]

Oyetunji and Anderson describe a PDS as the order of project phases, the people or groups involved, and the roles and responsibilities assigned to them.[3] Their definition highlights how the project moves through project stages (i.e., design, bid/selection, and construction), as well as the roles and responsibilities of each stakeholder at every stage of the project.

The Associated General Contractors of America (AGC) defines the project delivery method as "the comprehensive process of assigning the contractual responsibilities for designing and constructing a project."[4] AGC further addresses that a delivery system identifies the primary parties taking contractual responsibility for the project performance. This definition focuses on the legal and contract side of things, showing who is leading and in charge during each phase of the project.

When we compare different definitions, it becomes clear that a PDS is a complex idea. It involves organizing the project, developing the financing structure, deciding who does what, and managing contracts. Each definition gives us a different perspective, helping us better understand the PDS to achieve a project goal.

Table 1.1 Definitions of the Project Delivery System

	Definition of Project Delivery System
Miller et al. (2000)[a]	A system for organizing and financing construction process activities
Construction Industry Institute (CII) (2001)[b] and Oyetunji and Anderson (2006)[c]	A system that defines the sequence of project phases, parties involved, and assigned roles and responsibilities
The Associated General Contractors of America (AGC) (2004)[d]	A comprehensive process of assigning contractual responsibilities and a method for identifying the primary parties that are taking contractual responsibilities
This book	An organizational structure of how the project is procured and managed

[a] Miller, J. B., Garvin, M. J., Ibbs, W.C., and Mahoney, S. E. (2000). Toward a New Paradigm: Simultaneous Use of Multiple Project Delivery Methods. *Journal of Management in Engineering*, 16(3), 58–67.
[b] Construction Industry Institute (CII). (2001). Project Delivery and Contract Strategy. Publication RR165-12, Austin, TX.
[c] Oyetunji, A. A. and Anderson, S. D. (2006). Relative Effectiveness of Project Delivery and Contract Strategies. *Journal of Construction Engineering and Management*, 132(1), 3–13.
[d] Associated General Contractors of America (AGC) (2004). *Project Delivery Systems for Construction*. AGC, Arlington, VA.

Our Definition in This Book

As you can see from the various definitions we've discussed, many definitions often highlight how different stakeholders interact and how their tasks are divided to achieve the project's goals. This foundational understanding is crucial, but we can refine this definition further to make it more comprehensive and valuable for our study.

This book defines the PDS as "an organizational structure (or framework) of how the project is procured and managed." As you will see, the framework or organizational structure should define (1) project delivery methods (roles and responsibilities are assigned), (2) contract type (i.e., compensation structure), and (3) selection method (i.e., how your service providers are selected).

Brief History of Construction PDS

To better understand a construction PDS, it's helpful to look at how it has changed over time. Learning about its history gives you useful insight into how the PDS works in the construction industry.

Can you imagine how long PDSs have been in existence? The history of these systems is as ancient as the history of buildings and construction.

Then, consider early documents that recognize the builder's responsibility in construction. The Code of Hammurabi (1795–1750 B.C.) is among the earliest, acknowledging the builder's responsibility for design and construction to protect the owner's interests.

Master Builder

It is interesting to see how the idea of the master builder has evolved. The ancient legal document, the Code of Hammurabi, gave us an example that recognized the builder's duty to manage the project, comprehensively protecting the owner's interests. This early recognition of the builder's dual role marks the beginning of a long tradition of an integrated approach where a single individual or entity was responsible for the entire project phases (i.e., design and construction).

Much later, the Roman architect and engineer Vitruvius described a single individual called "Arkitekton" as responsible for the design and construction. The master builder approach carried on through medieval times and led to the construction of many well-known buildings. Master builders created the design, supervised the workers, and oversaw the entire construction process. In many ways, they were early examples of what we now call design-build (DB) or integrated project delivery (IPD).

The master builder approach, however, had its opponents who believed that the architect's role was different from the builder's work. Indeed, the master builder approach was initially criticized in 15th-century Italy by architects such as Leon Battista Alberti, who believed that the architect is the one who should provide simple sketches of a building; the builder is the one who conducts detailed design and construction. This separation of roles changed how people viewed project delivery, eventually leading to the specialized roles you see in construction today. By looking at the history of master builders and how their roles evolved, one can gain a better understanding of how today's PDSs developed.

Separation of Design and Construction

Can you guess when and why the separation of design and construction practices began? Despite some criticisms mentioned above, the separation of design and construction practices did not happen until the 19th century, after the spread of the Industrial Revolution all over Europe and the US. New types of buildings, such as complex factories, required designers with more design expertise but not contractors with the same expertise. This need for specialization led to professional organizations (e.g., American Institute of Architects), which helped establish a distinction between designers and builders.

In the 20th century, the US government played a significant role in further spreading this idea. Public contract laws in the US and other laws by individual states mandated the separation of design from construction to protect the public from corruption. Public laws even required the public owners to select contractors for public projects solely based on the price, while selecting designers based on quality. These laws implicitly assigned design professionals the role of the owner's agent and prohibited them from having any financial relationship with builders. The outcome was the design-bid-build (DBB) approach.

These legal requirements further solidified the separation of design and construction practices, implicitly assigning design professionals the role of the owner's agent. These laws prohibited any contractual relationship between designers and builders, ensuring that designers acted solely in the owner's best interest. This led to the widespread adoption of the DBB approach. The DBB method aimed to protect the public from corruption by ensuring transparency and accountability in the construction process.

Merging Design with Construction

Let's look at why the industry started to have problems with the traditional DBB method and how integrating two functions (i.e., design and construction) has helped solve some of those significant issues. Though DBB was suitable for specialization in each phase of the project, DBB led to typical construction problems such as claims, disputes, and delays for owners in the following decades. As building systems became more complex, architects and engineers needed to work closely with builders (i.e., contractors) to check the constructability and other construction issues during the design phase.

The Construction Management-at-Risk (CMAR) method was developed to address this issue. It allows the owner to bring a general contractor – the so-called GC/CM – into the project early so they can share their construction expertise with the design team. This collaboration helps develop designs that are not only technically sound but also practical and cost-effective to build. By working together during the design phase, the team can avoid delays, claims, and disputes common in traditional DBB projects – resulting in smoother project delivery. However, there is no contractual relationship between the designer and the general contractor in CMAR.

Meanwhile, another project delivery process gained popularity because it met owners' desire to assign both design and construction responsibility to a single entity. In DB, which began with private practice, the design-builder takes all the risks in designing and constructing the project.

This approach had the potential to reduce the cost and duration of projects and minimize the conflicts between project parties.

As the industry progressed beyond CMAR and DB, it introduced even more collaborative methods. One of the most advanced is IPD. In the IPD approach, all key stakeholders – such as the owner, designer, general contractor, and major trade partners – collaborate as one team from the start of the design phase under a single contract. This unique collaborative environment helps minimize project uncertainty, leading to desired project outcomes in terms of schedule, cost, quality, and relationship. Everyone involved shares responsibility for the project's outcome, creating integration and shared goals.

Understanding how each delivery method – DBB, CMAR, DB, and IPD – has evolved helps you see the purpose and advantages of each. This background sets the stage for a deeper exploration of PDSs in the following chapters.

Designing a PDS

When discussing PDSs, we challenge the conventional perspective. Unlike many current textbooks that view project delivery as merely an administrative process governed by rules or features, we propose a different viewpoint. This book positions the PDS as a design process to develop an organizational structure for project delivery.

So, why does this matter? Think about it: developing an organizational structure that leads to the owner's goals and expectations isn't as simple as selecting from a list of project delivery methods and contracting types. Instead, project delivery processes should be adapted to the specific contexts that address the unique characteristics of projects, including the owner's requirements. By viewing the PDS as a design product, the organizational structure would be equipped to help project participants achieve the best outcomes in both design and construction.

In this section, we will address three main aspects: identifying the components of the PDS, understanding the project context as input variables in the design process, and discussing design criteria in achieving successful project outcomes. By approaching the PDS as a design process, we can create more effective and efficient PDSs that are better suited to meet the unique demands of each project. This perspective allows for a more flexible and adaptive approach, ensuring that the PDS is a well-designed strategy tailored to the specific needs of the project.

Three Components of the PDS as Design Output

A PDS encompasses components related to an organizational structure regarding the design and construction of a project. Such elements include defining the project scope, organizing the teams (e.g., designers and builders), defining the commercial interests of each team, and sequencing the project phases. While different PDSs may vary in their approaches and structures, they all share some fundamental elements.

So, what exactly are the components of a PDS? Though many elements could exist, this book uses a simplified view of the PDS. We define that the PDS consists of three main components (see Figure 1.1):

- Project Delivery Method: This component defines the roles and responsibilities assigned to the stakeholders involved in the project. For example, the design and construction roles are separated in a DBB method, whereas in a DBB method, one entity is in charge of both design and construction.
- Contract Type: This component defines how the service providers (like designers and constructors) get reimbursed for their work. This could be through a fixed price contract, where the fixed price is agreed upon before work starts, or a reimbursable contract, where the service provider is paid for actual costs incurred plus a fee.
- Selection Method: This component defines how the service providers are chosen. They might be selected based on the lowest bid, their qualifications, or a combination of factors.

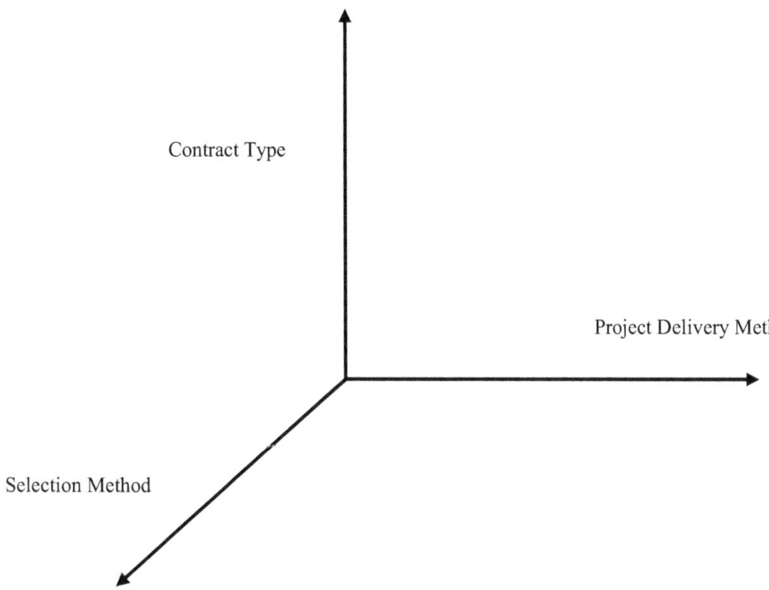

Figure 1.1 Components of the Project Delivery System.

By understanding these key parts, you'll be better prepared to see how different PDSs function and how they help meet project goals. We'll refer consistently to these components throughout the book to explain the various strategies used in project delivery.

Project Context as Input Variables in PDS Design

If you're an owner designing a PDS, there are three key components you need to determine: the project delivery method, the contract type, and the selection method. But how do you decide on these components? You need to consider the contexts that influence the design of your PDS. These contexts serve as input variables that guide your decisions, as shown in Figure 1.2.

The first input variable is the project characteristics. This involves the size, type, complexity, and level of uncertainty of the project. For instance, building a tiny house may require a different PDS than constructing a large complex building. Each project comes with its unique context and challenges. One critical factor among the project characteristics is the level of uncertainty. If a project is not well-defined, your approach to PDS design should be different than one when a project is clearly defined.

The project requirements are the second input variable for the PDS design. What are the project-specific requirements? They include budget, schedule, and quality expectations. For example, a project that needs to be completed quickly will have different requirements than one with a more flexible timeline. Imagine you have a school renovation project that needs to be finished before the new school year begins. The tight deadline will influence the PDS you design, possibly pushing the construction schedule that allows for overlapping design and construction phases to save time.

The last input variable is the owner and market conditions. This involves the owner's project management capability, market conditions,

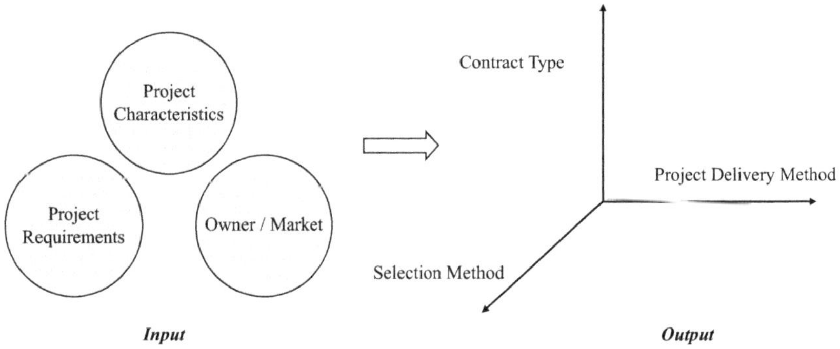

Figure 1.2 Project Delivery System as a Design Process.

and service provider availability. An owner with a team of experienced project specialists might approach project delivery differently than an owner with limited resources. The availability of qualified service can be an issue to be considered when designing the PDS. For example, even if an owner prefers the DB method, what if the market lacks service providers with DB experience? In such cases, the owner must consider not only their capabilities but also the availability of qualified service providers in the market.

When considering project characteristics, requirements, and owner and market conditions, you'll be better equipped to design a project framework that meets the unique needs of your project. For example, a more collaborative delivery system that allows for early involvement of specialty trades and key suppliers might be suitable if your project is highly complex and uncertain.

Design Criteria

As we covered earlier, a PDS is the structure used to organize how a project is bought and managed. You've also learned about the key inputs and outputs involved. Now, the question is: What design criteria should you use when designing your PDS? There are three main design criteria to consider when making that decision:

- The owner's willingness to take on project risks
- The level of integration between project participants
- How well the interests of all parties are aligned

Owner's Attitude Toward Project Risks

In many cases, the owner's attitude toward various risks serves as a critical design criterion when the owner designs his PDS: project delivery method, contract type (i.e., compensation structure), and selection method. For example, suppose the owner wants to avoid the risk of coordination between design and construction for the project. In that case, the owner may prefer a DB where the owner is not directly involved in the coordination between the designer and the contractor.

Two major challenges in construction project management are financial risks and schedule risks. These risks often come from two sources:

1. Productivity issues, where the actual productivity falls short of what was expected, and
2. Design and planning issues, where the actual work differs from the original design or plan.

The risks associated with productivity issues are mainly managed by choosing the compensation structure, while design and planning issues are addressed by selecting the project delivery method. This book focuses primarily on the owner's willingness to take on design-related risks because this choice affects how much control the owner wants over the design process.

But what if you want to make your project more integrated? To do this, you must design your PDS by adjusting the level of organizational integration and aligning interests to pursue the project goal best. You might be wondering if every project needs an integrated approach. The answer is no. However, since you're reading this book, you're likely interested in integration (after all, the book does emphasize an integrated approach). So, why is there a need for project integration? The need for project integration mainly stems from the inefficiencies in the fragmented approach of traditional PDSs. The construction industry faces significant challenges, such as increased costs, timelines, and adversarial relationships. Many of these issues are believed to originate from a fragmented approach.

So, what exactly is project integration? In the construction context, project integration has been defined as "where different organizations with different goals, needs, and cultures merge into a single cohesive and mutually supporting unit with the collaborative alignment of processes and cultures."[5] The degree of integration achieved for delivering a construction project is subject to applying contractual, organizational, and technological mechanisms. Simply put, stakeholders with different interests work together while sharing information to achieve a common goal.

Our efforts to design a PDS for project integration focus on two main criteria. First, the framework needs to ensure that different stakeholders are working together. Second, the PDS needs to align with the commercial interests of stakeholders. When everyone has aligned incentives, they are more likely to work together toward the project's success. Let's take a closer look at each of these design criteria.

Organizational Integration

Organizational integration is all about bringing all project parties to the table and facilitating their collaboration throughout the project. This means the project team works together among parties and shares information freely. However, achieving this level of integration requires the appropriate project framework (i.e., PDS). Organizational integration in PDSs usually occurs in the design phase. In the traditional DBB method, the level of integration is quite low because it prevents the contractor's team from getting involved in the design phase. This means the expertise of the contractors isn't leveraged during the design process, which can lead to inefficiencies and missed opportunities for improvement.

So, how can we adjust our PDS to achieve organizational integration? There are two main ways. First, organization integration can be achieved by assigning roles and responsibilities to each stakeholder in a way that allows them to collaborate with each other. For instance, if the general contractor's role includes preconstruction activities, they should be involved in the design phase. The project delivery method typically defines these roles and responsibilities, which in turn guide the level of integration. Second, organizational integration can be promoted by choosing the right project team with the right chemistry with their partners. This means that the selection process and criteria are crucial. You need to pick team members who are qualified and compatible with other team members (i.e., other stakeholders).

As you can see, organizational integration is key to making your project integrated. It requires a system where all parties are allowed to work together, share information freely, and have clearly defined roles and responsibilities. By carefully considering the level of integration in your PDS and selecting the right team, you can create a project framework that leads to organizational integration.

Alignment of Interests

Let's talk about the concept of interest alignment in project integration. This might sound a bit abstract, but the interest alignment is imperative in working collaboratively. Now, we all accept that all organizations and people behave in ways that maximize their profits and interests. There's no blame for such behavior – it's just how things work. However, this means that aligning the interests to have a common goal is crucial for achieving project team integration.

Poor performance in traditional contracting often stems from a misalignment of commercial interests between owners and contractors, or even among contractors themselves. Typically, specialty contractors focus on completing their part of the project and might not be interested in the project performance once their portion is done. They're paid for their services and have little incentive to seek the most cost-effective solutions for the overall project.

On the flip side, owners see the project as a whole and are looking for the best solutions in terms of cost, time, and quality. This difference in perspective can lead to a misalignment of commercial objectives. Owners want the project to be completed efficiently and economically, while contractors might focus more on completing their specific tasks and getting paid.

In a PDS, interest needs to be confined to commercial interests. So, how do we achieve interest alignment in a PDS? PDS is about adjusting the pricing or compensation structure and how each team gets paid. Remember that all organizations and people naturally behave in ways that maximize their own profits and interests. Therefore, you need to develop a pricing or compensation structure: how your service provider (or supplier) gets paid in a way that promotes collaboration.

Chapter Summary

1 The PDS is an organizational structure (or framework) of how the project is procured and managed.
2 The framework or organizational structure should define (1) project delivery methods (roles and responsibilities are assigned), (2) contract type (i.e., compensation structure), and (3) selection method (i.e., how your service providers are selected).
3 The separation of design and construction began in the 19th century due to the Industrial Revolution's demand for specialized building design. This shift, reinforced by professional organizations and legal requirements, established a distinction between designers and builders to specialize in expertise and prevent corruption.
4 The traditional DBB method, characterized as a fragmented approach, led to delays, disputes, and inefficiencies as projects became more complex and uncertain. To overcome the deficiencies of a fragmented approach, integrated project delivery methods, such as CMAR, DB, and IPD, have emerged, promoting early contractor involvement. The most collaborative of all, IPD, brings all major stakeholders, including key specialty trades, together under one contract, improving project outcomes through shared responsibility and aligned goals.
5 Three input variables guide your PDS design: the project characteristics, the project requirements, and the owner and market conditions.
6 There are three main design criteria to consider when making that decision: the owner's willingness to take on project risks, the level of integration between project participants, and the level of interest alignment.

Review Questions

1 How is the PDS different from the project delivery method?
2 List three components of the PDS in the construction industry.
3 Why do you think the level of project uncertainty impacts your design of the PDS?

4 Give three examples of the project delivery method. Which project delivery method prevents the contractor from collaborating with the designer in the design phase?
5 List two design criteria for project integration in the PDS design.
6 Under what circumstances might the project integration approach in the PDS be needed?
7 Project Description: The port of Vancouver needs an additional sizable paved staging area to support the loading and unloading of shipping containers from ships. The only available site near the cargo ship piers was previously used for producing creosote-impregnated railroad ties and is thus contaminated. Because the extent of contamination is not known, the port authority has decided to use a unit price contract for the cleanup of the site and the construction of the paved staging area. The scope of work involves the removal of contaminated soil from the 1000-foot by 800-foot site, backfill with compacted select material, and the surface paved with asphalt. The entire site is to be fenced with an eight-foot chain-link fence and lighted for night operations.

   Please describe the project characteristics as input variables for the PDS design. Note that your response should focus solely on the project characteristics and should not include the project requirements or the owner/market conditions, although they are also part of the input variables.

Notes

1 Miller, J. B., Garvin, M. J., William Ibbs, C., and Mahoney, S. E. (2000). Toward a New Paradigm: Simultaneous Use of Multiple Project Delivery Methods. *Journal of Management in Engineering*, 16(3), 58–67.
2 Construction Industry Institute (CII) (2001). Project Delivery and Contract Strategy. Publication RR165-12, Austin, TX.
3 Oyetunji, A. A. and Anderson, S. D. (2006). Relative Effectiveness of Project Delivery and Contract Strategies. *Journal of Construction Engineering and Management*, 132(1), 3–13.
4 Associated General Contractors of America (AGC) (2004). *Project Delivery Systems for Construction*. AGC, Arlington, VA.
5 Baiden, B. K. and Price, A. D. F. (2011). The Effect of Integration on Project Delivery Team Effectiveness. *International Journal of Project Management*, 29(2), 129–136.

Bibliography

Associated General Contractors of America (AGC) (2004). *Project Delivery Systems for Construction*. AGC, Arlington, VA.

Baiden, B. K. and Price, A. D. F. (2011). The Effect of Integration on Project Delivery Team Effectiveness. *International Journal of Project Management*, 29(2), 129–136.

Beard, J. L., Loulakis, M. C., and E. C. Wundram. 2001. *Design-Build: Planning through Development*. McGraw-Hill Professional, New York, NY.
Colvin, A. J. S. and Boswell, W. R. (2007). The Problem of Action and Interest Alignment: Beyond Job Requirements and Incentive Compensation. *Human Resource Management Review*, 17, 38–51.
Construction Industry Institute (CII). (1997). *Team Alignment during Pre-Project Planning of Capital Facilities*. Publication RR113-12, Austin, TX.
Construction Industry Institute (CII) (2001). *Project Delivery and Contract Strategy*. Publication RR165-12, Austin, TX.
Gottschalg, O. and Zollo, M. (2007). Interest Alignment and Competitive Advantage. *Journal of Academy of Management Review*, 32(2), 418–437.
Mearig, T. (2004). *Project Delivery Method Handbook*., State of Alaska Department of Education & Early Development, Juneau, AK.
Miller, J., Garvin, M. J., Ibbs, W., and Mahoney, S. (2000). Toward A New Paradigm: Simultaneous Use of Multiple Project Delivery Methods. *Journal of Management in Engineering*, 16(3), 58–67.
Oyetunji, A. and Anderson, S. (2006). Relative Effectiveness of Project Delivery and Contract Strategies. *Journal of Construction Engineering and Management*, 132(1), 3–13.

Part 2

Selection Method and Contract

Chapter 2

Selection Method in the Project Delivery Systems

Now, we are to start the second chapter. In this chapter, we will discuss one of the critical components of a project delivery system: how to choose your service providers for your project. Suppose that you're selecting a design-builder to remodel your house. You wouldn't just pick your design builder based on their bid price; you'd choose the company with the right skills and experiences, who works well with you, and who understands your goals and intentions. You may have some specific value that you want to use when selecting your design builder. As the owner, it's your responsibility to decide how you choose your service providers. Depending on the scope of services and the project delivery method, this selection process might involve a more complex procedure than just comparing their price. This selection method is often called the "procurement option."

When we define a project delivery system as an organizational structure, it's clear that organizing your project team is critical. Choosing each team member is vital because it impacts what kind of organization you'll be working with throughout the project. The selection method is not just an administrative procedure; it's the foundation for your organizational structure.

Team selection is about choosing the right members (or service providers) to deliver your project. Think of it as assembling a puzzle where each piece has to fit perfectly. This process allows you to have competent stakeholders involved in your project. In this chapter, we'll explore the different ways to choose the right team for your project. First, we'll look at how to select your designer. Then, we will address various selection methods to choose your construction contractor (s), such as Low Bid, Best Value, and Qualification-Based Selection (QBS). Each method has unique attributes, and we'll discuss which works best with different project contexts.

Selection of Designer

An architect is a person or an organization that provides design services for you, the owner. Hiring an architect should save you time and money

DOI: 10.1201/9781003628842-4

in turning what you want into architectural design conforming to the building codes and regulations, while streamlining the building process and providing an accurate picture of how the project will turn out, even before starting excavation. Design is a process that converts what you want into design drawings and specification documents. Therefore, an architect or design firm should communicate closely with the owner throughout the design phase. You may want to select a designer or design firm with extensive experience in a similar type of project and knowledge of construction methods.

Architectural services should be considered part of the project cost, typically just under 8%–10% of your budget. You pay for the ability to see many different aspects of a project – such as the owner's needs, material and method options, spatial constraints, and other possibilities and constraints – through professional eyes. If you are a public owner and your project is funded from federal funds, you should be under the Brooks Act when selecting a design firm or A/E (architect/engineer). Let's investigate the Brooks Act first.

The Famous Brooks Act[1] and QBS

Since the design team or A/E functions as an organization providing professional services, the selection process for the design team in public projects is governed by the Brooks Act. Similarly, when an agency construction management firm is involved, the Brooks Act also governs its selection process, provided the project is federally funded. This book does not address the selection process for agency construction management firms, as it is identical to that of design firms.

The Brooks Act (Public Law 92-582), enacted on October 18, 1972, establishes the procurement process by which professional service providers such as A/Es are selected for service contracts with government agencies (i.e., the public owners). The Brooks Act establishes a QBS process in which contracts for the professional service provider are negotiated based on demonstrated competence and qualification for the type of professional services required at a fair and reasonable price. Under QBS procurement procedures, price quotations are not a consideration in the selection process.

This QBS process, as established by the Brooks Act, has long been enthusiastically supported by every professional A/E society. There are seven basic steps involved in pursuing federal design work under QBS:

- Public solicitation for architectural and engineering services
- Submission of a statement of qualifications (SOQ) and supplemental accounts of capability to design specific projects for which public announcements were made

- Evaluation of both the general and project-specific statements
- Development of a shortlist of at least three submitting firms to interview with them
- Interviews with the firms
- Ranking of at least three of the most qualified firms
- Negotiation with the top-ranked firm.

It's important to note that while QBS procedures are legally mandated, agencies have some flexibility to implement minor adjustments as long as they comply with the guidelines established by the Federal Acquisition Regulation (FAR). Let's now review the general procedures for QBS.

Public Solicitation

QBS mandates that opportunities for design contracts must be publicly announced. Public owners post these opportunities on the Governmentwide Point of Entry, a federal online platform known as SAM.gov. The GPE contains information on proposed government procurements and contract awards.

Procurement actions valued at $25,000 or more must be published in the GPE.[2] Additionally, the GPE identifies awarded contracts when the contract amount exceeds $25,000 for civilian agencies or $100,000 for the Department of Defense. However, the GPE does not list procurements that are:

- Classified for reasons of national security
- For perishable items
- Placed under existing contracts
- Made only from foreign sources
- Not to be given advance publicity, as determined by the Small Business Administration

These notices in the GPE provide the location and scope of a project and may also contain such information as:

- Estimated design contract award range
- Project schedule and the date and time limit for receiving replies
- Categories of evaluation criteria and weight factors
- Any requirements for supplemental information

Statements of Qualification

Suppose an A/E firm wants to be considered for design service contracts with federal agencies. In that case, they must submit two key statements of qualifications to each agency they wish to work with. The first is Standard Form 254[3] (SF 254), known as the Architect-Engineer and Related Services Questionnaire. This form provides general information about the firm, including its size, capabilities, personnel, and past experiences. This form can also be updated and resubmitted whenever necessary. The A/E firm can submit this form simultaneously as the required project-specific form is submitted.

The next SOQ a firm should submit is the Standard Form 255 (SF 255), known as the Architect-Engineer and Related Services Questionnaire for Specific Project. Following the review of the notices in the CBD, if an A/E firm wants to be considered for a specific project listed in it, it must submit Standard Form 255. This form must be submitted in response to a particular solicitation and, when completed, contains the data relative to the project.

When a project is announced in the GPE, agencies typically do not notify firms directly, even if those firms have already submitted a Standard Form 254 (SF 254). Instead, each project announcement in the GPE includes a call for submissions, asking interested firms to provide both an SF 254 and an SF 255, along with any additional information specified in the notice. If a firm already has an up-to-date SF 254 on file with the agency, it does not need to submit the form again. However, the firm must still submit an SF 255, which demonstrates its qualifications specifically for the project in question.

The forms contain detailed instructions on how to fill them out. For instance, the SF 254 form explicitly says that you shouldn't include extra materials like brochures or photos unless they're asked for. On the other hand, the SF 255 form allows firms to add things like pictures or other graphics if it helps show their capabilities for the project.

Evaluation of Statements

The owner or the owner agency's architectural and engineering evaluation boards usually conduct the selection process. These boards comprise members with experience in architecture, engineering, construction, and government-related acquisition. The board members are typically chosen from the agency's or other agencies' professional employees. Sometimes, private practitioners or university faculty members can also be part of these boards if agency procedures allow it. However, when private practitioners sit on an evaluation board, neither they nor their firms can get a design contract awarded to avoid conflicts of interest.

The evaluation boards then review the qualifications statements, including Standard Forms 254 and 255. They must assess these qualifications based on the criteria listed in the Commerce Business Daily (CBD) notice. For instance, the criteria might include the firm's professional qualifications and experience with a specific type of project, the expertise and qualifications of the staff assigned to the project, the location of the firm's main office and its consultants, and an analysis of the firm's current workload. To help you understand better, I'll give you two examples of evaluation criteria.

The following evaluation criteria were used by a state highway department in evaluating the prospective designers' statements of qualification for a highway bridge:

- Applicable experience of the respondent team (includes project manager experience and similar bridge project experience) – 30 points
- Demonstration of ability and expertise to perform the job – 40 points
- Coordination experience with other organizations like DOT and FHWA – 20 points
- References – 10 points

The University of Washington used the following evaluation criteria in evaluating prospective designers' statements of qualification for a campus project:

- Relevant experience of the firm – 25 points
- Ability to manage schedule and budget – 15 points
- Project management approach – 10 points
- Permitting and other regulatory knowledge – 5 points
- Sustainable design experience – 5 points
- Staff experience and availability – 20 points
- Staffing plan – 10 points
- Past performance in utilization of small business entities, disadvantaged business enterprises, and minority and women-owned enterprises, as well as outreach plan for this project – 10 points

Development of a Shortlist

After evaluating the statements of qualifications, the boards prepare reports recommending which firms should be shortlisted. These reports rank at least three firms that need to move forward and discuss the project in more detail with them. The boards are not restricted in the number of firms they can select for these interviews. The final number is up to the board's discretion.

Interviews/Discussions with Firms

The interviews typically focus on discussing project concepts and evaluating the usefulness of different methods for providing the required services.

Before the interview, some agencies may send detailed selection criteria and additional information about the project to the firms that have been shortlisted. It's important to note that under the QBS system, the A/E firms are not supposed to develop any design products or propose their price as part of the competition for the project.

These interviews are usually held at the agency's office. However, in some cases, interviews might be conducted over the phone or via an online meeting platform like Zoom. The interviews are generally brief, lasting between 30 and 60 minutes.

Ranking of the "Top Three" Firms

After the interviews, the boards' reports must be prepared and submitted to the head of the agency or a designated representative. The reports list at least three firms considered the most qualified to provide the required services in order of competence. This step is regarded as the final selection of the competing firms. Suppose the firm chosen as the top preference is not the one recommended as the most highly qualified by the evaluation board. In that case, the agency head must provide a written explanation for this decision. The head of the agency, or their designated representative, is not allowed to add any new firms to the final report.

Negotiation with the Top-Ranked Firm

When the agency head makes the final selection, the contracting officer is authorized to begin negotiations with the top-ranked firm. The negotiations are conducted according to the procedures outlined in the FAR. Typically, the firm will be asked to submit a fee proposal that breaks down its direct and indirect costs, which serves as the starting point for the negotiation. If the proposed design fee is over $100,000, the fee proposal must be evaluated before moving forward with the negotiations.

If both parties can't agree on a fee within a reasonable time, the contracting officer will end talks with the top-ranked firm and move on to the second-ranked firm. This process continues until a satisfactory agreement is reached. If, by chance, negotiations fall through with all the selected firms, the officer will revisit the list of firms ranked by competence and qualifications and start the negotiation process again until a deal is made and a contract is awarded. It is worth noting that it is rare for negotiations not to succeed with the top-ranked firm.

The 6% Fee Limitation on Federal Design Contracts

Since 1939, federal construction agencies have been legally required to cap the fee paid to an architect or engineer at 6% of the estimated construction cost. Currently, at least four statutes enforce these fee limits for all civilian and military construction agencies, except for the U.S. Department of State.

Federal agencies have interpreted this 6% fee limit as applying only to the portion of the fee that covers the creation and delivery of "designs, plans, drawings, and specifications." Therefore, they don't apply the 6% limit to costs associated with other services like field investigations, surveys, topographical work, construction inspections, and project administration – basically, any service that doesn't involve creating and delivering designs or specifications. Most federal agencies have a list of these exempt services as part of their supplement to the FAR.

If Not Under the Brooks Act

If you're in the private sector or working with non-government funds, you're not necessarily bound by the Brooks Act. However, you probably don't want to choose your designer based solely on price, right? That's why most project owners still go through a process similar to QBS, even though they have more flexibility in how they do it. Instead of using an open public solicitation like the CBD, the owner or his selection board can send a "Request for Proposal" (RFP) directly to selected or preferred design firms. These firms are usually chosen based on their experience with similar projects and their specific expertise.

After receiving the proposals, the owner's selection board evaluates the proposals and makes recommendations for the final selection. The board can conduct interviews with the design firms during the evaluation process. Once a firm is chosen, the owner will negotiate and finalize the design contract, including the fees with that firm.

Selection of a General Contractor, General Contractor/Construction Manager, or Design-Builder

In selecting a general contractor (GC), general contractor/construction manager (GC/CM), and design-builder, the owner should choose one of four methods: Low Bid, Best Value-Fee (BVF), Best Value-Total Costs (BVTC), and QBS. Figure 2.1 shows two key factors to consider when selecting a service provider: how much weight you place on the price and how complete the design is at the time of selection. On the horizontal axis, we see the importance of price during the selection process, while the vertical axis represents the completeness of the design. As you move from left to right, the weight of the price increases, and as you move from bottom to top, the design gets further developed when selecting a service provider.

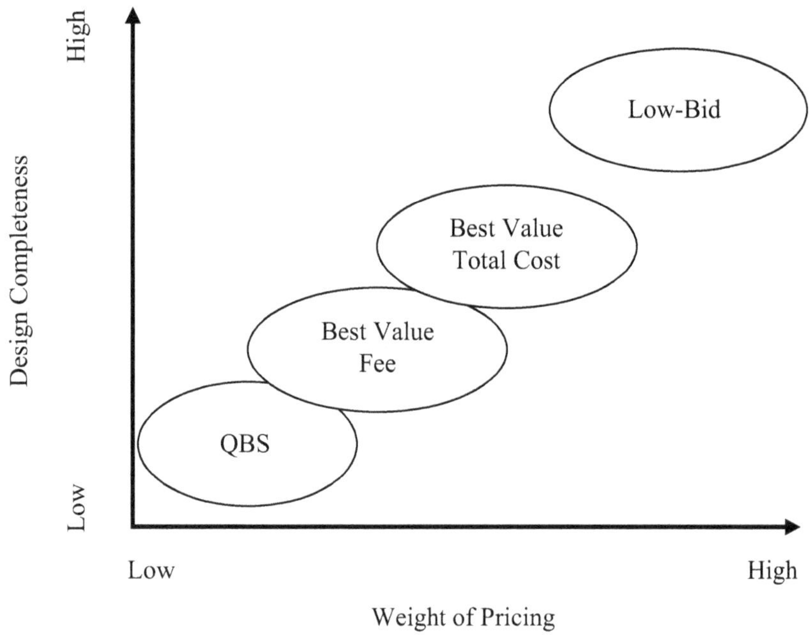

Figure 2.1 Criteria for Four Selection Methods.

A Low Bid approach is feasible when the design is fully developed, or the owner is confident that the design contingency level is near zero or at a minimum. Since this method focuses solely on price, it's common for the owner to pre-screen service providers, ensuring only qualified bidders can submit their prices. On the other hand, the BVF focuses on the fees because it's too early for contractors to estimate the total cost accurately. However, when the design is further developed, the BVTC considers both the total price and the qualifications or technical aspects of the proposals. QBS, on the other hand, considers only qualifications and technical elements without considering the price at all.

Low Bid

Low Bid procurement is defined as price competition, where price is a single factor in the final selection process. This method has many nicknames: hard bid, sealed bid, and competitive bid. As price is the only factor in the final selection, we need to discuss the price. What does the price include? In design-bid-build (DBB), the price in a bid document includes construction costs and a contractor's margin. In design-build (DB) projects, the price in a proposal includes construction costs and design costs in addition to a design-builder's margin.

Though price is a single factor in the final selection, most owners require the contractor's bid to be responsible and responsive. A responsive bid refers to a bid that meets all bid requirements. A responsible bid means that a bid should be prepared and submitted by a responsible contractor who is technically and financially capable. Many owners have a prequalification process to screen out non-responsible bidders.

Prequalification

Prequalification evaluates whether potential bidders have the skill, experience, sufficient financial resources, and capability to carry out a contract for the project. This prequalification process aims to evaluate a contractor and properly screen out unqualified contractors impartially.

A bidder should be regarded as non-responsive if they submit a bid on a construction project subject to prequalification for which it has not been accepted under the prequalification policy. The prequalification application form requires information on the ownership of the firm, the experience of the firm's personnel, any affiliations with other construction firms, bonding capacity, financial resources, the type of work performed by the firm, and other objective criteria rationally related to the particular contract to be awarded.

What information is to be included in the prequalification application? The required information on prequalification varies. However, most prequalification applications address the following items:

a Organizational Structure – The firm shall provide a list of all owners, officers, partners, or individuals authorized to represent or conduct business for or sign legal documents for the firm.
b Classification – The firm shall indicate the type(s) of work the firm's workforce and equipment regularly perform, licensure, and other pertinent information.
c Experience – The firm shall furnish a history documenting the experience with projects of similar size, scope, or complexity involving the type(s) of work for which prequalification is requested.
d Financial – Firms must provide at least one year of complete audited financial statements from the most current period available.
e Litigation/Claims – The firm must submit information regarding its litigation history, including litigation with owners and subcontractors.
f Timeliness – Firms must provide information on their success at completing projects on time, including the payment of liquidated damages.
g Capacity – Firms shall demonstrate sufficient bonding capacity, insurance, and resources for the project. Firms must provide relevant information on the personnel directly responsible for the work, including the office's location that will be primarily responsible for the job.

h Safety – Firms shall also demonstrate an acceptable safety history. Firms must provide copies of any complaints and safety violation reports from OSHA (Occupational Safety and Health Administration) or any other regulating agency associated with any construction project.
i Legal Authorization – The Applicant must provide a copy of their Contractor's License or a statement indicating they can acquire one before submitting a bid on a project. If an account is required, the Applicant shall identify the states in which they are licensed for this type of work.

Conditions for Low Bid

The low bid method allows the owner to leverage price competition among bidders. Since bidders need to be confident in their bid prices, the design should be completed or nearly completed to ensure minimal or no variations in the design at the time of procurement. Therefore, if the design is subject to significant changes at the time of procurement, the owner assumes a higher level of financial risk.

What If the Price Is Too Low?

What if the owner received a bid that was too low-priced? Is he happy with the price? When the owner puts a construction project out for a bid, he wants to get a fair and reasonable price for a high-quality job. Often, an agency may be legally obligated to abide by competitive bid laws that require the agency to choose the lowest-priced responsive, responsible bidder. But when a bid comes in too low – where the project requirements cannot realistically be met for the price offered – it can cause serious problems. Here are four consequences of an unrealistically low price.[4]

QUESTION OVER QUALITY

When people say, "You get what you pay for," they usually tell the listener that the low price will have some repercussions. Perhaps the bidder who submitted an impractical price is inexperienced. That would affect the quality of the work to be done. If the low price was purposeful, the contractor may be forced to swap out requested materials for cheaper alternatives or may have to find other methods to make it work. Either way, a bid that is far too low can result in a poorly constructed project.

PRICE FOR DISPUTE

When a bid comes in too low, it sets up the project for pricing problems down the line. The low price may be due to miscalculations, omissions, or deliberate misrepresentation. When the initial bid is unrealistically low, the final price can – and usually does – end up much higher. This is because additions and other modifications to the original scope of work must occur for the project to be completed. The change orders then cause even more headaches as they are identified. Negotiating the change order costs is frustrating for all parties involved, often with neither the owner nor the contractor convinced that a fair price has been reached.

WARNING FOR RELATIONSHIP

The friction of the pricing disputes and dissatisfaction with quality can negatively impact the relationship between contractor and owner. Your contractor is ready to work for any change order that increases the contract price from day one. The relationship between the owner and the contractor worsens if the number of change orders grows. This discord can affect the project performance during construction.

TIME DELAY

These consequences increase the likelihood of project delays and setbacks. In some cases, unresolved change orders can lead to legal disputes, which may result in work stoppages or delays. Such interruptions can cause productivity losses, additional damage, and increased costs.

Selection Process for Low Bid

Most owners use the pre-established procurement process. Most owners use the document called "invitation to bid" (ITB), which includes requests and information on the bid. The Low Bid selection usually consists of two processes: pre-bid conference and bid day.

Pre-Bid Conference

In many cases, the Low Bid process includes a pre-bid conference in which the owner and the A/E explain the attributes of the project and answer potential bidders' (i.e., contractors') questions about the ITB documents, including project scope and designs. Prospective bidders are permitted to request clarifications on the invitation for bids or RFPs by a stipulated date, and the pre-bid meeting is held within that period.

The process of pre-bid conferences is essential, especially for complex projects, to reduce ambiguity or uncertainty. The pre-bid conference is usually scheduled in the middle of the bid period. In many cases, participation at the pre-bid conference is mandatory, so all bidders should attend the pre-bid conference.

An agenda should be prepared to arrange how the pre-bid meeting will be carried out. A typical pre-bid meeting schedule follows:

- Opening remarks and introduction
- Presentation on the procurement aspect of the requirement
- Questions and answers on the procurement aspects
- Presentation on the technical aspects of the requirement
- Question and answers on the technical aspects
- Closing remarks

A representative of the owner's agency should present and discuss the entire solicitation document. A summary of each section of the ITB should be given, and bidders should be allowed to ask questions. The technical aspect of the requirement should also be discussed, offering bidders the technical details of the requirement.

Bid Day

Bid day is deadline-oriented. The term "bid deadline" refers to the date and time designated in the ITB as the last date and time for receipt of bids, as may be revised by addenda.[5] In public works, the deadline is firmly fixed. The ITB usually requires sealed bids. Additionally, the ITB requires bids to be submitted in a sealed envelope that is adequately addressed and endorsed. Oral, telephonic, electronic mail (e-mail), or facsimile Bids are usually invalid and not accepted.

After the fixed deadline (time) passes, the bid opening proceeds. Bid opening is a stage in the bidding process where the received bid envelopes are opened and examined by the ITB's advertiser or owner (Bid Agent). In an open or public bid opening, the bidders (and other parties) can witness the process and inspect the bids; only the owner's staff is present in a closed opening.

Competitive bids cannot be modified, withdrawn, or canceled after the deadline unless the owner consents. Such consent may be allowed under the terms outlined in the ITB documents if the bid contains an excusable mistake.

Then, what constitutes an excusable mistake? A written notice detailing the error and how it occurred must typically be submitted within a specified timeframe (usually within five working days after the bid opening). To qualify as an excusable mistake, the following conditions generally apply:

- The mistake makes the bid significantly different from what the bidder intended.
- The mistake occurred while filling out the bid and is not due to poor judgment, negligence in inspecting the project site, or failure to carefully review the drawings or specifications.

If a bidder requests to withdraw their bid, it is vital to immediately review their documentation to assess the nature of the mistake, how it happened, and the financial implications.

Best Value and QBS

Now that we've discussed the Low Bid method, let's move on to something more complex: the Best Value method and QBS. Unlike the straightforward Low Bid approach, these methods involve more complex evaluation procedures. In fact, it's common for the project owner to use a two-phased approach, which includes both a Request for Qualifications (RFQ) and an RFP. As shown in Figure 2.1, the Best Value method can be broken down into two types: BVF and BVTC. The difference lies in the pricing component in the proposal as well as the level of design completeness at the time of procurement. In the following sections, we'll dive deeper into each of these methods and explore when and how to use them effectively.

BVTC Method

The BVTC procurement method is a selection process used for construction services. The total construction costs and other non-cost factors are considered when evaluating proposals. This method is also known as the best value bid because it involves a competitive sealed proposal process.[6] The following criteria are used to further clarify the definition of a BVTC selection:

- Total Construction Cost (or design in addition to construction costs in the DB) is a weighted criterion for final contractor selection
- The final selection of a contractor is based on a weighting of the total construction costs and other criteria

The BVTC method, also known as the best value bid, can be applied in both the DBB and DB delivery methods. When using BVTC in DBB projects, the total cost refers to the total construction costs. However, the total cost in DB projects encompasses design and construction costs.

To employ the BVTC method in DB projects, it is essential for the owner to provide a clearly defined design guideline, allowing the contractor team to estimate their price in their proposal. Therefore, a more detailed and explicitly defined RFP is essential to implement the BVTC method effectively.

BVF Method

The BVF procurement method is a selection process where the fee is an essential factor in the final decision. This method considers the firm's fees and qualifications when making the final choice. The fees typically include general overhead and margins. In many cases, the contractor is also required to provide general conditions and preconstruction service fees in addition to the fee. This BVF approach is instrumental when the design is still in the early stage. At this early stage, the design details aren't developed enough for a contractor to develop a complete cost estimate, making BVF a good fit.[7]

The following criteria are used to further clarify the definition of a BVF selection:

- % Fee (or % Fee and General Conditions) is used as a weighted criterion for final contractor selection
- The final selection of a contractor is based on a weighting of fees and other criteria
- % design completeness is usually less than 20% of the design, not enough for a contractor to develop a construction cost estimate.

The BVF method is generally used in Construction Manager at Risk (CMAR), progressive DB, and Integrated Project Delivery (IPD) methods. It works well in these scenarios because it allows for more flexibility when the project details are still evolving. The fee proposal usually has two parts: a percentage based on the advertised maximum construction cost and a fixed amount for specific general conditions. In some cases, the RFP requires only a percentage fee, which includes all general overhead costs and profits, but not the general condition. When evaluating the fee, the board usually uses the normalized score: the tender that scores highest in the evaluation has its score inflated to 100% of the score (like ten points), and the scores of the other tenders are scaled up accordingly.

QBS Method

The QBS method is a procurement process where the primary criteria for selecting contractors are their qualifications and demonstrated competence. Unlike best-value methods that consider total cost or fees as a part of the selection criteria, QBS focuses entirely on the qualification and technical plan (i.e., project management, design development, and any project-related plan) that the service provider can bring to the project. Though the Brooks Act does not bind the selection process for a construction service provider, this QBS approach, as you can recall from earlier discussions, is designed to ensure that the project is executed by the most capable team, regardless of price considerations at the selection stage.

In the QBS process, there are no discussions of price or fees in the proposal during the selection phase. Instead, the evaluation is based on non-price factors such as the contractor's experience and work capacity. This method is particularly advantageous in complex or uncertain projects where the quality of work and the contractor's expertise in the design process are critical for the project's success. As shown in Figure 2.1, the QBS is appropriate if the design is usually less than 10% complete at the time of selection. By excluding price from the initial selection, QBS helps to foster a collaborative environment focused on delivering the best possible outcome.

This QBS method allows the contractor or their team to get involved in early design, which gives more opportunities to leverage the contractor's expertise, thereby improving design and planning quality as opposed to a limited contractor's role in the design process. Eventually, this attribute can reduce the risk of design changes or construction delays. Like BVF, QBS is widely used in various project delivery methods, including CMAR, DB, and IPD.

Selection Process Involving a Proposal Evaluation: Overview

The selection process is central to identifying the firm that will maximize the project owner's value. Crucial to a successful project is the project owner's responsibility first to establish a quality selection committee and then to allow adequate time and resources for the committee to do its work. The selection committee will determine the selection schedule, identify the evaluation criteria, release the solicitation, evaluate and rank the offerors based on their qualifications, and often select from these a shortlist comprising firms judged to be the best-qualified. Typically, the shortlisted firms are invited to submit technical proposals, which the selection committee will review and rank. The selection committee may conduct interviews, review reference checks, and take other steps to choose the offeror whose qualifications, combined with its price proposal, are judged to represent the most significant value to the project owner.

Most owners use two-phased selection to evaluate a proposal: an RFQ and an RFP. First, the owner issues an RFQ to identify and shortlist qualified firms. The SOQ, a response to RFQ, focuses on the contractors' qualifications, including their technical capabilities and past performance on similar projects. The second step involves a more detailed evaluation to further assess the shortlisted firms' qualifications and suitability for the project.

Some private owners may skip the RFQ process and proceed directly to the RFP stage for the BVF and QBS methods. In this scenario, the owner evaluates the proposals first and then selects the shortlisted firms for the interview process. This approach allows the owner to simplify and streamline the selection process while ensuring that the most qualified firm is chosen.

Selection Process #1: Choosing the Most Qualified Firms Through RFQ

The owner issues an RFQ. A construction firm is usually expected to respond to an RFQ solicitation with an SOQ, a written submittal that states the firm's qualifications. The RFQ usually requests various information regarding the proposer's qualifications:

- Basic company info
- Form of ownership
- Authorization/license in the state
- Personal capability
- Financial information
- Bonding capacity
- Safety information
- Experience of its kind
- Statement of why you should be selected

As seen in the list above, the RFQ will call for qualifications that fall into two categories: Pass/Fail (objective) and Scored (subjective). The "pass/fail" qualifications are required by objective and factual criteria, such as the minimum number of years in business or the minimum number of projects completed. These pass/fail criteria offer no room for interpretation and are prerequisites for further consideration. In contrast, "scored" qualifications are at least somewhat subjective and require evaluation to measure their worth or utility. An example might be the evaluation of a firm's relevant project experience. The owner evaluates submittals in response to an RFQ to determine a shortlist of firms that generally appear qualified for subsequent consideration on a specific project.

Selection Process #2: Choosing the Best Valued Firm Through RFP

The RFP contains evaluation criteria specific to the particular project under consideration. Invitations to respond with submittals to the RFP typically go only to a shortlist of firms prequalified through the prior RFQ process. The RFP in the Best Value Selection often involves two additional types of written submittals of proposals relevant to a specific project: (1) Technical proposal and (2) Price proposal.

The technical proposal is a construction company's written submittal demonstrating its capabilities and superior qualifications directly related to the project under consideration. The price proposal is a separate bid for the project's total construction cost. In requesting price proposals, the RFP must provide sufficient design information to allow proposers to submit bids for the project.

Scoring System in Evaluating Proposals

When the committee evaluates proposals, it should have a scoring system that justifies its decision. In the BVTC method, there are three scoring systems available in the market: (1) composite score system, (2) best value system, and (3) adjusted score system. For the BVF and QBS methods, the owner uses a composite score system since they cannot use either the best value system or the adjusted score system, both of which need total costs in their equation.[8]

Composite Score System

The technical review committee should review the qualitative proposals and establish a score for each qualitative proposal based on the factors, weighting, and process identified in the RFPs. The technical review committee should give a written composite score for each qualitative proposal. Table 2.1 presents that the technical review committee identified five factors and assigned a weight (maximum score) to each. The composite score is the sum of the scores of all factors. In Table 2.1, Proposer B is a winner since he has the highest score.

Best Value Score System

The technical review committee should calculate the technical score first. Then, the technical review committee calculates the best value score using the predetermined equation. While the composite score method gives a score to the price factor, the best value score method uses the price as it is. The best value score can be calculated as follows:

Table 2.1 Composite Score Example

	Statement of Qualification (1–25)	*Proposal Technical (1–50)*	*Interview (1–25)*	*Proposal Price (1–30)*	*Composite Score (1–130)*
Proposer A	24	46	21	30	121
Proposer B	25	48	22	28	123
Proposer C	23	48	24	26	121

Table 2.2 Best Value Score Example

	Technical Score (1–100)	Proposal Price	Best Value Score
Proposer A	92	$18,765,000	4.9
Proposer B	96	$16,450,000	5.8
Proposer C	96	$17,000,000	5.6

Table 2.3 Adjusted Price Example

	Technical Score (1–100)	Proposal Price	Adjusted Score
Proposer A	92	$18,765,000	203,967.4
Proposer B	96	$16,450,000	171,354.2
Proposer C	96	$17,000,000	177,083.3

- Best Value Score = Tech Score × $1 million/Price

As seen in Table 2.2, Proposer B is a winner since he has the highest score of 5.8.

Adjusted Score System

The technical review committee should calculate the technical score first. Then, the technical review committee calculates the adjusted score using the following equation. As in the best value score, the adjusted score method uses the price as it is. It should be noted that the proposer with the lowest score is the winner.

- Adjusted Price = Price/Score

As seen in Table 2.3, Proposer B is a winner since he has the lowest adjusted score.

Chapter Summary

1 The Brooks Act (Public Law 92-582) establishes a QBS process in which contracts for professional service providers, such as a designer, are negotiated based on demonstrated competence and qualifications.
2 To choose the A/E, most owners use QBS governed by the Brooks Act, which includes several processes: public solicitation, SOQ, evaluation of statements, interview process with shortlisted firms, and negotiation process.
3 On the other hand, selecting a general contractor, GC/CM, or DB has more options – you have four options to choose from: Low Bid, BVF, BVTC, and QBS.
4 The Low Bid is a popular selection method in public projects. Though price is a single selection factor in the Low Bid method, most owners

require the contractor's bid to be responsible and responsive. Since a bidder should be confident in their bid price, the design should be completed or nearly completed so that there is no or minimal variation for the design given to bidders at the time of procurement.

5 Best Value has two selection methods: BVF and BVTC. The BVTC can be applied in both the DBB and DB delivery methods. When using BVTC in a DBB project, the total cost refers to the total construction costs. In DB projects, however, the total cost encompasses both design and construction costs.
6 BVF is used when the design details aren't developed enough for a contractor to create a full cost estimate. The BVF method is used in CMAR, progressive DB, and IPD methods. It works well in these scenarios because it allows for more flexibility when the project details are still evolving.
7 QBS considers entirely the qualification and technical plan that the service provider can bring to the project, while price is not considered in the selection period. QBS allows the owner to select a contractor or contractor team early, usually less than 10% of the design complete. QBS is widely used in various project delivery methods, including CMAR, DB, and IPD.
8 BV and QBS methods usually use a two-phased selection process: two-phased selection to evaluate a proposal: an RFQ and an RFP. First, the owner issues an RFQ to identify and shortlist qualified firms. The SOQ, a response to RFQ, focuses on the contractors' qualifications, including their technical capabilities and past performance on similar projects. The second step (RFP & proposal) evaluates a proposal, a response to an RFP, which is a more detailed evaluation process to assess further the shortlisted firms' qualifications and suitability for the project.
9 When evaluating a proposal, there are three scoring methods: composite score system, best value score system, and adjusted score system. While the BVTF method can employ all three scoring systems, the BVF and QBS methods need to use only a composite score system since they cannot use either the best value system or the adjusted score system, both of which need total costs in their equation.

Review Questions

1 What selection method is the most appropriate if the following conditions are given?

- An RFQ and an RFP are required
- 30% of the design is included in the RFP
- Each proposer is required to propose a total construction price

2 Three proposers submitted their proposals on a project responding to the RFP. The owner evaluated their technical proposal. Their technical score and proposed price are as follows:

	Technical Score (100)	*Proposed Price*
Proposer #1	90	$10,582,000
Proposer #2	85	$ 9,253,000
Proposer #3	92	$10,675,000

3 List any case you can think of when the lowest bid is not awarded because it is not qualified to be a responsive bid.
4 Which federal law mandates using QBS for selecting designers or A/E services?
5 What scoring system must be used when evaluating proposals in a QBS project?
6 What selection methods typically use an RFQ/RFP process? (Name all selection methods that meet the given condition).
7 You are developing an acquisition plan to construct a new water treatment plant.

7.1 What criteria would you use to select the design firm for the project? Name two criteria and provide the reasons.
7.2 How would you weight the criteria selected in question 2? Would all elements of the selection criteria be weighted equally? What basis did you use to select the weight for each element of the selection criteria?

Notes

1 ACEC (American council of engineering companies) home page. www.asce.org.
2 FAR 5.101. https://www.acquisition.gov/far/5.101.
3 https://procurement.sc.gov/files/sf254.pdf.
4 www.RSMeans.com.
5 Excerpted from www.constructionworkzone.com.
6 Excerpted from "best practices for use of best value selection" (www.agc.org).
7 Excerpted from "best practices for use of best value selection" (www.agc.org).
8 https://law.justia.com/codes/indiana/2010/title5/ar30/ch7.html.

Bibliography

Alleman, D., Antoine A., Gransberg, D., and Molenaar, K. (2017). A Comparison of Qualifications-Based Selection and Best-Value Procurement for Construction Manager–General Contractor Highway Construction. *Transportation Research Record*, 2630(1), 59–67.

Anderson, S. D. and Oyetunji, A. (2003). Selection Procedure for Project Delivery and Contract Strategy, ASCE Construction Institute, *Construction Research Congress Proceedings*, Oahu, HI, March 19–21.

Associated General Contractors of America (AGC) (2004). *Project Delivery Systems for Construction.* AGC, Arlington, VA.

Gransberg, D. D., & Shane, J. S. (2013). Defining Best Value for Construction Manager/General Contractor Projects: The CMGC Learning Curve. *Journal of Management in Engineering*, 31(4), 04014060.

Mearig, T. (2004). *Project Delivery Method Handbook.*, State of Alaska Department of Education & Early Development, Juneau, AK.

Sanboskani, H. and El Asmar, M. (2025). Qualifications-Driven Procurement Selection Criteria for Design–Build and Construction Manager at Risk Firms. *Journal of Structural Design and Construction Practice*, 30(2), 04025010.

Yu, W., Wang, W., Chang, H., and Cheng, T. (2024). Novel A + Q Method for Improving Conventional Lowest Bid Contractor Selection Based on Post-Performance Records. *KSCE Journal of Civil Engineering*, 28, 4800–4812.

Chapter 3

Compensation Structure

In managing construction projects, a contract isn't just a formal binding agreement; it's the roadmap guiding the project owner and the service provider (e.g., contractor) through the terms and conditions. Each construction contract spells out the job details, timelines, and financial specifics, including payments, penalties, and the rights and obligations agreed upon by both parties. Among the components of the contract, critical is the compensation structure.

Construction contracts can be categorized depending on how a service provider gets disbursement for their service. The compensation structure defines the amount of the contract (i.e., how much a service provider will get for their service) and the terms of payment (i.e., how a service provider gets paid). It isn't just about the dollar amounts; it's about setting expectations and defining how you get paid. Therefore, the compensation structure impacts the commercial interests of a service provider, thereby affecting how a service provider behaves. Remember that it's the compensation setup that can really influence a contractor's approach to a project.

You know that the compensation structure involves more than just the contract amount. The compensation structure is also related to how information is shared or how much the owner can access the cost information. The level of shared information indicates transparency and trust between the owner and the service provider. The more information is shared, the more a climate of trust is fostered. Trust is the backbone of any integrated organization or project, as it enhances Collaboration and reduces conflicts that might arise. Therefore, understanding the intricacies of the compensation structure is critical, not only because it defines the financial transactions between the owner and the service provider but also because it influences the dynamics of their relationship.

In this regard, the compensation structure is one of the mechanisms through which the owner addresses how much risk the owner is willing to take on. In other words, the compensation structure transfers risk to the

DOI: 10.1201/9781003628842-5

contractor through various compensation terms and conditions: the baseline amount of compensation, the maximum amount of compensation, the information required for compensation, and the payment conditions.

It is noted that the compensation structure (or the type of contract) is not strictly tied to the project delivery method, though there is a typical match between a compensation structure and a project delivery method. Instead, the owner intends to determine the level of information sharing and risk transfer, thereby impacting the level of interest alignment.

There are various kinds of construction contracts used within the industry. They will vary from project to project. In this chapter, we will review the five types of construction contracts or compensation mechanisms as follows:

- Lump Sum Contract
- Unit Price Contract
- Cost-Plus Contract
- Guaranteed Maximum Price (GMP) Contract
- Estimated Maximum Price (EMP) Contract

In each type of contract, we will address the risks of each contract type from the owner's perspective, the contractor's perspective, the level of interest alignment, and the level of information transparency.

Lump Sum Contract

The most common contract type in the construction industry is a Lump Sum contract as a compensation structure. When diving into the Lump Sum contracts in construction, you might ask, "What exactly does this involve?" Well, it's one of the most straightforward contract types you'll encounter in the industry. So, what is the Lump Sum contract? Simply put, a Lump Sum contract is a fixed-price agreement for project delivery irrespective of actual costs. This price covers all services provided by the contractor or service provider, whether for construction only or for both design and construction. For example, the Lump Sum in Design-Bid-Build projects encompasses the total cost for all construction-related activities, including overheads and the contractor's margin. On the other hand, in Design-Build projects, the lump sum extends to cover the entire design process plus construction activities.

Under a Lump Sum agreement, the contractor is responsible for the proper job execution and should supply their means and methods to complete the work, conforming to the construction documents (i.e., the drawings and specifications). This fixed price, known as the Lump Sum amount, is derived from the contractor's bid, in which the contractor should consider all resource consumptions, overhead costs, expected profits, and contingency for the uncertainty.

When a contractor determines its bid price, the amount of overhead and margin calculated under a Lump Sum contract will vary but reflect how the contractor perceives the potential risks of the specific project. However, estimating a substantial overhead cost can lead the contractor to present higher construction costs, reducing the possibility of getting awarded. In addition, the overhead cost and margin also reflect the market condition – how competitive the market is.

Then, let's explore the attributes of a Lump Sum contract. The number one attribute is its simplicity as a compensation structure. Lump Sum contracts are designed to simplify the administrative aspects in terms of payment. One primary reason for choosing a Lump Sum contract is the reduction in contract administration costs and efforts related to quantity verification and measurement for the progress payment. With the total compensation agreed upon, there is no need for quantity checks or cost verification, which can significantly reduce overhead. It can reduce the administrative burden or overhead costs. A direct consequence of eliminating detailed quantity measures or cost verification is that project management staff, including inspectors, can shift their focus from quantity to quality. Project engineers can focus on more productive jobs rather than dealing with cost verification; inspectors can concentrate more on ensuring the work meets the requirements in the construction documents.

From an economic motivation perspective, Lump Sum contracts lock in financial compensation, providing a fixed revenue for service providers or contractors, regardless of cost or expense. This compensation structure motivates contractors to minimize costs. Furthermore, contractors are provided with flexibility in choosing the construction materials and methods. Such flexibility can benefit the contractor but may also pose risks to the owner. The following section will examine the circumstances under which an owner may choose to use a Lump Sum contract and the potential risks associated with that decision.

When Does the Owner Want to Use a Lump Sum?

Now we understand the advantages of the Lump Sum contract, which includes eliminating the need for detailed quantity measurements and cost verification, and it motivates the contractor to reduce its costs. Is eliminating detailed cost measures the sole reason an owner might prefer a Lump Sum compensation structure? This simplification overlooks the complex issues related to risk transfer. Let's explore how risks are allocated.

A Lump Sum contract is an excellent compensation structure if the requested work is well-defined and the construction documents are complete. The owner believes there is no or minimal change to the construction document. With no or minimal changes, the Lump Sum agreement will reduce owner risk, and the contractor has greater control over profit expectations because the contractor has the privilege to choose its own construction methods.

When agreed upon by parties, the stipulated sum contract might contain specific unit prices for items in uncertain quantities. For example, if the soil condition is quite uncertain, the owner may use the unit price method for earthwork (e.g., excavation) in the Lump Sum contract. Why does the unit price method need to be included in the Lump Sum contract? It is because the contractor takes too many financial risks.

Risk Allocation in the Lump Sum Contract

What risks are assigned to the owner and the contractor in the Lump Sum agreement? Primarily, the owner transfers the financial risk related to productivity by locking in the total project cost before the start of construction. Therefore, the contractor takes risks in terms of its productivity, prices for labor and materials, and supply chain variations unless there is a discrepancy between actual site conditions and the stipulated conditions in the construction documents. Even if actual productivity turns out lower than the expected levels used during the bidding process, the owner is only responsible for paying the agreed-upon amount, irrespective of the contractor's actual costs.

It looks like the contractor takes most of the project risks. Does the owner take any risks? Another significant risk involves uncertainty in the scope or quantity of work. In the Lump Sum contract, the owner bears the risk of the project definition, including the work scope. That being said, the contractor can request a change order if there are any changes in design or discrepancies between the actual site conditions and those described in the contract documents. For example, the excavation method can change based on unforeseen soil conditions. In such cases, the contractor can request a change order to account for the additional costs due to these differing soil conditions.

On a positive note, a Lump Sum contract encourages the contractor to explore innovative construction methods to reduce costs as long as their proposed method conforms to the contract documents. However, this cost-saving motivation may raise concerns for the owner regarding the quality of work. The owner may suspect the contractor might use inexpensive, lower-quality materials or products, provided they meet the minimum requirements.

So, how can the owner address and mitigate this potential quality risk? In Design-Bid-Build projects, the owner often prescribes detailed product selection requirements to limit the contractor's flexibility. By doing so, the owner restricts the contractor's flexibility in selecting means and methods, which can reduce the contractor's creativity to improve productivity. In Design-Build projects, on the other hand, the owner often focuses on controlling the performance criteria of the request for proposal rather than prescribing detailed materials or methods since the design-builder controls the design.

The Level of Interest Alignment and the Level of Transparency

Since the contractor agrees to provide all services for the stipulated Lump Sum, the contractor does not have to share any cost information with the owner. A Lump Sum contract is not open-book, and access to the contractor's or design-builder's cost information is limited. From the contractor's perspective, how much the contractor spends to complete the job is none of the owner's business. If the contractor saves money through his efforts to improve productivity, for example, the savings go toward the contractor's profits. Therefore, the level of transparency on cost information is deficient.

Suppose the general contractor also uses the Lump Sum contract in his subcontract. Then, each subcontractor tries to minimize its own costs since the subcontractor's revenue is already locked with the fixed amount. Therefore, you cannot expect your subcontractors to collaborate with each other to achieve the project goal. The Lump Sum contract, in many cases, brings about a silo mind. A silo mentality is a reluctance to share information with other stakeholders in the project. This attitude is seen as reducing the organization's efficiency and, at worst, contributing to a damaged organizational culture. At least, the Lump Sum contract is far from being aligned with commercial interests among stakeholders unless there is an additional contractual vehicle, such as incentives, to align stakeholders' interests.

Unit Price Contract

Under a unit price contract, a contractor is paid for the actual quantity of each line item performed as measured in the field during construction. Each unit price includes all labor, material, equipment, overhead, and profit attributable to that scope of work.

In this type of contract, both parties (i.e., the Owner and the Contractor) share the risks. The price for each task is calculated as follows:

- Price for a Specific Task = [Quantity of Work] × [Unit Price]

Unit price contracts involve the contractor determining a specific unit price on each bid item. The contractor takes the risk of productivity because the contractor pays the difference in case the actual unit price exceeds the contracted unit price. On the other hand, the owner determines the quantity in unit price at the time of procurement. The owner agrees to adjust the quantity if the actual quantity differs from the quantity the owner provides at the time of procurement. The owner takes the risk of quantity uncertainty since the owner pays the price for additional work quantity.

A unit price contract has benefits for both the owner and the contractor. Owners benefit because they are no longer responsible for any productivity changes. In contrast, contractors benefit because they do not have to worry about uncertainty regarding work quantity or inaccurate quantity take-off for specific tasks.

When Does the Owner Want to Use the Unit Price?

This type of contract is commonly used on construction jobs with the work quantity in question. It is frequently found in public civil works projects involving underground work. Highway and dam projects, of which the quantity is uncertain when bidding, are typical Unit Price contract projects. Since each task should be represented by unit work, work in the Unit Price contract should be defined by specific scope items such as footage of piping, square yards of base, or pavement (see Table 3.1).

It is not unusual to use unit price items within other contracts, such as a Lump Sum contract, when some works are not clearly defined or quantified when bidding. You may find cases where unit pricing may be incorporated into the pricing for specific portions of the project. For example, a Lump Sum contract may be put in place with a unit price stipulated for earthwork, such as excavating weathered rocks per cubic yard. It is the owner's burden to verify the quantity of work in question. In many cases, the owner needs to hire a quantity surveyor to verify the quantity of work.

Table 3.1 Bid Document Example (Unit Price)

Item No.	*Plan Quantity*	*Item Description*	*Price Per Unit (USD)*	*Total Amount (USD)*
1	Lump Sum	Mobilization		
2	186 C.Y.	Roadway Excavation incl. Haul		
3	82 C.Y	Embankment Compaction		
4	4 C.Y.	Quarry Spalls (Drainage)		
5	1 Each	Catch Basin		

Risk Allocation in the Unit Price Contracts

A distinct risk-sharing mechanism exists between the owner and contractor in Unit Price contracts. As previously discussed, the owner assumes the risk related to the quantity of work. The owner must estimate and specify the quantities during bidding before the contractor sets the unit price. Since the total contract value is calculated by multiplying the unit price by the quantity of work, the owner may be required to pay more than the initially agreed contract amount if the actual quantities exceed those outlined in the contract. To mitigate this risk, the owner often employs a quantity surveyor to measure and verify the quantities. As a result, the total compensation amount is not finalized until the project is complete.

On the other hand, the contractor assumes the risk associated with the unit price, which reflects the productivity or cost per unit of work. In this context, the contractor is responsible for any variances in productivity. If actual productivity falls short of what was anticipated during the bidding process, the contractor bears the financial impact of those variances.

Variation-in-Quantity Contract Clauses

In construction projects, the actual quantities often deviate from those in the unit price contract. The contractor may benefit from increased revenue when the quantities exceed the contracted quantity. However, when the quantity slightly deviates from the bid document, the change order process can impose administrative burdens on both the owner and the contractor. Contracting parties often address such issues by including a variation-in-quantity provision in unit-price contracts, which helps adjust the terms reasonably when quantities fluctuate significantly.

The variation-in-quantity provision allows the owner or contractor to request a quantity adjustment (i.e., change order) only if the quantity differs by more than a threshold, usually between 10% and 20%. In addition to avoiding administration burden, the provision protects both parties: contractors are shielded from losses due to reduced quantity within the threshold, while owners are safeguarded when quantities increase within the threshold. Please note that the variation-in-quantity clause is typically applied at the end of the contract when final quantities are confirmed.

Example. Variation-in-Quantity Clauses

Suppose a highway construction project where the contractor is supposed to excavate 10,000 cubic yards of soil. The contract specifies a unit price of $50 per cubic yard of excavation. The contract has a provision that

Table 3.2 Example, Variation-in-Quantity Case

	Item No.	*Plan Quantity*	*Actual Quantity*	*Item Description*	*Price Per Unit (USD)*	*Total Amount (USD)*	*Note*
Contract	##	10,000 C.Y.		Excavation, Soil	$50.00	$500,000.00	
Case 1	##	10,000 C.Y.	12,000 C.Y	Excavation, Soil	$50.00	$600,000.00	Change Order
Case 2	##	10,000 C.Y.	9,000 C.Y	Excavation, Soil	$50.00	$500,000.00	No Change Order

allows both parties to request a quantity adjustment when actual quantities differ by more than 15%. By the project's end, the actual excavation quantity is 12,000 m^3 – 20% more than initially estimated.

In this case, the variation-in-quantity provision allows the contractor to request the change order and adjust the quantity and contract price for the additional excavation. Therefore, the contractor is entitled to receive $600,000 instead of $500,000 because the variance goes beyond the threshold. On the other hand, if the actual quantity of excavation had only been 9000 cubic yards – 10% less than expected – the contractor could still receive the same amount of $500,000 instead of $450,000 since the actual variation is less than the threshold (see Table 3.2). This adjustment would help offset the contractor's reduced opportunity to cover fixed costs and overhead through a smaller volume of work.

The Level of Interest Alignment and the Level of Transparency

Regarding information transparency, the contractor does not want to share the cost information with the owner, which is similar to the Lump Sum contract. This is because their unit price is locked up when contracting. However, the contractor should share quantity information with the owner because the owner takes financial responsibility for the quantity changes. Except for quantity information, the level of information transparency is still low in unit price contract projects.

Since their unit price is locked up at the time of the contract, the level of interest alignment is still low, similar to a Lump Sum contract. You cannot expect your contractors under a unit price contract to collaborate to achieve the project's goal. The unit price contract also brings about silo minds, which will lead to sub-optimal use of their resources. Like a Lump Sum agreement, the unit price contract itself does not contribute to aligned interests among stakeholders. You might want to bring some incentive clauses into your contract to align the interests.

Cost-Plus Contract

Cost-plus contracts,[1] or cost-reimbursable contracts, are used when the scope of work is not fully defined. In this arrangement, all allowable project-related costs are reimbursed by the owner, with an additional fee to cover general overhead and profits. This fee can take various forms: a fixed amount or a percentage of the project costs. Project overhead (POH) costs can either be reimbursed or treated as a separate line item in the cost proposal, depending on the agreement between the parties. Some owners may require a lump sum amount for POH costs or general conditions, with only direct construction costs (such as labor, materials, equipment, and subcontractors) being reimbursed. Cost-Plus contracts, however, place greater financial risk on the owner.

Cost-Plus contracts are typically based on an agreed-upon fee structure since final construction costs are unknown until a project is complete. Owners often provide an estimated (or target) construction cost in the solicitation documents and request the contractor to propose a fee, which may also include a Lump Sum for preconstruction services. These contracts are often used when the construction contract is awarded before the design is barely developed, which allows the contractor to begin work while the design continues to evolve. This flexibility reduces project delivery time but also requires more oversight from the owner, who is more active in project controls.

There are several types of cost-plus contracts based on how the fee is structured:

- Cost-Plus Fixed Fee: A fixed amount that remains constant unless the scope of work changes.
- Cost-Plus Percentage Fee: A percentage of the actual construction cost, which can incentivize a contractor to increase costs.
- Cost-Plus Incentive Fee: The fee is tied to achieving cost targets, with reductions if costs exceed targets and increases if costs are lower.
- Cost-Plus Award Fee: A base fee combined with an additional performance-based fee evaluated periodically based on criteria such as work coordination, quality, and safety.

If the owner wants to establish limits on how much the contractor will be billing, the owner should seek a GMP.

When Does the Owner Want to Use Cost-Plus?

Owners typically use Cost-Plus contracts when the project is not clearly defined or when project conditions require flexibility. This contract type is commonly adopted when construction must begin quickly, even before the design is fully developed. In emergency or disaster recovery projects requiring immediate action, cost-plus contracts provide an efficient and flexible approach to commence work.

Cost-plus contracts are also preferred in the private sector when the owner works with a trusted contractor with whom they have established a long-term relationship. In some cases, especially for confidential projects, the owner may not wish to share information beyond a trusted contractor, making this contract ideal. Owners may also select this contract type when they want to participate actively in all major construction decisions.

Risk Allocation in Cost-Plus Contracts

Cost-plus contracts allocate significant financial risk to the owner. Without a well-managed scope of work, the project's final cost remains uncertain until the project is complete. This contract type may not always be cost-effective, exposing the owner to potential financial risk. This is especially true if the owner lacks knowledgeable personnel for project control. In this regard, owners often prefer to work with contractors they trust or have established long-term relationships with.

Owners must also invest more time and resources into project control than in Lump Sum contracts. Additional project controls, such as work authorization directives, can help mitigate some financial risks. For example, the owner may divide the project into multiple phases, each authorized up to a specific cost threshold. Should costs exceed the threshold, the contractor must notify the owner. While these control measures are helpful, they can impose an administrative burden on the owner.

The Level of Interest Alignment and the Level of Transparency

A Cost-Plus contract requires the contractor to use an open-book accounting system where cost information is accessible. Open-book accounting is a method of procuring work under which contractors are reimbursed based on transparent records of their incurred costs. Do you think open-book accounting promotes collaboration?

Collaboration and trust are key to success. Typically, overheads are expressed in the form of a percentage of the total specific project expenditure. The agreed profit needs to be determined at the outset based on the amount of risk. In a transparent organization, work is driven by collaboration, not competition. Stakeholders and their employees freely share knowledge with each other for the team's benefit rather than hoarding information for individual benefit. In this regard, a cost-plus contract helps align the interest level, though all cost risks are transferred to the owner.

Guaranteed Maximum Price

Owners may want to protect themselves if they are concerned about their financial risks in cost-plus contracts. A GMP contract is a cost-reimbursed contract for a price guaranteed not to exceed a stated maximum price (i.e., the price cap). The key feature of a GMP contract is the ceiling price (GMP), which sets a maximum limit on the total compensation. Suppose the actual costs of the project exceed the GMP. In that case, the contractor is responsible for the gap. As we saw in the previous section on cost-plus contracts, the owner bears significant financial risks in a cost-plus arrangement.

In contrast, GMP offers financial protection to the owner by ensuring that the project does not exceed a pre-agreed budget.

However, the GMP does not entirely shield the owner from financial risks. Although it may seem that the contractor bears the risk of cost overruns, the contractor is not responsible for costs arising from approved change orders or other allowable adjustments under the contract. This means that the GMP, like a Lump Sum contract, is subject to modifications through change orders. Therefore, the owner must have a change order control system to manage these adjustments effectively.

GMP Components and Calculations

GMP usually includes four components: (1) the contractor's general conditions, (2) the cost of work (COW) (cost of works), (3) the contractor's contingency, and (4) the contractor's fee. The COW refers to the direct construction costs going into the final construction deliverables. The labor, material, and subcontractor costs are included in COW. The contractor's general condition usually refers to project-specific overhead costs, such as project management and survey costs. The contractor's fee covers the contractor's general overhead and profits. Most GMP contracts include a contingency fund, which is used to cover unexpected expenses. This fund is particularly critical when the design is not completed when the contract is signed. The contingency helps manage uncertainties and prevents cost overruns from directly pushing the project beyond the GMP. Any leftovers from contingency funds (i.e., unused contingency) usually revert to the owner.

Many owners prefer to use the contractor's fee as a Lump Sum. The percentage fee is typically used for the selection phase and for estimating the interim numbers. However, those percentage fees change to the fixed fee when the GMP contract is made. Why do the owners use the Lump Sum for the contractor's fee? The fixed fee can change the contractor's behavior. Suppose the contractor manages to reduce their COW, which is excellent news. Right? However, the contractor's fee would be reduced if the fee is contracted as a percentage of the COW. Therefore, a fixed fee would incentivize the contractor to reduce their COW. On the other hand, the fee would increase if the GMP increases by more than 10% due to the change order.

If the actual final costs fall below the GMP, any savings (i.e., the difference between the GMP and the final costs) are typically shared between the owner and the contractor according to a predetermined ratio. For example, if the final project cost is $10 million less than the GMP and the contract specifies an 80:20 split, the owner would receive $8 million, while the contractor would take $2 million. This arrangement incentivizes the contractor to keep costs down, as both parties benefit from efficient project completion.

Example. Calculation of Total Compensation in the GMP Contract

Suppose the GMP is $115 million, which includes:

- Not-to-exceed amount $100 million for the COW and general conditions,
- $10 million as the contractor's fee (Lump Sum), and
- Not-to-exceed amount $5 million as contingency.

In addition, the contract stipulates a 70:30 savings split between the owner and the contractor if the actual final costs are lower than the GMP. Now, let's consider three potential scenarios when the project is complete:

1 Final COW is $95 million, with $5 million of contingency used:
2 In this case, the contractor's fee is $10 million (Lump Sum), and the total disbursement to the contractor would be $110.0 million ($95 million for COW + $10.0 million for the fee + $5 million contingency). Since the total actual cost is $5.0 million less than the GMP, 30% of the savings ($5.0 million) is allocated to the contractor, amounting to $1.5 million. Therefore, the total disbursement to the contractor would be $111.5 million ($110.0 million + $1.5 million in savings).
3 Final COW is $100 million, with $5 million of contingency used.
4 In this scenario, the contractor's fee remains $10 million, and the total disbursement would be $115 million ($100 million for COW + $10 million for the fee + $5 million contingency).
5 Since the final cost matches the GMP, no additional savings are distributed.
6 Final COW is $110 million, with $5 million of contingency used:
7 Here, the reimbursable COW is capped at $100 million, and the contractor's fee remains $10 million. Despite the actual COW exceeding $110 million, the contractor is only entitled to a total disbursement of $115 million due to the GMP cap. In this case, the contractor must absorb the $11 million shortfall (including $10 million of excess COW and a $1 million fee overage), making it the contractor's financial responsibility.

When Does the Owner Want to Use GMP?

Construction attorney Michelle Schaap, of Chiesa Shahinian and Giantomasi, told Construction Dive that deciding when to use a GMP depends mainly on the status of the plans.[2]

> If you don't have a complete set of plans, then you generally won't do a Lump Sum, which is why you might instead opt for the cost-plus,

> preferably with a guaranteed maximum price, for the obvious reason that if you do the cost-plus without the guaranteed maximum price, then it's just costs and costs and costs,

In short, GMP contracts are preferred when the owner needs upfront input from the contractor and wants cost assurance, even with design contingency. There are several contexts in which the owner may consider GMP for the compensation structure for their projects. Some key contexts where a GMP contract might be appropriate include:

The first context involves projects requiring construction to begin before the design is fully complete or fast-tracked. Since construction cannot start until a construction contract is made, GMP contracts are often used when the design is not fully complete at the time of contracting. Since the detailed scope may still evolve with some design contingency level, the GMP provides a cost ceiling, giving the owner cost certainty while allowing flexibility for design development. In many GMP agreements, some contingency is offered to address such design uncertainty when the GMP contract is made.

The second context involves projects with an owner's need for cost certainty in a cost-plus case. In projects where cost overruns would be problematic for the owner's budget, GMP contracts assure that the total project cost will not exceed a pre-agreed ceiling unless change orders are formally approved. This gives the owner financial control while allowing some flexibility. For example, a city government may use a GMP contract with some level of contingency for a public infrastructure project, such as a new transit center. The GMP ensures that even if there are issues such as construction productivity (i.e., actual costs exceeding the expected costs), the total project cost won't surpass a specified maximum, helping the city maintain its budget.

The third case is where the owner wants to leverage a collaborative relationship with the contractor. Given that GMP contracts operate on an open-book accounting system, they inherently assume a foundation of trust and cooperation between both parties. In addition, GMP contracts also foster collaboration between the owner and contractor. The contractor has an incentive to manage costs efficiently since savings can be shared, aligning their interests with the owner's goal. For example, an IT company constructing a new corporate headquarters may use a GMP contract to foster a collaborative relationship with the contractor. It is no wonder many big tech companies have recently used GMP contracts to build their campuses. The contractor is incentivized to find cost-saving measures without sacrificing quality, knowing that part of the savings will benefit both parties. The owner also expects the contractor to share its expertise in construction during the design phase.

Risk Allocation in the GMP

Although GMP contracts are a kind of cost-plus agreement, they significantly shift financial risks from the owner to the contractor. A GMP contract protects the owner from financial risks with a ceiling on the total compensation. Then, does a GMP contract have any incentive for cost savings? Compared to Lump-Sum contracts, GMP contracts offer a different approach to cost savings. In Lump Sum agreements, the contractor retains 100% of the savings if the actual construction costs are lower than the agreed-upon price (i.e., the Lump Sum). However, the owner does not benefit from the cost savings in the Lump Sum contract. On the other hand, any savings under the GMP contracts are typically shared between the owner and the contractor based on a predefined split ratio.

Despite these advantages to the owner, GMP contracts pose certain risks for the owner. One notable challenge is that the GMP is usually finalized before the design is fully complete. This can lead to price inflation as the contractor may inflate costs to account for design uncertainties. Additionally, since GMP contracts are typically negotiated with a single contractor in the middle of the design phase, the owner may lose purchasing leverage or risk overpaying if the GMP is set higher than the market price. This lack of price competition leaves the owner vulnerable to accepting an inflated GMP proposal even though the owner can convert the project to a Lump Sum contract. Effective cost control (e.g., change order control) and collaboration with the contractors are essential to keep the project within the agreed maximum price.

The Level of Interest Alignment and the Level of Transparency

GMP contracts create unique dynamics of interest for both owners and contractors. As discussed, how risks are allocated in a GMP contract leads stakeholders to act in a way that protects their interests. During the design phase, the contractor tends to increase the GMP, factoring in potential financial risks and uncertainties. Considering the contractor's financial risks, such protection (i.e., adding buffers to their estimated costs) seems necessary when the design is incomplete. With a higher GMP, contractors protect themselves from absorbing potential overruns exceeding the GMP.

On the other hand, owners seek to limit the GMP to remain below their internal budget. During the design phase, they negotiate the GMP to protect their financial risks. Owners often set a GMP that allows for contingencies within their budget parameters. While contractors attempt to safeguard themselves by proposing a higher GMP, the owner must carefully manage these negotiations to avoid the risk of overpaying. Once the construction phase begins, both parties are tied to this agreed maximum price.

In practice, the contractor's motivation shifts during construction. On the one hand, they are incentivized to keep costs under the GMP, particularly if savings are shared between the owner and contractor. The larger the contractor's share in this ratio, the stronger their incentive to reduce construction costs and improve productivity. On the other hand, they are discouraged from keeping the actual costs below the GMP if the contractor does not receive any portion of the savings. Instead, contractors may focus on meeting the cap rather than reducing costs. Therefore, structuring an appropriate split ratio is essential to align the contractor's interests.

Regarding transparency, GMP contracts use open-book accounting, which promotes transparency. Open-book accounting allows the owner to access the contractor's cost data, including labor, materials, and other project-related expenses. This level of transparency fosters collaboration and builds trust between the owner and contractor. For this reason, many owners prefer to work with contractors with whom they have an established relationship, as trust and open communication are critical to the success of GMP contracts.

Estimated Maximum Price

As seen in GMP contracts, contractors often tend to inflate the GMP to protect themselves from financial risks. This may create tension between the owner and the contractor: the owners may suspect that the proposed GMP price is inflated or market competitive. If owners begin to doubt whether they are getting a fair price, such doubt could impact the relationship and even trust between the parties. To address these concerns, some owners have explored alternative compensation structures where they share more financial risks.[3]

An EMP contract structure offers a different approach to compensation than the GMP model. EMP represents the sum of design costs, construction costs, and a negotiated team contingency and fee. Among the components of EMP, the team's fees are at risk as their amount depends on actual expenses. What sets EMP apart is its collaborative nature and the unique risk-sharing structure it imposes on the project team. Under this model, the contractor, designers, and other key stakeholders agree to put 100% of their profits if the actual costs exceed the EMP. In return, the owner guarantees to pay the full cost of the work (COW) – including direct project costs and POH – regardless of the final expenses. If the actual costs exceed the COW, the excess amount is deducted from the fee.

Example. Calculation of Total Compensation in the EMP Contract

Let's take an example. Suppose that EMP is \$503 million, which includes \$450 million of COW (including the General Conditions), \$23 million of contingency, \$30 million of fee or at-risk amount with a split of 60–40 (owner: project team) if there are savings. Note that the \$30 million fee or at-risk amount does not decrease even if the COW decreases. There are four scenarios when the project is complete:

1 Final COW is \$440 million, with \$23 million contingency used.
2 Suppose the actual COW, including the project's direct costs and POH costs, is \$440 million. Although the actual COW is less than \$450, the fee, based on contract terms, should be \$30 million. This brings the total price to \$493 million (\$440 + \$23 + \$30). As a result, there is a project savings of \$10 million (\$503 million – \$493 million). With a 60–40 split, the contractor's share of the savings is \$4 million, making the total compensation to the project team \$497 million (\$493 million + \$3 million).
3 Final COW is \$450 million, with \$23 million contingency used:
  Suppose the actual COW, including the project's General Conditions, is \$450 million. The fee on the COW amounts to \$30 million with \$23 million contingency, bringing the total price to \$503 million. There are no savings, as the final price equals the EMP.
4 Final COW is \$470 million, with \$23 million contingency used:
  Suppose the actual COW, including the project's General Conditions, is \$470 million, which exceeds the COW (\$450 million) in the EMP. The contingency of \$23 million is used. Then, the difference of \$20 million absorbs the at-risk fee. As a result, the total compensation to the project team would be \$503 million (\$470 million COW + \$23 million Contingency + \$10 million fee).
5 Final COW is \$490 million, with \$23 million contingency used:
  Suppose the actual COW, including the project's General Conditions, is \$490 million, which exceeds the COW by \$40 million set in the EMP. As a result, the actual COW completely absorbs the \$30 million fee set in the EMP. Since the owner fully reimburses 100% of the COW, the total compensation to the project team would be \$513 million (\$490 million COW + 23 million Contingency + \$0 fee).

In scenario (4), the owner must pay an additional \$10 million beyond the EMP, representing their share of the financial pain. Meanwhile, the project team receives no general overhead or profit, reflecting their portion of the pain-sharing.

When Does the Owner Want to Use EMP?

An EMP contract operates in a similar context to a GMP contract. Both are typically used in an integrated project delivery approach. Like the GMP contract, the EMP contract is effective when the design process needs input from the contractor upfront, especially in projects where construction complexities or uncertainties benefit from the contractor's or trade contractor's expertise during the design process.

Owners typically choose an EMP contract when there is a strong, trusted relationship with the contractor's team. This contract structure works best when both parties have a history of successful collaboration and mutual respect. The contractor team is more likely to agree to an EMP arrangement if they believe investing their loyalty and effort will lead to a continued partnership with the owner. The success of an EMP contract depends not just on financial terms but also on the shared understanding that both parties are committed to the project's success.

Risks Associated with the EMP

The key difference between GMP and EMP exists in how the financial risk is allocated. While GMP contracts significantly burden the contractor, the EMP contract has a pain-gain risk-sharing mechanism between the owner and the contractor. Under the EMP, both the owner and the contractor share in the financial outcomes of the project. This shared financial risk allocation structure fosters collaboration in keeping projects under budget.

For the contractor, the major risk lies in the potential loss of profit and general overhead if the COW exceeds the contracted amount. Under the EMP model, the contractor's fee is diminished if the project's actual costs exceed the COW and the contractor's contingency. Sometimes, the contractor may end up with no fee (i.e., zero profits and general overhead). This creates a balance where the contractor is motivated to control costs but is protected from financial loss, unlike in a GMP contract, where they bear a greater portion of the financial risk. On the other hand, owners face the primary financial risk of reimbursing the full COW if it exceeds the EMP.

Despite the risks involved, EMP contracts offer both parties mutual benefits from cost savings. When the project comes in under budget, both parties share the financial benefits or savings. Typically, the contractor receives a more generous share of these savings than under GMP contracts. This incentivizes the contractor to improve their productivity.

The Level of Interest Alignment and the Level of Transparency

Like GMP and cost-plus contracts, the EMP contract uses open-book accounting, where cost information is accessible to the owner. This cost transparency helps build trust between the parties, allowing for a collaborative project environment. Compared to the GMP contract, the EMP contract balances risk-sharing between the owner and the contractor, aligning their interests throughout the project. Both parties share the financial outcomes (i.e., cost overruns or savings), which implies that they share an interest in keeping the project within budget.

The contractor is incentivized to control costs without sacrificing quality. At the same time, the owner is involved in key operational decisions (e.g., submittal approval) to help the contractor control its costs. Given an EMP contract's collaborative and transparent nature, the owner must engage in key project decisions with the contractor team. This level of collaboration is essential for successful cost management. When both parties have confidence in each other's intentions, the transparency of cost information and the pain-gain sharing mechanism can lead to successful project delivery, nurturing their relationship.

Case Study: The Port of Vancouver

Project: The Port Staging Area
Location: Vancouver, WA
Contract Type: Unit Price

Description of Procurement

The port of Vancouver needs an additional sizable, paved staging area to support the loading and unloading of shipping containers from ships. The only available site near the cargo ship piers was previously used for producing creosote-impregnated railroad ties and is thus contaminated. Because the extent of contamination is unknown, the port authority has decided to use a unit price contract for the cleanup of the site and the construction of the paved staging area. The scope of work involves removing contaminated soil from the 1000-foot by 800-foot site, backfilling with compacted select material, and paving the surface with asphalt. The entire site will be fenced with an 8-foot chain-link fence and lighted for night operations.

You have been asked to develop the bid form for the construction contract for the project. The following items are to be included in the bid form: (1) excavation, (2) compaction, (3) refilling/backfilling, (4) additional treatment required to improve the soil suitability, (5) use of imported soil for fill, (6) soil testing, (7) concreting of the substrate for asphalting, (8) asphalting, (9) fencing, (10) lighting, and (11) mobilization.

Question 1.1. What work item (s) do you recommend be listed as Unit-Price items on the bid? Why did you select these items?

Question 1.2. What areas of work do you recommend be listed as Lump Sum items on the bid? Why did you select these items?

Question 2. The port authority has decided to use a two-step procurement method for this project. The first step is to pre-qualify the bidders, who will be invited to submit bids during the second step. What criteria do you recommend the port authority use to pre-qualify the bidders? Why did you select these criteria?

Question 3. What methods do you suggest that the port authority adopt during the execution of the contract to determine the exact quantities of work required for each unit price item that you identified in Question 1.1 above?

Chapter Summary

1 The construction projects use various contracts defining compensation methodology, dominating the commercial interests: Lump Sum, Unit Price, Cost-Plus, GMP, and EMP.
2 In the Lump Sum contract, the contractor agrees to complete the project for a fixed price, bearing the risk if costs exceed the estimate but keeping any savings. This compensation structure motivates contractors to minimize costs. Furthermore, contractors are provided with flexibility in choosing the construction materials and methods.
3 A Lump Sum contract is an appropriate compensation structure if the requested work is well-defined and the construction documents are complete. With no or minimal changes, the lump-sum agreement will reduce owner risk while the contractor has greater control over profit expectations. In the lump-sum contract, the owner transfers the financial risk related to productivity by locking in the total project cost. On the other hand, the owner bears the risk of the work scope and definition.
4 A Lump Sum contract is not open-book, and access to the contractor's or design-builder's cost information is limited. Therefore, the level of transparency on cost information is deficient. The Lump Sum contract,

in many cases, brings about a silo mind. A silo mentality is a reluctance to share information with other stakeholders in the project.

5 In Unit Price contracts, the owner pays the contractor based on the actual quantity of work done at a set unit price. Owners benefit because they are no longer responsible for any productivity changes. In contrast, contractors benefit because they do not have to worry about uncertainty regarding work quantity.

6 The Unit Price contract is frequently found in public civil works projects involving underground work, where the work quantity is uncertain when bidding. The Unit Price contract is characterized by a shared risk. The owner assumes the risk related to the quantity of work. On the other hand, the contractor assumes the risk associated with the unit price, which reflects the productivity or cost per unit of work

7 In the Unit Price contract, the contractor does not want to share the cost information with the owner, which is similar to the Lump Sum contract. However, the contractor should share quantity information with the owner.

8 Cost-Plus Contracts are the compensation structure where the contractor is reimbursed for all costs plus an additional fee for profit, encouraging transparency and trust, but placing most financial risk on the owner.

9 Owners typically use cost-plus contracts when the project is not clearly defined or when project conditions require flexibility. Cost-Plus contracts allocate significant financial risk to the owner. Owners must also invest more time and resources into project control than in Lump Sum contracts. A cost-plus contract requires the contractor to use an open-book accounting system where cost information is accessible.

10 GMP Contracts are similar to cost-plus but with a cap on the maximum price, where the contractor bears the risk of any cost overruns beyond this cap. However, the GMP does not entirely shield the owner from financial risks. Like a Lump Sum contract, the GMP is subject to modifications through change orders.

11 GMP contracts are preferred when the owner needs upfront input from the contractor and wants cost assurance, even with design contingency. Compared to Lump Sum contracts, GMP contracts offer a different approach to cost savings. Any savings under the GMP contracts are typically shared between the owner and the contractor based on a predefined split ratio.

12 In the EMP contracts, the integrated team (i.e., contractor, designers, and other key stakeholders) agrees to put 100% of their profits if the actual costs exceed the EMP. In return, the owner guarantees to pay the full COW regardless of the final expenses.

13 Like the GMP contract, the EMP contract is effective when the design process needs input from the contractor upfront, especially in projects where construction complexities or uncertainties benefit from the contractor's or trade contractor's expertise during the design process.

Review Questions

1 What project delivery methods typically use a Lump Sum contract?
2 How does GMP differ from traditional Cost-Plus contracts?
3 The City of Romson plans to construct a new state-of-the-art water treatment plant to comply with revised state water quality requirements and serve a rapidly growing population. The city's director of public works is considering procuring a construction contractor for the project using either a lump-sum contract or a cost-plus contract.

   3.1 What advantages & risks might the city realize by selecting a Lump Sum contract as opposed to a Cost-Plus contract?
   3.2 What advantages & risks might the city realize by selecting a Cost-Plus contract as opposed to a lump-sum contract?
   3.3 Which alternative fee method would you recommend for structuring the fee for the water treatment plant if a Cost-Plus contract is to be used? Why did you make this selection?

Notes

1 Cost-Plus Contract in this book refers to the cost-plus contract without a cap. Therefore, GMP is not in this category because GMP is Cost-plus with a Cap.
2 www.constructiondive.com.
3 Darrington, J.W. and Lichtig, W.A. (2010). Rethinking the "G" in GMP: Why Estimated Maximum Price Contracts Make Sense on Collaborative Projects. *The Construction Lawyer*, 30(2), 29–39.

Bibliography

Construction Industry Institute (CII) (2001). *Project Delivery and Contract Strategy*. Publication RR165-12, Austin, TX.

Darrington, J.W. and Lichtig, W.A. (2010). Rethinking the "G" in GMP: Why Estimated Maximum Price Contracts Make Sense on Collaborative Projects. *The Construction Lawyer*, 30(2), 29–39.

Gordon, C. M. (1994). Choosing an Appropriate Construction Contracting Method. *Journal of Construction Engineering and Management*, 120(1), 196–210.

Ogundare, T. I., Kassa, R., Maali, O., Lines, B., Smithwick, J., and Sullivan, K. (2024). Current Practices in the Development of Guaranteed Maximum Price

(GMP) Contracts. *Proceedings of the Construction Research Congress 2024*, 146–154.

Palihakkara, A. and Perera, B. (2022). Risk Management in Guaranteed Maximum Price Contracts. *Smart and Sustainable Built Environment*, 11(2), 368–386.

Rahman, M. M. and Kumaraswamy, M. M. (2004). Contracting Relationship Trends and Transitions. *Journal of Management in Engineering*, 20(4), 147–161.

Tran, D., Brihac, A., Nguyen, L., and Kwak, Y. (2018). Project Cost Implications of Competitive Guaranteed Maximum Price Contracts. *Journal of Management in Engineering*, 34(2), 05018001.

Part 3

Project Delivery Method

Chapter 4

Design-Bid-Build (DBB)

The first project delivery method we will explore in this chapter is Design-Bid-Build (DBB). DBB is a conventional project delivery method where the owner develops a distinct contractual relationship with a designer and a contractor. The DBB process initiates when the owner engages a designer to create a design document (i.e., design drawings and construction documents). Once the design is completed, the owner solicits bids from multiple contractors through a price-competitive selection process to carry out the construction work conforming to the design and construction documents. The contractor is usually selected based on either a Low Bid or a Best Value Total Cost approach.

DBB follows a fragmented procurement process, as it separates the design phase from the construction phase. Each phase is carried out sequentially and independently, impacting project stakeholders' roles, responsibilities, and relationships. Historically, the DBB delivery method is widely used, especially in public projects. However, it was not history's first delivery method to manage construction projects. As discussed in Chapter 1, the industry used a more integrated approach, with a master builder managing both the design and construction phases, before taking the fragmented approach.

The 19th-century Industrial Revolution increased the growing separation between design and construction processes. With increased income levels, people began seeking diverse forms of housing. New, complex building types require specialized design expertise and management skills in managing the construction process. Professional organizations such as the American Institute of Architects (AIA) also helped separate the design community from the construction community. This specialization helped improve their capability but contributed to the increased fragmentation in the industry.

Furthermore, the Miller Act (1935) reinforced this separation by requiring general contractors (GCs) on government projects to provide surety bonds. As many design firms did not have enough financial resources to

DOI: 10.1201/9781003628842-7

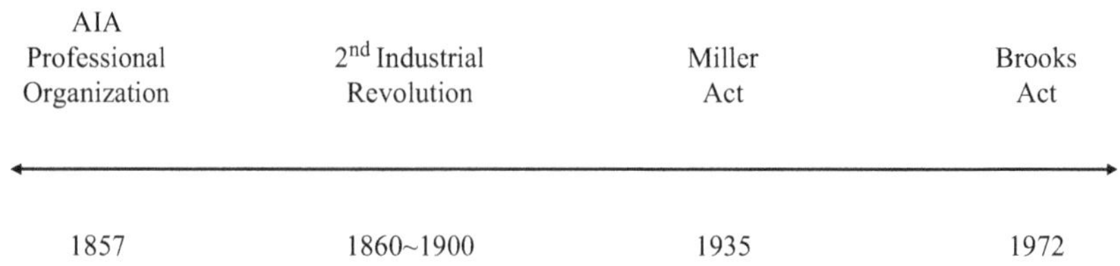

Figure 4.1 Historical Events Promoting the Fragmentation.

meet these bonding requirements, their role became restricted to design. At the same time, contractors took on the business risks associated with construction. Later, the Brooks Act, which requires public owners to choose designers using QBS, affirmed this separation by establishing distinctions between the selection criteria for designers and contractors. Figure 4.1 shows the timeline of the relevant events that fostered the fragmented approach.

Although DBB has been criticized for its fragmented nature, it is still widely used, particularly in the public sector. Understanding the characteristics and processes of the DBB projects is essential for managing the DBB projects as either an owner or a contractor. In this chapter, we will examine the key features of the DBB delivery method, the major processes in each project phase, and risk allocations among the project stakeholders.

Major Attributes of the DBB

Spearin Doctrine

As we've discussed, the DBB method separates the design, bidding, and construction phases, implying that each phase is executed independently. The design team develops the plans responding to an owner's needs, and then the contractor steps in to build a facility according to those plans. This means that the contractor is responsible for executing the construction, including quality control (QC), but isn't involved in making design decisions.

But what happens if the contractor discovers a flaw in the design or specification during construction? Or what if problems arise during construction because of a design mistake? Who's responsible for that? Is it the contractor, since they're in charge of the construction process? Or is it the design team, since they developed the plans? Or could it even be the owner's responsibility? This is where the Spearin Doctrine comes into play, and it's an essential aspect of the DBB method that you should understand.

Before the 19th century, contractors typically bore most project risks associated with construction unless their contract explicitly stated otherwise or when uncontrollable natural disasters (i.e., an Act of God[1]) made it impossible for contractors to complete the project. It was often said, "Where one agrees to do, for a fixed sum, he will not be excused or become entitled to additional compensation because unforeseen difficulties are encountered."

However, over time, courts began to recognize exceptions to this rule. One of the most significant exceptions is the owner's implied warranty of the plans and specifications, which was established in the *United States v. Spearin*.[2] In this case, the U.S. Supreme Court ruled that if a contractor is required to build according to the plans and specifications provided by the owner, the contractor isn't responsible for any defects in those plans. Essentially, the owner guarantees that the design is complete and error-free. This owner's responsibility isn't waived by clauses requiring contractors to visit the site or check the plans – they are still protected from the consequences of design flaws.

In the DBB projects, the owner takes on the risk of any defects or errors in the construction documents (i.e., the drawings and specifications). In other words, if there's an issue with the design, it's the owner, not the contractor, who is responsible for the outcomes. In simple terms, the Spearin Doctrine protects the contractor when they stick to the plans and specifications provided by the owner. For example, the contractor brings additional resources because of the design omissions in the electrical drawings. In that case, the contractor isn't liable for any loss or damage caused by those defects or any deviation between the construction documents and the actual conditions.

However, even though the owner assumes the risk for design errors or omissions, they can still seek legal recourse against the design professionals (i.e., the architect or engineer) if they believe the design errors resulted from the designer's negligence. However, here's the thing: that kind of dispute between the owner and the designer is separate from the delivery method. Such a dispute between the owner and the designer falls outside the scope of the Spearin Doctrine itself.

Now, is the Spearin Doctrine absolute? Unfortunately for contractors, there are exceptions. In 2007, the Ohio Supreme Court shook up the construction industry by narrowing the application of the Spearin Doctrine. In *Dugan & Meyers Construction Co. v. Ohio Dept. of Administrative Services*,[3] the trial court initially sided with the contractor under the Spearin Doctrine due to damages caused by excessive design changes. However, the Ohio Supreme Court reversed the ruling when the case was appealed. They decided that the Spearin Doctrine didn't apply to delays caused by design changes. Instead, the court pointed to specific contract clauses, such as "no damages for delay" and "written requests for time extension," to justify their decision. What does it imply for the industry, especially contractors? The Spearin Doctrine doesn't invalidate an express contractual provision.[4]

Sequential Engineering

One of the key features of the DBB method is the separation between product design (done by the designer) and process design (led by the contractor). In this approach, each stage of the project – product design, process design, and construction – is completed in sequence, with the next stage only beginning once the previous one is finished. This is what's known as sequential engineering. In this system, information flows in only one direction: from product design to process design. However, the process design does not influence or provide feedback on the product design.

Now, what does this imply in practice? Because the contractor is not involved in the design phase, knowledge about the construction process or materials isn't often thoroughly investigated during the design phase. Once construction begins, such issues can lead to undesirable outcomes (e.g., design changes). Let's take an example. You are working on a building construction project that involves installing custom-designed lighting fixtures. When your electrical trade starts planning how to install them, they realize there's not enough space for the crew to install the fixture. As a result, a design change might be needed, resulting in additional costs and time to the project. As projects become more complex and uncertain, the drawbacks of sequential engineering become more evident. Without the contractor's input during the design phase, the risk of costly design changes and inefficiencies during construction increases.

Lump Sum Contract and Vulnerability to Changes

In the DBB method, projects typically use a lump sum contract where the contractor agrees to complete the project at a fixed price (i.e., the lump sum). The contractor bears the risk of any cost increases while performing the scope of work in the contract. However, this fixed price can increase if the owner approves a change order.

Now, you might wonder when a change order would come into play. Change orders are approved only if the actual conditions differ from those specified in the bid or construction documents. This happens because, in the DBB method, the bidding process takes place only after the design is 100% complete. However, there's the catch: even though the design is supposed to be complete, there's always a chance that something in the construction process doesn't match the plan (i.e., drawing and specification). That's where the change order comes in. This book introduces a graph that illustrates the relationship between the vulnerability to change orders and the level of design details in the design documents when the contract is made.

In DBB, the design is entirely or almost completed when the owner selects their contractor, which might sound like a good thing, but it actually makes the project more vulnerable to change orders. The more detailed the design, the more likely any deviation from the plan (e.g., unexpected site conditions or minor design errors) could require a costly change order. Moreover, the sequential engineering, which the DBB method employs, implies that process design (e.g., construction methods and materials) is not fully considered during the product design phase. Since the process design isn't factored into the design phase, constructability or logistical challenges might not surface until after the design is complete. This disconnect can increase the likelihood of change orders during construction.

This change order vulnerability gets even worse when the DBB projects use the low bid selection method. Contractors who win the job with the lowest price often work on tight margins under a lump sum contract. As a result, the contractors have little room for profit unless a change order is approved. This setup encourages contractors to pursue change orders, further increasing the likelihood of change orders.

The DBB method is also commonly used for comparison in the Design-Build (DB) method. In DB, the vulnerability to change orders is much lower than in DBB because the contractor is involved in both the design and construction phases, allowing them to provide process design as one team when the designer develops product design, as shown in

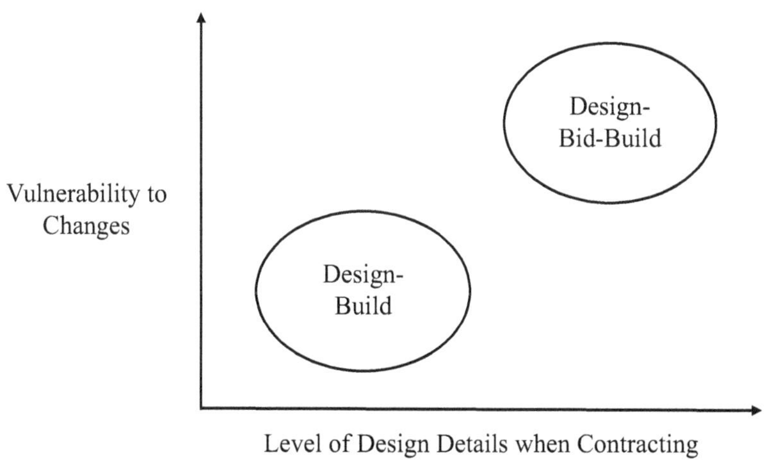

Figure 4.2 Vulnerability-to-Change Graph.

Figure 4.2. We'll dive deeper into DB in a later chapter. Still, for now, it's important to note that the level of detail in DBB design documents can make projects more susceptible to costly change orders once construction is underway.

Does DBB Fit Your Needs?

DBB is the most widely used project delivery method in the construction industry. The DBB structure clearly separates design from construction, with the owner holding separate contracts with the designer and the contractor. In most cases, procurement laws and regulatory frameworks are procedurally aligned with DBB. DBB remains the "default" choice for many projects unless project contexts and needs call for alternative delivery methods such as Construction Manager at Risk (CMAR) or DB. To determine whether DBB is the right fit, owners should assess the project context as well as the owner's needs. Generally, most owners would find a healthy pool of designers and contractors, as it is the default project delivery method. Therefore, market condition is typically not a limiting factor when considering DBB.

Project Context

If the project needs the contractor's early involvement, DBB is not your project. Typically, the design process requires the contractor's expertise when the project is complex and uncertain. In other words, DBB can be best suited for your project where the design concept and construction methods are conventional. If a project requires a contractor's input or involves technical challenges that benefit from contractor involvement early on, DBB may not be your best option.

The Owner's Needs

DBB offers owners complete control over the design prior to construction. This may be appealing to owners who want to thoroughly define the project scope and requirements before engaging a contractor. In most cases, the owners do not prioritize the project speed over cost savings through competition. Because construction cannot begin until the design is fully completed and competitively bid, DBB is not suitable for fast-track delivery. If speed is the owner's priority, other delivery methods like CMAR or DB may offer better alternatives. However, DBB usually provides a low-price contract due to market competition over the finalized design and specifications.

Project Delivery Phases

Design Phase

In a DBB project, the owner hires a professional designer, often called the Architect of Record (AOR), to provide design services. Typically, the owner uses a Qualification-Based Selection (QBS) to select their designer or AOR. The Brooks Act requires owners to use QBS to choose their design professionals for public projects. The designer or AOR collaborates closely with the owner and key stakeholders – such as user groups or maintenance teams – to understand and incorporate their needs into the design documents, including drawings and specifications. In addition, the designer may involve consulting engineers to handle the design of specialized systems, such as mechanical or electrical systems, much like a GC brings in subcontractors.

However, the DBB project's design process does not have the contractor's input. The process design is not fully considered since DBB uses a sequential engineering model. It's also worth noting that if a construction contractor is brought in for consultation during the design phase, they are usually excluded from participating in the bidding process for the construction work, further limiting their involvement in developing the design.

Bid Phase

In the DBB projects, the owner typically uses a **low-bid selection** process to choose the contractor. In this approach, contractors compete by offering the lowest price to win the project. The nature of the bid documents in DBB makes it well-suited for this method because, by the time the bidding process begins, the design is fully complete, and there's no design contingency to account for. This means that the contractor is able to estimate all necessary resources to develop an accurate estimate to complete the work described in a bid package.

So, what exactly is a bid package? A bid package is a set of documents issued by the owner to potential bidders (contractors) to help them prepare their bids. It typically includes an invitation to bid (ITB). An ITB includes details about a project, such as project scope, legal requirements, and the bid selection process. Since the bid package, including the ITB, is the only set of documents that contractors rely on during the bidding process, it must be clear, complete, and free from ambiguity. Any unclear or incomplete information in the bid package can come back to haunt the owner in the form of costly change orders once the project starts. A lack of clarity in the bid documents allows contractors to request changes later, which can drive up project costs.

If you are an owner or an owner's agency, you should also be aware of the Spearin Doctrine. For example, even if the bid package states that the contractor is responsible for verifying soil conditions, the owner's responsibility under the Spearin Doctrine cannot easily be waived with simple wording in the bid package. The owner is still accountable for the accuracy of the design and specifications provided to the contractor.

Sometimes, the owner may want to include alternates in the bid package. **Bid alternates** refer to specific components or features of the project that are not included in the base price but can be added depending on the project's budget. For example, suppose there is a project to build a multi-family housing development. The owner might include an alternate for a solar panel window system that increases the building's energy efficiency. Using alternates, the owner can adjust the project's scope to better fit the available budget, depending on how the bids come in.

Alternates can help manage project costs but may also create certain concerns for contractors. During bidding, contractors might ask for a prioritized list of alternates so that they can project which features the owner will most likely accept. This is especially important in public projects where the process must be transparent, but it can also be useful in private projects. Contractors are often concerned that if alternates are grouped in a certain way, it could give an unfair advantage to a specific contractor. However, in the private sector, this is less of an issue because the owner has more freedom to select the contractor they prefer or even award the project to a higher bidder if they choose. Ultimately, it's up to the architect and the owner to decide whether to provide a prioritized list of alternates.

Risk Allocation in DBB

DBB remains the most commonly used project delivery method for public construction projects. One of the key factors in the success of any DBB project, either as an owner or a contractor, is understanding how risks are allocated among the different stakeholders. In DBB, the owner, designer, and contractor each take on distinct roles and responsibilities. For example, the designer is responsible for developing the project design documents, while the contractor assumes the risks associated with construction execution. Each stakeholder needs to be held accountable for the risks to which they are best equipped to manage.

Design Errors

People often think designers should take all the responsibility for design errors in DBB projects. However, it is not that simple. In DBB projects, the design contract is completely distinct from the construction contract, complicating the accountability line. According to the Spearin Doctrine, the owner is primarily responsible for design errors. This means that before construction begins, the owner should guarantee that the design meets all project requirements, including building codes. The contractor is generally protected from liability if issues arise during construction due to design errors.

You may wonder what exactly the designer is responsible for in their design service. If the owner believes that design errors result from negligence – meaning the designer failed to meet the standard of care – they can pursue legal action against the design professionals. However, it's important to remember that proving a designer's negligence can be complicated and challenging.

The designer's responsibility is typically limited to the professional standard of care – a level of competence and diligence expected from a reasonably skilled professional. This standard has evolved over time and is often referred to as common law, based on past court decisions regarding the accepted norms of professional behavior. Most litigation regarding the architect's design error states, "an error in judgment is not necessarily evidence of negligence, for mistakes and miscalculations are incident to all of the business of life." In other words, perfection is not the standard of care for architectural practice.[5] This is the common law standard of care for architects, which is usually described as the level of skill and care employed by architects practicing in similar circumstances and geographical areas.

Cost Overruns

We need to consider two scenarios when discussing cost overrun risks in the DBB projects. The first scenario involves the case where the expected construction costs, once the design is complete, exceed the owner's budget. In such cases, the owner bears the risk, even if the designer initially promised that the project would stay within budget. Legally, architects are not held accountable for cost overruns related to design unless negligence is proven. A notable example is the case of *Beede vs. Coombs* (1896),[6] where an architect designed a house with a budget of $2500, but the final cost reached $2700. Although the architect promised to meet the client's budget, the court ruled that an architect must only exercise reasonable skill and ability. Please remember that architects do not guarantee that a project stays under budget.

The second scenario involves cost overruns related to productivity during construction. In most DBB projects, owners use a lump sum contract in which the contractor is responsible for any cost overruns unless there is an approved change order. This puts the risk of construction-related cost overruns squarely on the contractor. However, the contractor can choose their means and methods as long as they meet the requirements stipulated in the construction documents (drawings and specifications).

Change Order

As we've discussed, change orders are critical in the DBB projects because they directly impact the contract amount and the owner's budget. From the owner's perspective, it's imperative to remain vigilant about change orders throughout the project. Owners should be aware of two types of change orders, both of which can drive up the total project costs.

The first change order type occurs when there is a discrepancy between the actual site conditions and what is described in the bid package or construction documents. If the contractor encounters unexpected conditions – such as unforeseen soil conditions or inaccuracies in the design documents – they are likely to request a change order to adjust the cost of the work. This type of change order can significantly burden the owner because it results in additional costs beyond the lump sum contract amount.

The second type of change order occurs when the owner requests changes to the scope of work in the contract document. The owner can request modifications to design elements, such as adding new work elements during construction. In response, the contractor typically submits a change order to account for the additional resources needed for such changes.

Did you see why the owner should be careful about potential change orders? As you see in both cases, whether due to discrepancies or scope changes, the owner assumes the risk of change orders, as these change orders usually push the project beyond the original budget.

Schedule Overrun

In any project, the risk of a schedule overrun should be assumed by the stakeholder responsible for causing the delay, regardless of the project delivery method. In the DBB projects, the ITB typically specifies the allowed project duration. A tighter schedule can reduce the number of contractors willing to bid, as contractors may be concerned about schedule challenges.

If the contractor causes the schedule overrun, the contractor may be subject to liquidated damages. These are predefined damages agreed upon in the contract, which the owner can collect as compensation if the contractor fails to meet the specified schedule (i.e., contract due). Liquidated damages – sometimes referred to as liquidated and ascertained damages – are legally enforceable in the United States and serve as a financial penalty for delays. Therefore, contractors must carefully manage their construction schedule to avoid these penalties.

On the other hand, if the owner causes the schedule delay, the owner is at risk of a claim from the contractor. In this case, the contractor may seek compensation for the additional costs due to the delay, such as extended labor, equipment, or material costs. However, it is noted that the compensation due to the schedule delay should be stipulated in the contract. Both parties must carefully monitor the project's progress and schedule. In most cases, the specification includes guidelines and several steps to prevent unnecessary delays, as schedule overruns can have financial consequences for both the contractor and the owner.

Quality Problem

The architect assumes the risk related to quality assurance, while the contractor is accountable for QC. The architect/engineer (A/E) should be responsible for design quality: the design documents (drawings and specifications) meet the project's requirements. Speaking of window installation, for example, the architect might specify the minimum thermal performance and dimensions for the windows. Still, they typically don't specify the window's exact brand or model number. During the construction phase, the architect's role shifts to construction administration, which includes reviewing the contractor's work. The architect typically reviews submittals and responds to requests for information, if any, to help verify that the contractor complies with the approved plans and specifications. However, the design professionals are not directly responsible for day-to-day QC during construction.

In DBB, the contractor is responsible for QC. The QC covers the materials, workmanship, and deliverables. Continuing with the window installation example, the contractor chooses a window product that meets the design and specification requirements. Then, the contractor should turn in the submittal to get their choices approved. Once the windows are installed, the contractor is also responsible for inspecting the installation. Suppose any quality issues (e.g., faulty seals) arise during the inspection. In that case, the contractor must fix these issues. An independent third-party inspector is sometimes hired to perform quality checks in addition to a contractor-led inspection.

The Level of Integration

In this chapter, we have examined the main features and processes of the DBB method. We also addressed how various project risks are distributed to different project stakeholders. Now, let's explore the level of integration in DBB projects.

In project delivery systems, integration refers to how different stakeholders work collaboratively to leverage each other's expertise when necessary. Since most of a project's value is determined during the design and engineering phases, integrating the knowledge of design and construction professionals at critical points can help avoid issues later on. However, in the DBB projects, opportunities for such integration are restricted, as discussed in the sequential engineering section. For example, while the designer might consult with an external construction expert during the design phase, the GC, who leads the construction process, is prohibited from getting involved in the design process. This restriction results from DBB's sequential engineering, where process design (i.e., construction plan) is separated from product design. Due to the sequential engineering, the contractor's expertise and insights are restricted from influencing design decisions.

What kind of opportunities would be lost if the contractor's expertise is prohibited from influencing the design process? This limitation may imply that constructability considerations, logistical challenges, or cost-saving alternatives may not be factored into the design. For example, a contractor might suggest different equipment layouts to optimize installation and maintenance, but their input typically arrives too late to influence the initial design in DBB projects.

You would argue that a construction expert who provides consulting to the designer during the design process can serve as a GC in the construction phase of DBB projects. Does that work? From a fairness perspective, some government agencies argue that allowing a contractor who has consulted during the design phase to bid on the project could create conflicts of interest or give that contractor an unfair advantage. Therefore, public projects often exclude contractors involved in the design process from participating in the bidding, which further limits early-stage integration. This fragmented approach, a hallmark of sequential engineering, prevents the flow of valuable knowledge from contractors and suppliers into the initial design phase. As a result, projects delivered using DBB are often more prone to change orders during construction, as contractors and suppliers bring forward constructability issues or adjustments that weren't considered in the design.

Case Study[7]: Simpson Investment Campus

Project: Investment Bank Campus
Location: Bellevue, WA
Budget: $60,000,000 (design and construction)
Project Delivery System: DBB
Contract Type: Lump Sum

Background

Simpson Investment, headquartered in downtown Bellevue, decided to construct a second office building adjacent to its main headquarters to respond to the expanding demands. The project used a traditional DBB method, with NBR as the architect and UWCM as the GC. The project, valued at $60 million, was to be completed in 30 months under a Lump Sum contract. The bid documents included detailed design drawings and specifications. The bid document specified that the new building's boiler system should be compatible with the existing building's control systems, as both buildings were to be centrally managed from a single control room.

When they started testing the systems at the end of the construction phase,[8] they discovered major problems. The boiler prescribed in the design documents was undersized and incompatible with the control system in the main building. In short, the boiler failed to meet the operational requirements, leading to delays and significant changes.

The Issue

Simpson Investment claimed that UWCM, the GC, was responsible for verifying the technical specifications and their compatibility with the existing system. They pointed to the language of the bid documents that required the contractor to verify all design and operational parameters. UWCM, on the other hand, contended that it was not their duty to correct a design error from the bid documents prepared by NBR, the architect.

Question 01. Would you agree that DBB is the best delivery method for this case? Defend your response.

Question 02. Who should be in charge of the costs related to the faulty design and rework of the boiler? Provide the reasoning behind your decision.

Chapter Summary

1 The DBB method uses a fragmented procurement process, as it separates the design phase from the construction phase. Historically, the Miller Act (1935) reinforced this separation by requiring GCs on government projects to provide surety bonds. Later, the Brooks Act, which requires public owners to choose designers using QBS, affirmed this separation.

2 In the DBB projects, the owner takes on the risk of any defects or errors in the construction documents according to the Spearin Doctrine. The Spearin Doctrine protects the contractor when they stick to the plans and specifications.
3 The DBB projects follow sequential engineering, where product design (done by the designer) is separated from process design (led by the contractor). Because the contractor is not involved in the design phase, knowledge about the construction process or materials isn't often thoroughly investigated during the design phase.
4 In the DBB projects, the process design isn't factored into the design phase: constructability or logistical challenges might not surface until after the design is complete. This disconnect can increase the likelihood of change orders during construction. This change order vulnerability gets even worse when the DBB projects use the low bid selection method.
5 The DBB project delivery process is relatively straightforward. The owner uses a QBS to select their designer. However, the DBB project's design process does not have the contractor's input. Then, the owner typically uses a low-bid selection process to choose the contractor. The bid package includes instructions to bid, bid forms, contract forms, bond forms, and the project's drawings and specifications. Sometimes, the bid package includes alternates, components not included in the base price but that can be added depending on the project's budget.
6 In DBB, the owner, designer, and contractor each take on distinct roles and responsibilities. According to the Spearin Doctrine, the owner is primarily responsible for design errors. During the design phase, the owner bears the risk of expected costs exceeding the budget when design is complete, even if the designer promised that the project would stay within budget. Once the construction begins, the GC bears the risk of cost overruns related to productivity.
7 If the contractor encounters unexpected conditions, they are likely to request a change order to adjust the cost of the work. This type of change order burdens the owner. In any project, the risk of a schedule overrun should be assumed by the stakeholder responsible for causing the delay. When it comes to quality, the architect assumes the risk related to quality assurance, while the contractor is accountable for QC.
8 In the DBB projects, opportunities for such integration are restricted because of the sequential engineering. Due to the sequential engineering, the contractor's expertise and insights are restricted from influencing design decisions. This limitation may imply that constructability

considerations, logistical challenges, or cost-saving alternatives may not be factored into the design.

Review Questions

1 Please respond to the following questions.

1.1 From the owner's perspective, what is the most important aspect of the Spearin Doctrine in preparing bid documents?

1.2 Which of the following is the difference between the DBB and other project delivery methods? Choose only one choice.

a The contractor has incentives to lower total construction costs.
b The owner has control over the design process.
c The contract price cannot change during the construction period.
d The contractor is not involved in the design phase.

2 Below is a public project that used a DBB method. Please respond to the questions at the end of the scenario.

The City of Clyde Hill is under a court order to construct a tertiary sewage treatment plant to comply with state environmental regulations. The city hired a design firm to develop plans and specifications for the new treatment plant. The design firm developed an initial cost estimate for the project of $30 million. The city conducted a special election, and the voters narrowly approved the city's request for approval to sell $30 million in municipal bonds to cover the estimated construction costs.

The design is now complete, and the final cost estimate for construction is $35.8 million. You are hired by the city to provide advice on a procurement strategy for the project. The city decided to use a public Lump Sum contract for the project but is very concerned about the estimated construction cost.

2.1 What strategy can you recommend the city adopt to ensure there is adequate funding to award a contract at the conclusion of the bidding?

2.2 Do you think the DBB method helps align stakeholders' interests (i.e., GC, A/E, specialty trades)? Defend your answer.

3 Below is a typical K–12 school construction that used a DBB method. Please respond to the questions at the end of the scenario.

One of the vast metropolitan areas went through an influx of population seeking IT jobs. As a consequence, the city needed an additional elementary school. The city had a project budget of $5.55 million for design and construction services. The city had professionals in design and construction who had experience in similar projects. The city had only two years to complete the project. They thought it was very tight in terms of schedule. Since the town had no experience in GC/construction manager or DB, the city decided to go with the DBB method.

A local architect was hired for $410k and delivered a construction document in five months. The project was advertised and open for bidding for the statutorily required 30-day period. The city and the designer agreed to have two alternates in their bid. Sets of bid documents, including drawings, were picked up by ten GCs for a deposit of $200 per set. For some reason, only three bids were received on a bid day.

The city received three bids from local GCs, and their bids ranged from $495,000 to $535,000 without alternates. The proposed costs ranged as follows:

Total	*Contractors*	*Base Bid*	*Alternates*
A	$495,000	$57,000	$552,000
B	$512,000	$56,000	$568,000
C	$535,000	$55,000	$590,000

The city accepted only one alternate out of two because of the budget constraint. Finally, the city made a Lump Sum contract of $512,000 with Contractor A. The contract completion date was 14 months after the construction contract was made. The construction job kicked off as scheduled, but faced with several issues:

The geotechnical report included in the bid package did not provide accurate information. The filling material should be added to the original estimate, which caused the Contractor to pay an additional $50,000.

The contractor requested a change order for a special driveway pavement that was not covered in the construction document. The amount of the change order request for the waterproofing was $37,000.

Since the school should open as scheduled, the Owner had to accept the change order request. Finally, the project was completed as scheduled.

3.1 What selection method did the city use?
3.2 What is the purpose of alternates in the bid package?
3.3 Do you think that the project was appropriate for DBB? If you don't agree, what would the best delivery method? Why?
3.4 Why do you think some of the contractors dropped out during the bidding process?
3.5 Did the Spearin Doctrine apply to this project? Where?

Notes

1 www.sgrlaw.com.
2 *Spearin*, 248 U.S. at 136.
3 https://law.justia.com/cases/ohio/supreme-court-of-ohio/2007/2007-ohio-1687.html.
4 Bestpracticesconstructionlaw.com.
5 https://www.aia.org/resource-center/standard-care-should-i-care. *Coombs v. Beede*, 89 Me. 187, 36 A. 104 (Me. 1896).
6 https://www.aia.org/resource-center/standard-care-should-i-care.
7 This case study is a hypothetical scenario inspired by the author's professional experience. While it reflects real. project contexts, all company names and data have been modified for illustrative purposes only.
8 It is called a commissioning process.

Bibliography

Almuhannadi, M., and Ghareeb, A. (2024). Enhancing Design-Bid-Build Project Delivery: A Comprehensive Review and Framework for Contractor Selection and Project Optimization in the Construction Industry, *Organization, Technology and Management in Construction: an International Journal*, 16(1), 63–80.

Anderson, S. D. and Oyetunji, A. (2003). Selection Procedure for Project Delivery and Contract Strategy, ASCE Construction Institute. *Construction Research Congress* 2003, Oahu, HI, March 19–21.

Associated General Contractors of America (AGC) (2004). *Project Delivery Systems for Construction.* AGC, Arlington, VA.

Construction Industry Institute (CII). (2001). *Project Delivery and Contract Strategy*. Publication RR165-12, Austin, TX.

Miller, J. B., Garvin, M. J., William Ibbs, C., and Mahoney, S. E. (2000). Toward a New Paradigm: Simultaneous Use of Multiple Project Delivery Methods. *Journal of Management in Engineering*, 16(3), 58–67.

Touran, A., Gransberg, D. D., Molenaar, K. R., Ghavamifar, K., Mason, D. J., and Fithian, L. A. (2009). *A Guidbook for the Evaluation of Project Delivery Methods.* Transit Cooperative Research Program, Washington, DC.

Chapter 5

Construction Management at Risk (CMAR)

In this chapter, we will explore the Construction Management at Risk (CMAR) delivery method, a collaborative delivery approach. In California, CMAR is usually called CM/GC (Construction Manager/General Contractor), while in Washington, it is called the GC/CM (General Contractor/Construction Manager) method. Though the terminology may differ, the core principles remain unchanged: the project owner engages a contractor early in the design phase to provide the contractor's input, which can improve the design quality and reflect the constructability, leading to better project outcomes.

The Design-Bid-Build (DBB) delivery method has been the dominant project delivery method, especially for public projects. However, DBB's organizational structure often led to challenges, including legal claims, change orders, and project delays for private and public owners. As the construction industry evolved, the building systems got complex, demanding collaboration between the designer and the construction contractor in the design phase. Suppose a contractor identifies a design that does not account for constructability during construction. In that case, a design change may be necessitated, leading to costs and extended project duration.

Owners and industry leaders began to question the limitations of DBB's sequential engineering model, which restricts the involvement of construction contractors during the design phase. They recognized that engaging contractors earlier could improve project outcomes. In addition, the industry began rethinking the competitive bidding model focused on low bids: the Low-Bid approach was seen as incompatible with the need for early contractor input. The industry also questioned the adequacy of traditional compensation structures (i.e., Lump Sum).

The industry was finding ways to use contractors' expertise in the design process. This shift in mindset paved the way for the CMAR method, which promotes an organizational structure where the contractor or CM/GC, specialized in construction means and methods, contributes to design development (DD) by working with the designer.

DOI: 10.1201/9781003628842-8

The CMAR projects, called CM/GC or GC/CM, provide distinct features. During the preconstruction (or design) phase, the contractor serves as a construction manager, providing preconstruction services such as design feasibility studies, cost estimation, and scheduling. Once the project moves into the construction phase, this same contractor assumes the role of general contractor, overseeing the construction work and managing trade contracts.

In this chapter, we will delve into the characteristics of the CMAR delivery method, discussing the main processes of each project phase. We will also address how risks are allocated among the project stakeholders. By understanding these elements, readers will realize how CMAR can foster collaboration, while they can address some of the limitations of the CMAR delivery method.

Major Attributes of the CMAR

Two-Phase Delivery Process

In the CM/GC delivery method, the owner makes two separate contracts: one with the designer and another with the CM/GC. The contract with the CM/GC is divided into two distinct agreements in different phases: one for preconstruction services and the other for construction services.

In the design phase, the CM/GC collaborates closely with the designer and the owner. During this phase, the CM/GC's preconstruction services include, but are not limited to, constructability review, cost estimating, and project scheduling. This early involvement helps align the project's product design with process design, eventually reducing the likelihood of surprises or change orders during construction (see Figure 5.1).

Once the design progresses to a point where the design contingency level is reduced to an acceptable range, the contractor and project owner begin negotiations on the construction contract price (i.e., Guaranteed Maximum Price [GMP]). If both parties agree on the GMP, the project moves into the construction phase, where the CM/GC "changes hats" and takes

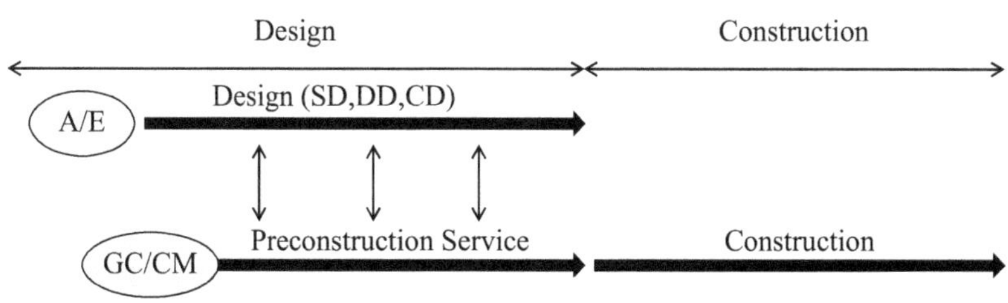

Figure 5.1 CMAR Process (Conceptual Flow).

on the role of the general contractor. Although the CMAR method follows a two-phase delivery process, the timing of the CM/GC's involvement can influence their role and the project's outcomes. Typically, there are three scenarios regarding when the CM/GC (or GC/CM) might be selected and brought into the project (see Figure 5.2):

Early Involvement Before Designer Selection

In this scenario, the CM/GC is engaged even before the designer is chosen. This is particularly evident when the owner – often a private project owner – has established strong relationships with certain CM/GC firms through prior working experience. The CM/GC usually helps the owner choose a designer whose expertise aligns with the project and with whom they have good working chemistry. Once a designer is selected, the CM/GC provides early input on methods and materials that will impact the design process. Since the CM/GC gets involved early, the project benefits from increased flexibility and expanded opportunities in design compared to the other two scenarios. For example, the CM/GC might propose innovative construction methods or materials that the designer may not be familiar with, promoting the project's efficiency and potential cost savings.

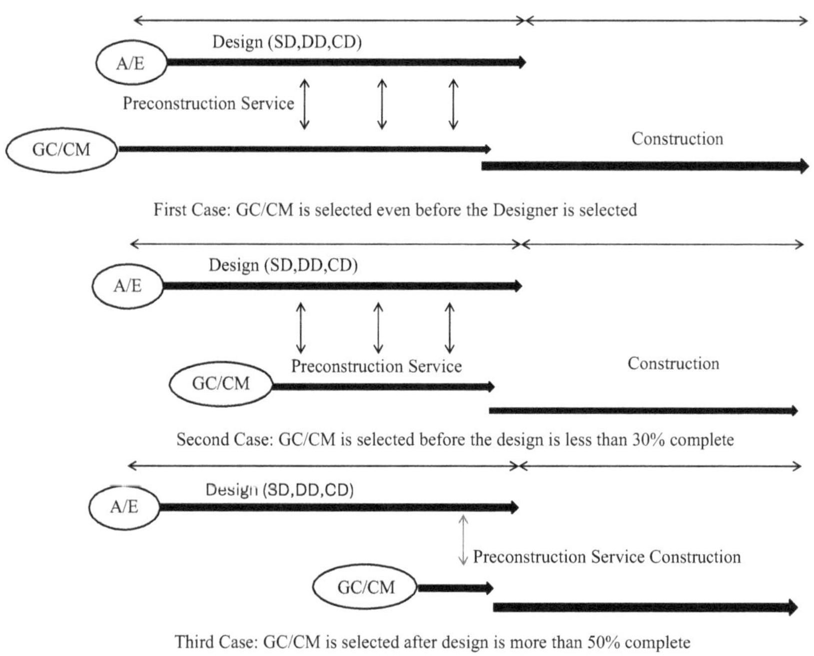

Figure 5.2 Three Different Cases of CM/GC's Involvement.

Selection When Design Is up to 30% Complete

This is the most common approach. When the design is at an early stage (no more than 30% complete), the designer can help the owner select a CM/GC who provides preconstruction service (e.g., constructability review). At this point, the CM/GC has the chance to contribute to design decisions during DD, bringing valuable insights into construction methods, material choices, and constructability. Although their influence is somewhat limited compared to the first scenario, the CM/GC can still significantly impact the project by advising on constructability and identifying potential risks.

Selection When Design Is Over 50% Complete

In this scenario, the CM/GC is brought on board after the design is already substantially developed, similar to the traditional DBB approach. With the design primarily complete, the CM/GC has limited or no opportunity to influence design decisions, which restricts the CM/GC's role in the design process. In this scenario, the CM/GC's role during the design phase is limited to reviewing design documents before the construction documents (CDs) are finalized. As a result, many professionals consider this case less representative of the CM/GC method.

In general, owners prefer to engage the CM/GC early, ideally when the design is no more than 30% complete. This timing allows for selection based on either qualifications or a best-value fee proposal, enabling the owner to choose a CM/GC who brings expertise to the project.

Contract Types: GMP as a Governing Compensation Structure

Usually, the CMAR (or CM/GC, GC/CM) delivery method goes with a GMP contract. As you learned in Chapter 2, a GMP contract is a cost-reimbursed contract where the owner contracts with a service provider to perform a fixed scope of work in exchange for a guaranteed price not exceeding a stated maximum price. The key feature of a GMP contract is the ceiling price (GMP), which sets a maximum limit on the total compensation the contractor can receive. Suppose the actual costs of the project exceed the GMP. In that case, the contractor is responsible for covering the overrun unless the GMP is adjusted through a formal change order approved by the owner.

GMP usually includes four components: (1) the contractor's general conditions, (2) the cost of work (COW), (3) the contractor's fee, and (4) the contractor's contingency, as shown in Figure 5.3. The COW refers to the direct construction costs going into the final construction delivery. The labor, material, and subcontractor costs are included in COW. The

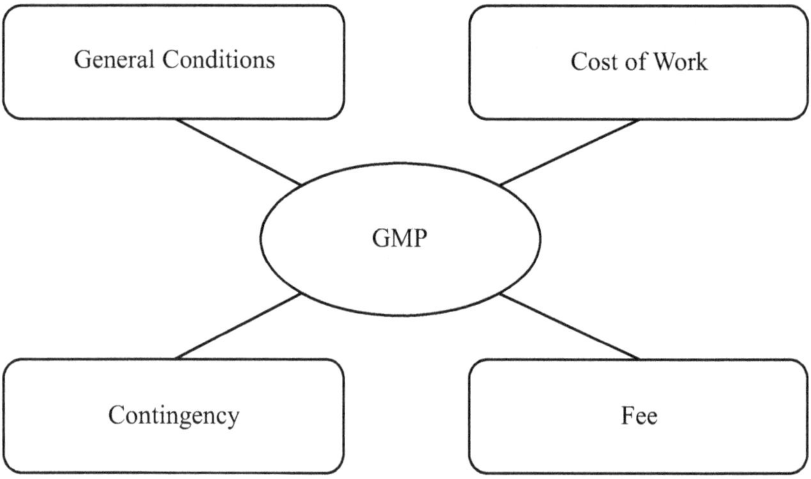

Figure 5.3 GMP Components.

owner usually requires the contractor to provide the COW organized by trade. The contractor's general condition refers to project-specific overhead costs, such as supervision and survey costs. The contractor's fee covers the Contractor's general overhead and profits. Most GMP contracts include a contingency fund, which is used to cover unexpected expenses during construction. The contingency helps manage uncertainties and prevents cost overruns from directly impacting the contractor's profits or pushing the project beyond the GMP. Any leftovers from unused contingency funds usually revert to the owner.

If the actual final costs come below the GMP, how do you deal with any savings (i.e., the difference between the GMP and the final costs)? The owner does not have to share any savings, but most owners are usually willing to share some savings to incentivize the contractor to work under budget. For example, if the final project cost is $10 million less than the GMP with an 80:20 split, the owner would receive $8 million of the savings, while the contractor would take $2 million. This shared savings arrangement incentivizes the contractor to keep costs down, as both parties benefit from efficient project completion.

Since the GMP contract has a chance of change orders, we need to take a look at how compensation changes in the case of a change order. There would be three types of change orders in the GMP contract: (1) the change order that does not increase or decrease the COW, (2) the change order that increases the COW, and (3) the change order that decreases the COW. For the change orders, the fees should be determined according to the relevant contract clauses.

Though the GMP has been used in the CMAR projects to facilitate the collaboration, there is some research[1] that claims that the GMP contract has some drawbacks: (1) GMP encourages adversarial behavior, protecting their margin, and (2) contractors are not encouraged to increase their risks, such as innovative technologies.

While a GMP contract is the most common compensation structure in CMAR projects, there are some cases where lump sum or cost-plus agreements may also be used. In rare cases, the CMAR project can be structured as a cost-plus contract. As in a GMP contract, this structure still gives the owner access to the contractor's cost information. However, without a cost cap, there is a significant financial risk for the owner if costs rise unexpectedly. For this reason, cost-plus contracts are used only when the project needs to start quickly, bypassing the GMP negotiation process in the preconstruction phase.

Though uncommon, some CMAR projects may use a Lump Sum contract if both parties agree. For a Lump Sum contract to be effective, the design must be nearly 100% complete, with minimal or no design contingency remaining. This ensures an accurate final price but eliminates the opportunity for fast-tracking (i.e., starting construction before design completion). Therefore, Lump Sum contracts in the CMAR projects are incompatible with fast-track schedules.

Another situation in which a Lump Sum might be used is if the owner and CM/GC cannot reach an agreement on a GMP. In this case, the owner may convert the project to a DBB approach, soliciting competitive bids and awarding the project to the lowest bidder under a Lump Sum contract. Since the GMP structure is the predominant contract type in CMAR projects, the discussions in this chapter will primarily assume that a GMP contract governs a CMAR project.

Fast Track

Fast track is a project delivery strategy that enables construction to start before the design is complete. In a fast-tracked project, the designer provides some completed design packages (e.g., earthwork and structural design packages) to the construction contractor as they are developed, allowing construction to begin while other design packages (e.g., MEP design packages) are still in progress. This overlap between design and construction phases shortens the overall project timeline, which is often desirable for construction project owners seeking quicker project delivery.

The fast-tracking of the project is therefore achieved through overlap between the design and construction phases. The overlap between the design and construction phases can be realized by involving the construction contractor in the design phase. For fast-tracking to succeed, two key conditions must be in place. First, the contractor responsible for construction (often the general contractor) must be involved during the design phase. This early involvement allows close coordination between the designer and contractor, which is essential for managing the overlap effectively. Second, the construction contract must be made before the design is finalized. In other words, the contract terms and conditions, such as the contract price, should be agreed upon during the design phase. This allows construction to proceed without waiting for complete design documents.

In traditional DBB delivery, fast-tracking is challenging because DBB does not meet these conditions. In DBB, the contractor is only brought in after the design is 100% complete, meaning there's no opportunity for overlapping phases. Therefore, there is no chance of overlapping between the design and construction phases. However, these conditions are met with the CMAR method. The CMAR model enables early contractor involvement and allows for a flexible approach to contracting and scheduling, making fast-tracking a practical strategy. In the CMAR delivery method, the GMP is typically negotiated and finalized before CDs are fully complete. It is noted that the contractor cannot begin construction until the GMP is established.

During the design phase, the project is divided into multiple work packages, such as foundation and structural packages. Most civil and architectural design packages, except MEP packages, are usually completed before the GMP is finalized. Suppose the GMP is set early, for example, when the design is only about 50% complete. In that case, construction can begin sooner, with a more significant overlap between the design and construction phases, allowing for an aggressive fast-track approach, as shown in Figure 5.4. On the other hand, if the GMP is finalized when the CDs are nearly complete, the project may lose fast-tracking opportunities. Therefore, in the CMAR projects, the timing of GMP finalization largely determines the extent to which a fast-track strategy can be implemented.

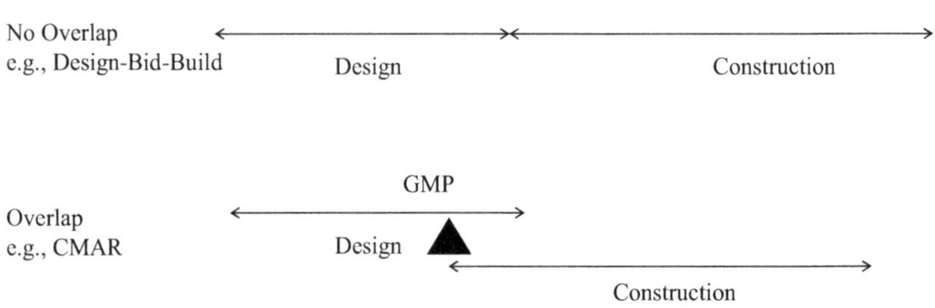

Figure 5.4 Fast Track and Overlap Between Design and Construction.

Concurrent Engineering

Concurrent engineering is an integrated approach to the design and preconstruction process in which product and process design are developed simultaneously rather than sequentially. Unlike the traditional DBB method – where the designer completes the product design first, and then the contractor devises a process for construction – concurrent engineering involves both disciplines working together from the outset. This requires a multi-disciplinary project team, bringing together designers, contractors, key trade specialists, and other key stakeholders early in the design process to consider how design choices may impact later stages of construction.

In the CMAR delivery method, concurrent engineering becomes especially effective when the contractor, or CM/GC, is engaged early in the design phase, ideally before the design reaches 30% completion. With the CM/GC involved at this stage, they can invite trade contractors and major suppliers to participate in the design discussions. For example, an HVAC (Heating, Ventilation, and Air Conditioning) trade contractor may be consulted during the DD stage of a large commercial building to advise on the MEP system layout, which can prevent redesigns or spatial conflicts during installation. Similarly, a structural steel fabricator may review structural plans to suggest design adjustments that improve constructability. Such preconstruction service provided by trade contractors or suppliers is called "design assist."

This early collaboration allows for the simultaneous development of the product and process designs. While the designer works on architectural and structural plans, the contractor's team can develop detailed strategies for project sequencing, scheduling, and logistics. For example, if the designer specifies a unique curtain wall system, the CM/GC and facade subcontractor can assess the feasibility of the design, develop installation methods and logistics plans, and identify any potential construction risks while the design is still evolving. Throughout such a process, the CM/GC and subcontractor can provide feedback to help the designer refine the curtain wall design. This kind of iterative design process, where product and process design inform each other, shows the collaborative nature of concurrent engineering.

As you see, the CMAR method supports concurrent engineering, facilitating a coordinated approach to design and preconstruction. This approach can reduce the likelihood of rework and improve cost control, as key construction processes are defined and planned alongside the design. However, the effectiveness of concurrent engineering depends on how early the CM/GC is brought into the design phase and how early the CM/GC brings its key trade partners into the process. The earlier they're involved, the more effectively the team can leverage concurrent engineering to benefit the project.

Spearin Doctrine

As you learned in Chapter 4, the Spearin doctrine defines the responsibility of design errors. According to the Spearin Doctrine, the implied warranty of design adequacy protects contractors on traditional DBB construction projects. The Spearin doctrine provides that a contractor who is "bound to construct or build according to plans and specifications prepared by the owner ... will not be responsible for the consequences of defects in the plans and specifications."[2] Whether the Spearin doctrine applies outside the traditional DBB context remains an open question. Here, we can discuss whether the Spearin doctrine is applied to the context of CMAR.

Since we are discussing the responsibility for design errors, it would be appropriate to clarify who leads the design process and holds the responsibility for it. In CMAR projects, the owner typically hires an architect to design the project, just as in traditional DBB projects. However, early in the design process, the owner also brings in a CM/GC to act as a construction manager during the design or preconstruction phase.

During the design (preconstruction) phase, the CM/GC collaborates with the designer and the owner. They provide preconstruction services, such as preliminary scheduling, constructability study, interim cost estimates, and value engineering. Unlike in DBB, where the owner and their architect or engineer have exclusive control over the project's plans and specifications, CMAR involves the CM/GC in the design process. The CM/GC plays an active role in developing the plans and specifications, collaborating with the designer until the CDs are finalized.

Suppose that the CM/GC provides feedback on an HVAC design draft. Based on the input, the architect/engineer (A/E) modifies the design with the approval of the owner's representative or agency. But what if this revised design later turns out to be flawed? Who do you think leads the design decision processes? Who bears responsibility for the error – the owner, the A/E, or the CM/GC who contributed feedback? And does the Spearin doctrine, which usually holds the owner responsible for design adequacy, still apply in this collaborative setting?

A notable case in Massachusetts addressed the applicability of the Spearin doctrine to CMAR contracts.[3] The Spearin doctrine, which traditionally implies that the owner warrants the adequacy of the design provided to the contractor, was examined in this CMAR context. The court upheld that holding separate contracts for design and construction does not negate the Spearin doctrine. However, the CM/GC's involvement in the design phase introduces a layer of complexity regarding responsibility for design errors or omissions. The court held that the Spearin doctrine still applies to CMAR projects. According to the court decision, the CM/GC's duty to provide input into the design (i.e., preconstruction service) does not eliminate the owner's implied warranty of the plans and specifications.

In particular, the extent of the CM/GC's participation in the design and any contractual delegation of design responsibilities to the CM/GC can impact the court's interpretation of liability under the implied warranty. Additionally, the owner's control over design decisions may influence this determination. Generally, the greater the CM/GC's design responsibilities under the contract, the more likely they will share responsibility for design-related issues when courts assess the owner's liability under the Spearin doctrine. Considering the owner still controls the design process in many CMAR projects, the Spearin doctrine is still applied to most CMAR projects.

Does CMAR Fit Your Needs?

Selecting the appropriate project delivery method is critical for any public or private owner. Suppose the owner seeks any project delivery method other than the traditional DBB. In that case, the owner should be confident that their choice of the project delivery method is aligned with the project context, the owner's needs, and the market environment. The following considerations will help determine whether CMAR is the right fit for your project.

Project Context

The owner should investigate the context of the project to ensure that it is in line with the CMAR. It is noted that specific project characteristics are better suited for the CMAR model. According to Washington State's procurement guidelines,[4] CMAR is recommended when:

- The project requires complex scheduling or coordination.
- Construction will take place in an occupied facility that must remain operational.
- The general contractor's involvement during the design stage is critical to the project's success.
- The project involves a technically complex work environment.
- The project requires specialized work on a historically significant building.

These conditions reflect a context where CMAR's collaborative structure, particularly the general contractor's participation during design, is needed to reduce risk and improve productivity.

The Owner's Needs and Ability

The owner's role in a CMAR project differs from other delivery methods, especially Design-Build (DB). A key distinction from DB lies in the control over the design process and design liability. Under the Spearin Doctrine, the owner in a CMAR contract is generally responsible for the design and specification. In this regard, owners should actively participate in the DD and coordination between the designer and the contractor (GC/CM).

If the project delivery speed is critical for the owner, CMAR should support a fast-track approach, where construction can begin before design is fully complete. This allows schedule compression, compared to traditional DBB, where construction follows a fully completed design and bid process.

More importantly, the owner must be capable of leading the CMAR project. It involves not only legal capability but also technical capability. The owner should have professional staff who are able to coordinate the design and construction process between the designer and the contractor. The owner should mainly be able to manage the GMP structure, which involves fee negotiation and open-book accounting. In addition, most states view CMAR as an alternative to DBB, which is generally accepted as the standard delivery method. Therefore, public owners in many states must obtain approval to use CMAR for specific projects. Private owners, however, are typically free to use CMAR without going through a qualification process.

In Washington state,[5] for example, a public entity must apply for certification to use CMAR. Once certified, the entity may use CMAR for eligible projects over the next three years without needing additional committee approval for each project.

Suppose a public university in Washington State plans to renovate an occupied research facility with complex HVAC and electrical systems. Because the building must remain operational during the renovation and the systems are highly technical, the CMAR method is generally recommended. The university would acquire certification by demonstrating that it meets the required qualifications and that the project aligns with the state's criteria. This certification would allow the university to use CMAR for this project and other eligible projects over the next three years.

Can You Bring the Competent CM/GC?

Even when the project and owner are well-suited to CMAR, the market needs to support this delivery method. To make CMAR projects successful, the project should have (1) experienced general contractors who have served as CM/GCs (e.g., preconstruction services) and (2) experienced mechanical and electrical contractors (ECs) who have served as design-assist service providers. In addition, the owner needs to have some trust or relationship with the CM/GC because the GMP contract structure needs transparency.

Project Delivery Process

As discussed in the previous section, there are three timings at which a CM/GC (or GC/CM) might be brought into a project: (1) before the designer is selected; (2) when the design is no more than 30% complete; and (3) when the design is more than 50% complete. Since the second scenario – where the CM/GC is engaged when the design is around 20%–30% complete – is the most prevalent, this section will focus on the design, selection, and construction process, assuming the second scenario.

Selection Process

The owner hires a professional designer (or Architect on Record) to provide a design service. Similar to traditional DBB delivery, the owner usually uses a qualification-based selection (QBS) to select a professional designer. The Owner's agency should use QBS to choose a designer for a public project because of the Brooks Act.

Once the designer is selected, the owner selects the CM/GC with the designer's assistance. The owner typically chooses one of two methods for selecting the CM/GC: QBS or the Best Value-Fee method.

When evaluating CM/GC firms, the owner needs to check beyond basic qualifications as a general contractor. The selected CM/GC is expected to have specific skills and experience in areas critical to the success of a CMAR project, including:

- Developing comprehensive project schedules
- Preparing accurate construction cost estimates
- Conducting value engineering and life-cycle cost analyses
- Collaborating with subcontractors to generate cost-effective options and assess market and labor conditions
- Familiarity with the local labor and subcontracting market
- Analyzing and comparing alternative designs
- Demonstrating expertise in construction methods and techniques relevant to the project type
- Performing constructability reviews to identify potential design issues before construction
- Conducting interdisciplinary plan checks and reviewing design and CDs
- Coordinating and communicating effectively with various project team disciplines

Table 5.1 Example Selection Criteria in the CMAR Project

Item	*Max Score*
Responsiveness of submittal	10
References	10
Experience, past performances	20
Experience in healthcare facilities	
Experience in higher education projects	
Proposed personnel	10
Proposed approach	10
Past relationship with the owner and design team	20
Proposed fee	10
Interview	10
Total	100

These qualifications enable the CM/GC to provide valuable input during the design phase, improve project coordination, and effectively manage cost and schedule risks. In the selection process, the owner usually follows a two-phase procurement approach: (1) shortlisting the top-qualified firms and (2) selecting the best-qualified firm using selection criteria and evaluation. The owner should define the selection criteria carefully, assigning specific weight to each criterion.

Table 5.1 provides an example of selection criteria that could be used in a CMAR method. This example assumes a best-value approach, which considers both qualifications and fees. Generally, the fee component is less than 30% of the total score, emphasizing qualifications over cost. It's also common to include an interview process for finalists, allowing the owner to assess each firm's approach and compatibility with the owner and the project context.

Design Process

Suppose the CM/GC is selected during the Schematic Design phase. In this case, the designer leads the design process, collaborating with the CM/GC and taking input from the owner. Even in a CMAR project, the owner maintains active involvement and oversight in the design process by closely coordinating with the designer and providing feedback.

The CM/GC regularly updates the cost estimates as the design evolves. The accuracy of these interim cost estimates largely depends on the CM/GC's expertise, including their cost database, as many design variables are still being defined. Some owners request the CM/GC to submit cost estimates at specific design milestones (e.g., 50%, 75% design complete). During this phase, the CM/GC also prepares a preliminary construction schedule and a packaging strategy, working with the design team.

One of the key components in preconstruction is value engineering, a structured approach to improving the design to reduce costs without compromising quality or functionality. Value engineering sessions involve all design teams to achieve design goals at the lowest total cost over the project's lifecycle (including operational and maintenance costs).

In addition to providing constructability feedback, interim cost estimates, and scheduling updates, the CM/GC also leads discussions on risk management. These discussions involve identifying potential risks, developing mitigation strategies, and assigning responsibility for each identified risk, ensuring that proactive measures are in place.

Before the project moves into the CD phase, the CM/GC engages key trade contractors (e.g., MEP subcontractors) to provide "design assist" services. This involvement enables trade contractors to review the design, offer estimates and scheduling insights for their specific trades, and provide feedback that may improve constructability. The CM/GC also defines scope requirements by trade, laying the groundwork for procuring trade contractors. In fast-track projects, certain early construction activities – such as earthwork or structural framing – may begin even before the interior and MEP (mechanical, electrical, and plumbing) designs are fully completed.

During DD, the CM/GC is also responsible for identifying products with long lead times (typically those requiring more than six months lead time). Addressing these items early is critical to avoiding delays during construction. The CM/GC also plays a critical role in securing project stakeholders (i.e., specialty trades and suppliers). They oversee the bidding process, including evaluating proposals, often with the involvement of the design team and the owner. Together, the project team can decide on the best approach for subcontracts. For subcontracted work, the CM/GC typically obtains at least three bids per trade category and interviews at least two bidders. Although the CM/GC holds the subcontract with each trade contractor, the owner often remains actively involved in the selection process, given that subcontract costs directly impact the final project cost.

The CM/GC submits a GMP proposal with a list of assumptions based on the current design documents to the owner to finalize the GMP. This GMP proposal includes a contingency to cover uncertainties that may arise due to the level of design completeness. If the owner accepts the GMP, both parties enter the second phase of the contract – a GMP contract for construction services. It is noted that GMP should be fixed before construction begins. In other words, if the project team wants to use fast-track, they should finalize GMP early.

While the owner generally intends to enter into an open-book Cost of the Work plus Fee arrangement with a GMP with the selected CMAR, they usually retain the option to choose a different CMAR if necessary. For example, if the owner and CM/GC cannot agree on the GMP or if the owner determines that a different selection or bidding process would better serve their interests, they have the right to make that change.

Generally, the CM/GC is expected to provide the following key deliverables during the design phase:

- GMP proposal
- Material sourcing plan
- Subcontractor selection plan
- Risk management plan
- Construction schedule

Construction Process

Once the GMP contract for construction services is made, the CM/GC shifts roles to the general contractor, similar to a traditional general contractor in DBB projects. In this role, the CM/GC provides an experienced professional team, including field superintendents, to coordinate and oversee all construction activities. The CM/GC's responsibilities include, but are not limited to, managing each trade, controlling safety and quality, and addressing all construction-related tasks. Since much of the CM/GC's work in the construction phase mirrors that of a general contractor, this section will focus on cost management.

In CMAR delivery, cost management is a collaborative and "open-book" process in which the CM/GC shares cost information with the owner. In the light of a collaborative and transparent cost management process, the CM/GC needs to inform the owner of any cost issues as they arise, especially if they may lead to a change order. To effectively manage project costs, the CM/GC maintains a "log of pending cost issues," which tracks potential adjustments that could affect the final project cost. This log is typically updated monthly, providing the owner insight into financial impacts.

A cost issues log might include cost items such as design changes, material price fluctuations, and site conditions. Table 5.2 provides an example of a pending cost issue log, which the CM/GC shares with the owner for transparency and collaborative decision-making. With such a log maintained, the CM/GC makes sure the owner is aware of potential financial impacts early on, allowing for timely decisions to adjust the project scope or approve additional funds.

The open-book of CMAR also involves distinguishing between cost issues that fall under contingency versus those that qualify as change orders. This distinction would be critical because each category affects the GMP differently. Change orders adjust the GMP, addressing additional costs due to scope changes or circumstances such as soil conditions. In contrast, the contingency is part of the GMP reserved for unexpected issues

Table 5.2 Example Log of Pending Cost Issues in the CMAR Project

Issue ID	*Description*	*Cost Impact*	*Status*	*Comments*
1.	Design change in HVAC layout	$38,000	Under review	Awaiting engineer's feedback
2.	Steel price increase	$274,000	Approved	Steel price increased by 10% in Q2
3.	Additional site grading required	$165,000	Pending approval	Unexpected soil condition
4.	Change in façade materials	$365,000	Under discussion	Proposed switch to alternative material for durability

within the original scope that don't increase the GMP. While owners often prefer to categorize many pending issues under contingency to keep the GMP unchanged, contractors may advocate for classifying these as change orders, as they increase the GMP. For example, a project requires an unexpected upgrade to fireproofing in the steel framework. In that case, the owner might argue that contingency should cover it, while the contractor may consider it a change order. Negotiation and clear communication are needed to reach an agreement.

Risk Allocation in CMAR

CMAR is one of the popular project delivery methods for owners who want to pursue the integrated approach. This model allows for early contractor involvement in design and a fast-track approach. It also provides budget certainty through a GMP, which sets the project's cost ceiling. However, despite its advantages to the owner, CMAR involves risk allocation among the owner, designer, and CM/GC. Understanding which party assumes responsibility for specific areas is essential for minimizing disputes and effective project delivery.

Design Errors

As discussed in the "Spearin Doctrine" section, design errors would be challenging in CMAR projects due to their collaborative but not fully integrated context in the design phase. It's noted that while the CM/GC provides the preconstruction activities in the design process, the owner retains control over design decisions in collaboration with the A/E. Because of this retained control, the Spearin Doctrine – which holds that the owner warrants the accuracy of the design – may still apply (see the "Spearin Doctrine" section).

When you have such a case in your project, the extent of the CM/GC's involvement and influence over the design can affect their responsibility for design-related issues. To effectively manage design risks in CMAR projects, developing clear and explicit contractual language specifying the CM/GC's design-related responsibilities is essential. As practical advice, if you're acting as an owner's representative, it may be wise to assume that the Spearin Doctrine could apply to your CMAR project.

Cost Overruns

In a CMAR project with a GMP, the CM/GC assumes responsibility for any cost overruns exceeding the GMP unless these costs arise from owner-approved change orders. The CM/GC needs to manage all project costs within the agreed-upon GMP, including construction costs, general conditions, and general overhead. Suppose expenses exceed the GMP due to factors (e.g., labor productivity) within the pre-defined project scope. In that case, the CM/GC absorbs the additional expenses, thereby shielding the owner from unexpected cost increases.

The way project costs are managed (e.g., change orders) in CMAR projects is similar to that in traditional DBB projects. Since CMAR includes separate contracts for design and construction, the CM/GC is accountable for construction-related risks while the designer remains responsible for the quality and accuracy of the design. Suppose that material prices increase unexpectedly after the GMP is set. In that case, the CM/GC must bear this cost escalation unless the contract includes specific provisions for such price adjustments. On the other hand, the owner is typically cautious about any change orders, as these can increase the GMP (or ceiling price).

Let us explore the change order issue, as it is the central subject in cost management. If the owner initiates a scope change, such as upgrading finishes, this request qualifies as a change order, leading to the GMP increase. However, cost issues that arise from circumstances other than owner-directed scope changes or site conditions often pose more complex challenges. These issues can take time to resolve, as the owner may be reluctant to approve change orders. Instead, owners may prefer to classify these issues under the project's contingency, as discussed in the earlier section.

Schedule Delays

From project design through construction and close-out, the CM/GC plays a critical role in managing the project schedule to keep the project on the right track. The master schedule, typically approved by the owner at the beginning of construction, serves as the roadmap or reference when managing the construction schedule. Any schedule changes or updates should be based on the approved master schedule, usually in the form of a critical path method schedule. The CM/GC is responsible for meeting the project milestones outlined in the contract.

To manage the construction schedule, the CM/GC should track construction progress and identify variances between the expected and actual progress. When unfavorable variances (i.e., delays) occur, the CM/GC should take proper action with an updated schedule. These schedule updates are usually communicated with the owner regularly, typically on a monthly basis, to inform the owner of the early signs of potential schedule problems.

In CMAR contracts, schedule delays can have different implications depending on their causes. If delays result from the contractor's inefficiencies or productivity issues, the CM/GC should be responsible for the delays. They may end up paying liquidated damages, compensating the owner for time-related losses. However, if the owner causes delays, such as late approval processes, the CM/GC would request a schedule extension and, in some cases, additional compensation depending on the contract clauses.

For example, in a subway construction project, new restrictions were imposed due to complaints from nearby residents. Specifically, the permitted hours for material deliveries by truck were cut nearly in half, which limited the window for transporting materials. In this case, the CM/GC would document the delay, notify the owner, and negotiate an extension to the project schedule. Depending on the terms of the contract, the owner may grant a time extension but may not always provide additional financial compensation. Therefore, the CM/GC should understand their contractual terms and conditions in terms of schedule performance and delay compensation before construction begins.

Quality Problems

In CMAR projects, quality control during construction is primarily the responsibility of the CM/GC. The CM/GC must proactively manage quality on-site. Quality control includes setting up the quality control plan, supervising the on-site workmanship, and inspecting the completed tasks. At the start of construction, the CM/GC should submit a comprehensive quality control plan for the owner's approval. This plan should address the administrative procedures and technical standards to make sure that construction meets the quality requirements stipulated in the contract documents.

While the owner typically pays for any required testing and independent inspection services, the CM/GC must coordinate with these testing and inspection agencies. Suppose quality tests fail due to poor workmanship or materials provided by the CM/GC or its subcontractors. In that case, the CM/GC may be responsible for the costs of additional testing and corrective measures. Suppose a large commercial building project requires imported soil for site filling. As a result, the CM/GC must take corrective actions. The CM/GC is responsible for the cost of this additional testing and any necessary rework.

The Level of Integration

Throughout this chapter, we have explored the key characteristics and processes of the CMAR projects. We also addressed how various risks are allocated among stakeholders. Now, let's take a look at the level of integration in CMAR, compared to the DBB approach, which is often limited in terms of project integration.

As you can see, CMAR is a cooperative delivery method, but it is short of being fully integrated. In CMAR, the CM/GC is involved early in the design phase, allowing the contractor to collaborate with the owner and designer to offer preconstruction services (e.g., provide constructability study and propose innovative products or methods). The CM/GC may also bring trade partners and major suppliers into the DD phase to support effective value engineering and design review (known as "design-assist" services). However, despite these collaborative elements, CMAR doesn't achieve the same level of integration as more fully integrated approaches like DB or Integrated Project Delivery. Why do you think the CMAR projects have some limitations in terms of integration?

One main reason is the contractual structure in CMAR. The owner makes two separate contracts: one with the architect (A/E) for design services and another with the CM/GC for preconstruction and construction services. This contractual separation indicates that the architect and the CM/GC are not directly bound to work together contractually, which can limit their motivation to align fully on project goals. Although both parties participate in the design process, they may have less incentive to work as a team to achieve the best possible outcomes for the project.

Another limiting factor is the financial risk allocation in CMAR's GMP structure. In a CMAR project with a GMP, the CM/GC bears the financial risk for costs exceeding the GMP. The CM/GC may add a buffer to the GMP (i.e., inflate the price) to protect themselves. Additionally, this risk allocation often discourages the CM/GC from proposing new or innovative products and methods, as these tend to carry higher cost variability since they are usually not fully tested risks in construction projects. Since the CM/GC would have to absorb any unforeseen costs beyond the GMP, they may be less inclined to take unproven solutions that could improve project value or reduce costs.

Therefore, while CMAR offers a more cooperative approach than DBB by involving the contractor early and supporting design collaboration, it does not fully align all parties under a single integrated goal. The owner may find it challenging to achieve innovative or cost-saving solutions, given that the CM/GC is focused on managing financial risk.

Case Study: Biotech Building

Project: New biotechnology building at the Higher Education Institute
Location: Seattle, WA
Budget: $6,200,000 (design and construction)
Project Delivery System: CMAR
Contract Type: GMP Contract

Project Planning

The owner needed an education and research building for the growing biotechnology field, which has expanded rapidly recently. The Department of Biotechnology submitted a request for a new building. With prudent planning over eight months, the owner (a school's project delivery office) went over site selection, programming, and budgeting. The owner has professional staff to oversee the design and construction process. The proposed new building includes the renovation and connection with the existing structure. As with school projects, this project has a firm due date, so speed is essential. Since the project involves a complex biotechnology system that needs close coordination with the user, maintenance staff, and construction contractors, the owner feels that CMAR is the appropriate delivery method for this new project. The owner is concerned about the renovation and connection part of the project, as the as-built documents may not be accurate. A working budget for design and construction was set at $6,200,000 based on prospective needs, with a reserve fund of $600,000 (a reserve fund was not part of the $6,200,000 working budget).

Design Phase and Procurement Phase

The owner hired a local architect for $500,000, and the owner has worked closely with the designer. The owner proceeded with the selection process to choose the CM/GC using the best value fee method with advice from the designer. Three general contractors were invited to the final phase, which included proposal submission and an interview. Finally, the owner selected one general contractor and made a two-stage agreement with the firm. Details are as follows:

- The first phase of the contract (a lump sum of $250,000) is for preconstruction services.
- The second phase of the contract is for construction services.
- GMP is to be set after a design is 70% complete.
- The owner does not have to enter into the second phase of the contract if GMP exceeds the owner's budget.
- CM/GC and the owner agreed to adjust the scope of work if the GMP exceeds the owner's budget.
- The CM/GC agreed to fast-track the project and to take responsibility for updating the cost information during construction as the design is completed.

In the first phase of the contract, the owner participated in design coordination meetings in which the CM/GC provided an interim estimate, expected schedule updates, and a design review. The CM/GC submitted a GMP proposal of $5.8 million, including a contingency of $300,000, at about 70% design completion.

The parties negotiated the price and work scope, but did not come to an agreement. The owner was concerned about the schedule tightness.

Case Study Questions

Question 1. (Planning) Was the Owner's choice of project delivery method appropriate? Defend your answer.
Question 2. (Planning) If the owner selected a DB method, what would prevent the owner from implementing his plan?
Question 3. (Design and Procurement) Do you think the level of the owner's involvement in design seems appropriate?
Question 4. (Design and Procurement) Should the owner select the trade contractors, or should the CM/GC select the trade contractors?
Question 5. (Design and Procurement) If both parties cannot agree on GMP, what should the owner do as the next step to complete the project?

Chapter Summary

1 In the CM/GC delivery method, the owner makes two separate contracts: one with the designer and another with the CM/GC. The contract with the CM/GC itself is divided into two distinct agreements

in different phases: one for preconstruction services and the other for construction services.

2 Although the CMAR method follows a two-phased delivery process, the timing of the CM/GC's involvement can influence their role and the project's outcomes. Typically, there are three scenarios regarding when the CM/GC (or GC/CM) might be selected and brought into the project: (1) CM/GC's involvement even before the architect is selected, (2) CM/GC's involvement when the design is up to 30% complete, and (3) CM/GC's involvement after the design is more than 50% complete.
3 In general, the CMAR delivery method goes with a GMP contract. The key feature of a GMP contract is the ceiling price, which sets a maximum limit. If the actual final costs come below the GMP, most owners are usually willing to share some savings to incentivize the contractor to work under budget.
4 The CMAR model enables early contractor involvement and allows for a flexible approach to contracting and scheduling, making fast-tracking a practical strategy. However, the GMP should be finalized before the design is complete for the fast-track approach.
5 In CMAR projects, concurrent engineering becomes effective when the CM/GC is engaged early in the design phase. With the CM/GC involved at this stage, they can invite trade contractors and major suppliers to participate in the design discussions. This early collaboration allows for the simultaneous development of the product and process designs.
6 Considering the owner still controls the design process in many CMAR projects, the Spearin doctrine is still applied to most CMAR projects.
7 Once the designer is selected, the owner selects the CM/GC with the designer's assistance. The owner typically chooses one of two methods for selecting the CM/GC: QBS or the Best Value-Fee method. In the selection process, the owner follows a two-phase procurement approach: (1) shortlisting the top-qualified firms and (2) selecting the best-qualified firm using selection criteria and evaluation.
8 During the design phase, the CM/GC regularly updates the cost estimates as the design evolves. In addition, the CM/GC provides various preconstruction services: value engineering, procurement plan (e.g., subcontractor selection plan), risk management plan, and construction schedule. The CM/GC also submits a GMP proposal with a list of assumptions based on the design documents to the owner to finalize the GMP.
9 Once the GMP contract for construction services is made, the CM/GC shifts roles to the general contractor. The CM/GC's responsibilities during the construction phase include, but are not limited to, managing each trade, controlling safety and quality, and addressing all construction-related tasks. In the construction phase, cost management

is a collaborative and "open-book" process in which the CM/GC shares cost information with the owner. The CM/GC needs to inform the owner of any cost issues as they arise, especially if they may lead to a change order.

10 In CMAR projects, various risks are allocated to different stakeholders. For design errors, the Spearin Doctrine may still apply since the owner retains control over design decisions. You'd better develop explicit contractual language specifying the CM/GC's design-related responsibilities. Under the GMP contract, the CM/GC assumes responsibility for any cost overruns exceeding the GMP unless these costs arise from owner-approved change orders. Quality control in CMAR projects is primarily the responsibility of the CM/GC.

11 CMAR is generally considered a cooperative delivery method, but it is short of being fully integrated. The CM/GC may bring trade partners and major suppliers into the DD phase to support effective value engineering and design review. Despite these collaborative elements, CMAR doesn't achieve the same level of integration as more fully integrated approaches. The architect and the CM/GC are not directly bound to work together contractually, which can limit their motivation to align fully on project goals. In a GMP contract, the CM/GC bears the financial risk for costs exceeding the GMP. This risk allocation may discourage the CM/GC from proposing new or innovative products and methods.

Review Questions

1 Please respond to the following questions.

1.1 Which of the following is not characteristic of CMAR compared to DBB? Choose only one answer choice

a GMP is a prevailing method of contract
b CMAR is suitable for complex projects
c A Low Bid is usually used as a selection method
d Value engineering would be easily facilitated

1.2 What is the condition for the owner to leverage fast-track in the CM/GC project?

1.3 Which of the following is not a major reason to choose a CM/GC method? Choose only one choice

a The owner's ability to provide input on the selection of specialty trades
b The project can leverage the contractor's expertise in new technology and constructability

   c The owner can choose a CM/GC contractor based on their own criteria rather than on the low price
   d CM/GC lowers total construction costs compared to the tradi tional DBB method.

2 Do you think that CM/GC with GMP fosters innovation or adopting new technology? Why? Defend your answer.
3 Below is a short description of a new sewage treatment plant project. Please respond to the questions at the end of the scenario.

   The City of Highland Falls plans to construct a new sewage treatment plant. Since the city lacks a construction staff, they hired a CM/GC. The city had already hired a design firm to prepare design concepts for the new plant, but the concepts were incomplete. You are asked to recommend a procurement strategy.

   3.1 When do you recommend that the city hire a CM/GC? Why did you select this time?
   3.2 The city used a two-step procurement procedure for this project. The first step is to request prospective contractors to submit their qualifications. Based on the evaluation of the contractor's qualifications, the city will select the five top-ranked firms to submit their bids or proposals. What specific qualification information do you recommend the city require prospective contractors to submit in response to the request for qualifications (RFQs)?

4 A financial group plans to build a new local campus in Seattle, WA. The proposed project is a 15-story office building with a steel structure and a curtain wall facade. The owner initially engaged a local architectural firm to develop the design.

   Then, the owner brought TXX, a GC/CM, to provide preconstruction services when the design was around 70% complete. TXX's responsibilities included conducting a design review during the CD phase (e.g., constructability study), developing an interim schedule and estimate, and identifying project risks. TXX also needed to secure key subcontractors and suppliers to mitigate procurement risks, particularly for the steel structure and curtain wall. With the design approximately 90% complete, still in the CD phase, the owner and TXX entered into a GMP contract.

   4.1 What would be the CM/GC's primary role in the preconstruction phase?
   4.2 Why did the owner choose the CM/GC so late in the design process?
   4.3 Do you think the project could have benefited from a fast-track approach?

5 Below is a new hospital facility where a typical CM/GC delivery method was applied. Please respond to the questions at the end of the scenario.

A medical group planned to build a new hospital facility in Seattle, WA. Since a new facility will be located adjacent to the current hospital in use, tight and close coordination is required. In addition, the Owner wants to get his facility management team engaged in the design phase. The owner has its own project delivery group with its architects and project managers. The owner's project delivery group determined to utilize a CMAR project delivery method.

The owner hired an architect to prepare designs for the construction of a new hospital facility. When the designs were partially completed, the owner issued an RFQ and Request for Proposal to select CM/GC on the best value fee basis. Three contractors were invited to submit a proposal. The contract for preconstruction service was made. During the design phase, the designer held a weekly design coordination meeting. The design coordination meeting involves the designer, the owner's group, and the CM/GC. The CM/GC's senior estimator updated the budget (expected construction costs) in the meeting. The design team held two value engineering sessions in which every stakeholder brought their own idea of cost savings. In the middle of the DD phase, the CM/GC invited a local mechanical contractor and EC for design assist service.[6]

At the end of the DD phase, the owner and the CM/GC negotiated GMP, which included a 5% contingency. The CM/GC proposed a 10% contingency based on the uncertainty of design documents, but the Owner rejected and counteroffered 5%. Finally, they agreed to a 5% contingency. The CM/GC then entered into a series of subcontracts with trade contractors, including electrical work on the project. The CM/GC used a Lump Sum contract when he made subcontracts.

5.1 (Planning) Was the owner's choice of project delivery method appropriate? Defend your answer.

5.2 (Design and Procurement) Why do you think the CM/GC used a Lump Sum contract when they made subcontracts?

Notes

1 Darrington, J. W. and Lichtig, W.A. (2010). Rethinking the "G" in GMP: Why Estimated Maximum Price Contracts Make Sense on Collaborative Projects. *The Construction Lawyer*, 30(2).

2 McCree & Co. v. State, 91 N.W.2d 713, 714 (Minn. 1958); see also United States v. Spearin, 248 U.S. 132, 136–137 (1918).

3 *Coghlin Electrical Contractors, Inc. v. Gilbane Building Co., 472 Mass. 549, 36 N.E.3d 505 (2015).*
4 RCW 39.10.340, https://app.leg.wa.gov/rcw/default.aspx?cite=39.10.340.
5 RCW.39.10.260, https://app.leg.wa.gov/rcw/default.aspx?cite=39.10.260.
6 Design-assist is a term describing a method to support early involvement of mechanical and electrical contractors. It is a process that allows the mechanical and electrical contractors to assist the design team through consultation with out taking responsibility for the design. They provide information to the design team on details, fabrication, installation, and potential suppliers.

Bibliography

Anderson, S. D. and Oyetunji, A. (2003). Selection Procedure for Project Delivery and Contract Strategy, ASCE Construction Institute. *Construction Research Congress* 2003, Oahu, HI, March 19–21.

Associated General Contractors of America (AGC) (2004). *Project Delivery Systems for Construction.* AGC, Arlington, VA.

Construction Industry Institute (CII) (2001). *Project Delivery and Contract Strategy.* Publication RR165-12, Austin, TX.

Darrington, J.W. and Lichtig, W.A. (2010). Rethinking the “G” in GMP: Why Estimated Maximum Price Contracts Make Sense on Collaborative Projects. *The Construction Lawyer*, 30(2), 29–39.

Minchin, E., Ptschelinzew, L., Migliaccio, G., Gatti, U., Atkins, K., Warne, T., Hostetler, G., and Asiamah, S. (2014). *NCHRP Report 787: Guide for Design Management on Design-Build and Construction Manager/General Contractor Projects.* Transportation Research Board, Washington, DC.

Palihakkara, A. and Perera, B. (2022). Risk Management in Guaranteed Maximum Price Contracts. *Smart and Sustainable Built Environment*, 11(2), 368–386.

Chapter 6

Design Build (DB)

This chapter will explore the Design-Build (DB) project delivery method, an integrated approach where a single contract addresses design and construction services. This method stands in contrast to the traditional Design-Bid-Build (DBB) method, where the owner manages separate contracts for design and construction, resulting in a fragmented delivery process. While the Construction Manager at Risk (CMAR) approach promotes collaboration during the design process with the CM/GG's early involvement, the level of integration and collaboration in the CMAR should be limited due to its organizational structure.

In any construction project, the primary stakeholders include the owner, the designer (architect or engineer), and the contractor. The relationships between these participants, as well as the roles and responsibilities, are usually defined by the project delivery method. Typically, owners select a delivery method based on the level of control they want to retain over the design process and the risk they are willing to assume.

Did we mention single responsibility? What does the single responsibility mean in terms of project management? The DB method combines design and construction responsibilities in a single contract. A design-builder is accountable for both project phases. This approach offers several advantages, such as reduced project timelines (i.e., aggressive fast-track approach) and streamlined decision-making. In a typical DB setup, the owner defines the project requirements and then selects a design-builder capable of executing both the design and construction phases.

Within the DB model, there are several variations, allowing owners to choose a specific type of DB arrangement that aligns with their desired level of control over the design and risk tolerance. For instance, in a "bridging" DB model, the owner may hire a designer (often called a "bridge architect") to develop the design to approximately 30% completion before selecting a design-builder to finalize the design and carry

DOI: 10.1201/9781003628842-9

out construction. In conventional cases, the owner may set project requirements and then engage a design-builder to create a design that meets these specifications. This chapter will further explore the different types of DB models and the circumstances under which each might be most beneficial.

The evolution of DB reflects a growing need for collaboration and integration among the project stakeholders in project delivery. Historically, the separation of design and construction began in the mid-1800s as projects grew complex, requiring specialized expertise. Before this, projects were delivered by "master builders," who took responsibility for the entire process, from design to construction. DB returns to this "master-builder" concept in many ways, providing a single responsibility for the project.

While initially popular in the private sector, public owners were slower to adopt DB due to procurement laws that mandated the separation of design and construction services. However, the public sector's interest in DB projects grew over time, particularly after the U.S. Congress passed the National Defense Authorization Act of 1996 (also known as the Clinger-Cohen Act), which allowed federal agencies to use a two-phase selection process.[1] This legislation opened the door for broader use of DB in the public sector.

Today, DB is used for various projects, from commercial buildings to infrastructure projects like highways. This delivery method offers several advantages for owners:

- Single-source responsibility for both design and construction
- Early development of the project budget
- Reduced project delivery timelines
- Promoted collaboration

While DB offers numerous advantages, it may not suit all projects. Owners who wish to retain control over design details may find traditional DB challenging, as design or requirement modifications can become costly once the DB contract is made. In such cases, these owners might prefer a progressive Design-Build (PDB) approach, which allows them to retain more influence over the design process.

In this chapter, we will explore the types of DB models, the major attributes and delivery process of the DB delivery method, the cases where DB is effective, and risk allocation. By the end of this chapter, readers will have a comprehensive understanding of the DB method and will be ready to use the DB method as practitioners.

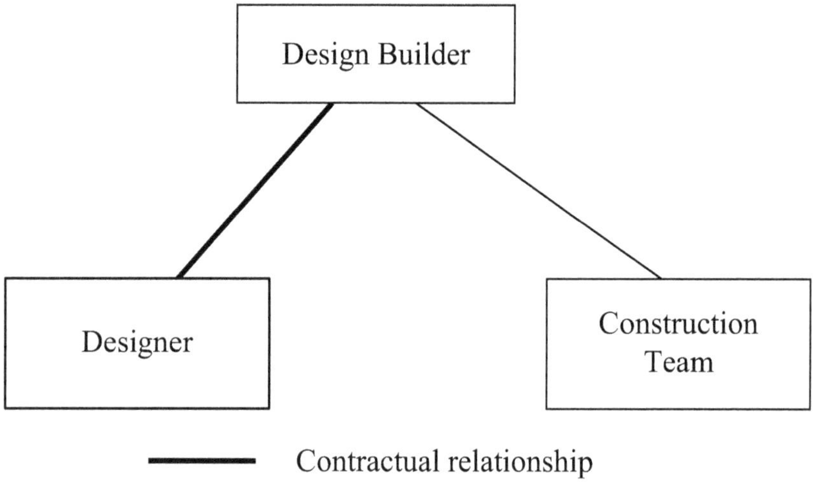

Figure 6.1 Contractor-Led Design-Build Organization.

Types of DB Methods

Classification Based on DB Entity

This section will address several ways to classify DB methods. The first classification approach is based on the DB team structure. We can categorize DB entities into four main types:

Contractor-Led Team

This is the most common DB structure, especially in heavy civil and building construction projects. In this model, a general contractor takes the lead, entering into a contract with the project owner for both design and construction services. The general contractor hires a design firm as a subcontractor to lead the design phase, as shown in Figure 6.1.

Designer-Led Team

In contrast to the contractor-led model, the design firm takes on the contract with the project owner and subcontracts the construction work to a contractor, as shown in Figure 6.2. Designer-led organizations are often spearheaded by engineering firms on industrial projects such as utility plants rather than architectural firms, as engineering firms tend to be better capitalized and less risk-averse in this context.

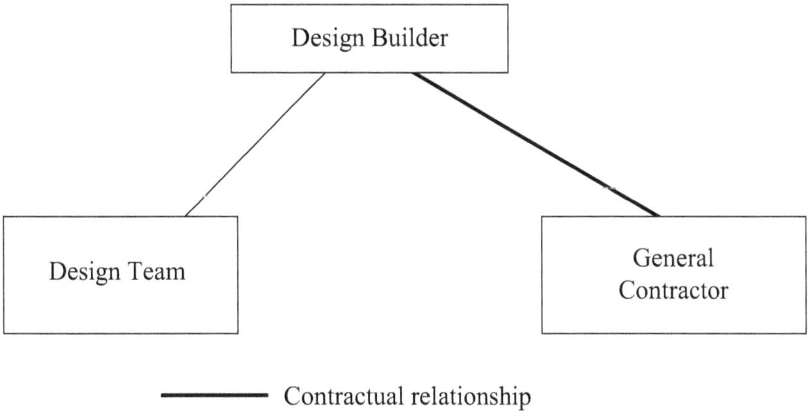

Figure 6.2 Designer-Led Design-Build Organization.

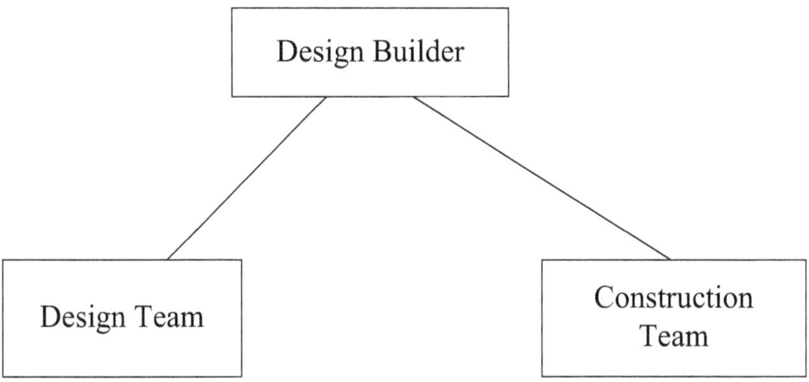

Figure 6.3 Integrated Design-Build Organization.

Integrated DB Firm

An integrated DB firm combines design and construction services under one roof, with licensed personnel capable of performing both roles, as shown in Figure 6.3. Such an integrated DB firm has a construction wing as well as a design wing. This type of organization provides a high level of coordination, as both design and construction functions are managed by a single entity, often leading to efficiencies in collaboration and communication.

Joint Venture

A joint venture involves multiple companies teaming up to deliver a project. This structure resembles an integrated DB firm, as it brings

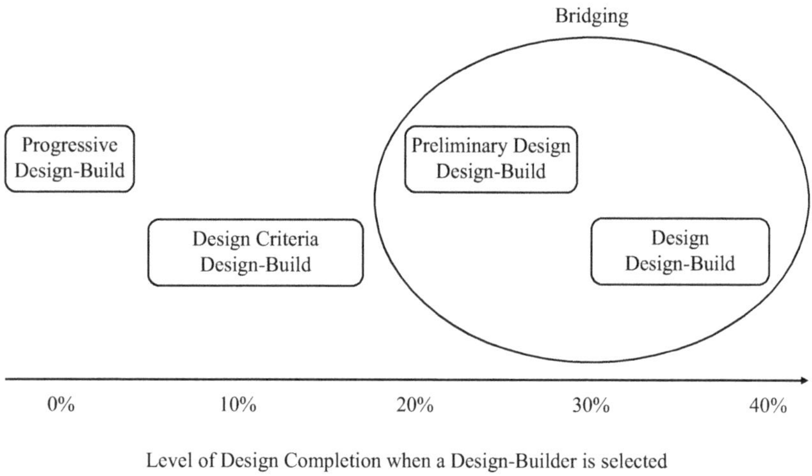

Figure 6.4 Design-Build Classification (the Level of Design Completion).

together the skills of both design and construction firms. Joint ventures can involve multiple construction firms and one or more design firms, and they may be formed either for a single project or for pursuing multiple projects.

Each DB structure provides unique advantages depending on the project requirements and the level of collaboration required. By understanding the characteristics of each, owners may prefer a specific type of design-builder that aligns with their particular needs. For example, the owner may prefer a contractor-led design-builder if the project seems more complex and requires expertise in site risk management.

Classification Based on the Level of Design Completeness

DB can also be classified according to the level of design completion at the time a design-builder is solicited. Figure 6.4 illustrates the different types of DB contracts based on the extent of design provided in the Request for Proposal (RFP).

Progressive DB

One of the owner's concerns with the DB approach is losing control over design development (DD). So, how can the owner retain their influence over the design process while still using DB?

PDB addresses this issue by involving the design-builder in the early design phase. Their early engagement enables the design-builder to develop the design while the owner still keeps communicating with the design-builder. In PDB, the design-builder works directly with the owner's team from the early design phase, providing the owner with an opportunity for their input and feedback. Unlike traditional DB, where the lump sum contract is dominant, the owners prefer the GMP contract for PDB projects. The GMP in PDB is typically established when about 40%–60% of the design is complete. This timing allows the owner to work closely with the design-builder to develop and refine the design, while allowing the design-builder to use a fast-track approach.

With PDB, the owner retains control over DD until the final price (i.e., GMP) is locked in. This approach allows for a collaborative design process, allowing the owner to influence design details while maintaining a single contract for both design and construction.

Design Criteria DB

In the Design Criteria DB approach, the owner typically engages knowledgeable professionals or design consulting firms to assist with facility planning, programming, and developing design criteria for the RFP documents. For effective design criteria development, it's essential to understand the owner's needs in terms of performance.

Design criteria generally outline the minimum performance requirements, with certain design limitations, allowing the design-builder to control the overall design process. This method enables design-builders to introduce innovative solutions that can improve project value for the owner.

Bridging

Many owners who employ a DB delivery method for the first time are concerned that they may lose control of the design; the delivered results may differ from what they want. In the DB method, "Bridging" is a project delivery approach where the owner retains more control over the initial design stages by hiring a "bridging architect" before hiring a design-builder. There are two types of bridging: (1) Preliminary Design DB and (2) Design DB.

Preliminary Design DB (Bridging)

Owners can develop a preliminary design before soliciting proposals from prospective design-builders. In this approach, the owner advances the design to a level of design completion, typically around 20%–25%, before issuing an RFP. This preliminary design work can be conducted by the owner's in-house staff or external design consultants called "bridging architects."

The preliminary design DB approach results in a more prescriptive RFP with well-defined drawings and specifications. In this method, the owner's designer (e.g., bridge architects) establishes the preliminary design, which the design-builder is then responsible for completing the design and construction. The bridging documents included in a DB RFP may consist of partially completed (25%–40%) architectural, mechanical, and electrical drawings. While these documents set the primary design direction, variations are generally discouraged in proposals. However, if a proposed modification can improve project value for the owner, some flexibility is permitted for the design-builder to suggest creative improvements to the preliminary design. While preliminary design DB allows for a certain level of innovation, the scope for creative input from the design-builder is more limited compared to other DB types like PDB or Design Criteria DB.

Design DB (Bridging)

When a bridging architect or designer develops more than 35%–40% of the architectural, mechanical, and electrical drawings before soliciting proposals, this process is referred to as design DB. Often called "Draw-Build," this approach involves highly detailed bridging documents in the RFP, significantly reducing the opportunity for the design-builder to introduce creative solutions.

In the Design DB, the bridging documents limit the design-builder's flexibility to create design solutions, as the design variables are already defined. Therefore, with limited scope for design input, the proposal evaluation usually assigns a substantial weight to price. This approach often shifts focus to cost, as the design elements are mostly pre-determined. Therefore, the owner may expect the design-builder to execute the established design within budget constraints. Due to its restrictive context, some professionals and researchers do not consider design DB a true form of DB, as it lacks the integrated design and construction development.

As discussed in the preceding sections, DB RFPs can vary widely in terms of design completeness. Some RFPs incorporate minimal design documentation, enabling design-builders to leverage their creativity and innovation. Others, like bridging, employ extensive "bridging" documents to have the owner's intention reflected in the design documents, giving the owner greater control over the design outcome.

Major Attributes of DB Project Delivery

Concurrent Engineering

As we learned in the previous chapters, concurrent engineering is a holistic approach to the design and preconstruction process, where product and process design progress simultaneously. In the CMGC delivery method, concurrent engineering is possible if the CMGC is engaged early, typically before 30% of the design is complete. However, because the CMGC's role in DD is limited due to their contractual relationship, their input on the process design (i.e., procurement or constructability) and their integration with the design process may be limited. By contrast, in the DB delivery method, the design and construction teams work together throughout the project lifecycle, promoting a more integrated approach to concurrent engineering.

For concurrent engineering to be effective in the DB model, the design-builder needs to engage key subcontractors as early as possible. Suppose that during the development of an HVAC (Heating, Ventilation, and Air Conditioning) system, the mechanical subcontractor, electrical subcontractor, designers, and general contractor are all seated together in the design phase. They can discuss structural and mechanical, electrical, and plumbing (MEP) design, as well as the logistics and installation methods for the HVAC system. This early collaboration allows the team to identify potential issues, such as space limitations for HVAC ductwork. Through such a concurrent engineering process, the team can avoid costly design changes or delays later in the project.

The level of concurrent engineering achievable depends on the type of DB project. For example, bridging methods, such as Preliminary Design DB or Design DB, may limit concurrent engineering opportunities because most design variables are determined before a design-builder is fully engaged. In contrast, models that involve the design-builder early in the process (e.g., Design Criteria DB or Progressive DB) provide more opportunities for concurrent engineering.

Single Responsibility and Spearin Doctrine

In the DB delivery method, single responsibility indicates that a single entity (i.e., the design-builder) is contracted to provide both design (architecture and engineering) and construction services, regardless of the form of the design-builder (e.g., a contractor-led design-builder). In a traditional project delivery approach, the owner typically hires a designer (i.e., an architect and/or engineer) to develop the project's drawings and specifications. Once these are complete, the owner contracts with a general contractor through competitive bidding or negotiation for the construction service. This creates two separate contracts, requiring the owner to coordinate between the designer and contractor, addressing issues that arise between the two functions (i.e., the design and the construction).

In contrast, in the DB projects, the owner interacts with just one entity that oversees both design and construction. This single contract structure places full accountability on the design-builder for managing project performance, including quality and safety. For the owner, this results in simplified communication, faster response times, and expedited project delivery, as they no longer serve as a "middleman" coordinating between design and construction.

Do you think that single responsibility relieves the owner of the Spearin doctrine (i.e., the guarantee of the design document)? Traditionally, under the Spearin Doctrine, if the owner provides the design, they implicitly guarantee its completeness and suitability. In DBB projects, this doctrine means that if a contractor builds according to the owner-provided design and issues arise, the owner may bear responsibility for those issues.

However, the Spearin Doctrine's applicability in DB projects is more complex due to the single responsibility structure and different contexts depending on the type of DB:

Progressive DB

The owner retains influence over the design process. This level of involvement and collaboration during the design process makes it challenging to determine how the Spearin Doctrine is applied. The degree of the owner's responsibility may depend on their specific level of control and involvement in the design process, which can differ from project to project.

Design Criteria DB

A design-builder is responsible for design and construction, so the owner is not held accountable for design errors. The Spearin Doctrine generally does not apply because a design-builder assumes full accountability for the design unless the error results from the misleading or inaccurate contents in the RFP.

Bridging (e.g., Preliminary Design DB or Design DB)

In bridging methods, the owner provides partially or substantially completed design documents, such as the bridging documents, which outline key design aspects. Since these documents are owner-provided, the owner may still bear responsibility for their completeness and accuracy. The Spearin Doctrine could apply to the bridging documents, as a design-builder relies on these as a starting point and may not be held accountable for issues arising from the bridging documents.

While the DB method's single responsibility structure often limits the relevance of the Spearin Doctrine, certain DB variations (i.e., Bridging or Progressive DB) may involve owner-provided design elements. In these cases, the owner's liability under the Spearin Doctrine may still be effective, depending on their level of involvement and influence.

Fast-Track

Fast-track is a project delivery approach that allows construction to begin before the entire set of construction documents is complete. This approach is discussed in the CMAR chapter, where it's noted that although fast-tracking is possible in CMAR, it is limited by the need to establish the GMP before construction can begin. In CMAR projects, the GMP acts as a safeguard for the owner, so construction can't start without a finalized contract. After all, who would start building without an agreed contract amount?

However, the contract with the design-builder is usually made early in the design phase. Since the design-builder is responsible for design and construction, the project can move to the construction phase without waiting for a fully developed design or a separate construction contract. This allows for an easy transition from design to construction, even if the design of some systems (e.g., HVAC) is still being refined during construction. For example, suppose an HVAC system has yet to be fully detailed. Under the fast-track approach, civil works – such as site preparation, foundations, and structural framing – can begin while the HVAC design is not finalized. By the time the project reaches the stage where HVAC design is needed (e.g., selecting a supplier), the detailed design for the HVAC system should be ready.

The ability to fast-track is one of the key advantages of the DB delivery method, particularly for large public projects where time is a critical factor. By compressing the project timeline, fast-tracking minimizes disruption to the public, leading to savings on overhead and social costs. For these reasons, fast-tracking is often a primary consideration for public sector clients who prefer the DB method over traditional DBB or CMAR methods.

Lack of Owner's Control over Design

The owner typically has limited control over DD in the DB method. The design-builder has primary responsibility for design and construction. Once the contract is awarded, the owner has limited involvement in guiding the design, aside from specifying requirements in the RFP. This limited design control can raise concerns for owners, especially regarding design quality. The level of owner control varies depending on the specific type of DB approach:

Progressive DB

Although the design-builder is involved early, the owner retains more control over DD in this model. The owner collaborates with the design-builder throughout the design phase until the GMP is set. This ongoing involvement allows the owner to provide feedback and make adjustments during DD.

Design Criteria DB

It is a typical DB. The owner has limited control over the design. Therefore, the owner should carefully develop the RFP, where the design and performance criteria should be explicitly provided.

Bridging DB

This model gives the owner control by allowing them to engage a bridging architect to develop preliminary design documents before selecting the design-builder. For example, in Design DB (sometimes called

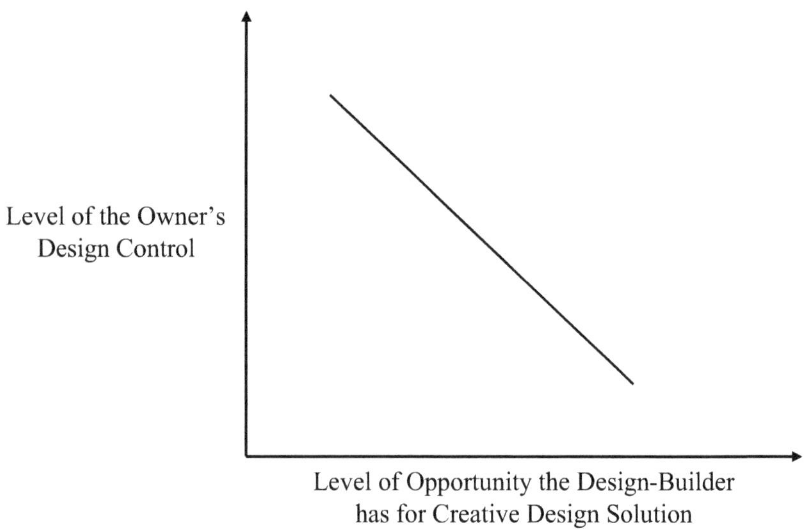

Figure 6.5 Relationship Between the Owner's Design Control and Creative Design Opportunity.

"Draw-Build"), the bridging architect may complete more than 35% of the design, giving the owner a more substantial influence on the initial design direction. However, once the DB team is engaged, further changes to the design are limited.

Generally speaking, the more control the owner has over design, the less opportunity or flexibility the design-builder has to propose creative or innovative solutions (Figure 6.5). In summary, while the DB method provides a single point of responsibility and limits the owner's control over design, owners must carefully consider how much design control they need.

Does DB Fit Your Needs?

In the planning stage, you, as an owner, should determine which delivery method is appropriate for your project. This section will explore how the owner can decide if the DB is the appropriate delivery method.

Define the Goals of the Project

The first step in the planning phase is for the owner to define the project's goals. The following questions are good starting points for defining the project goals.

- What does the owner intend to do with the project once it has been completed?
- Does the owner have special needs that may impact project design and construction?

Determine Owner's Ability

Before discussing the owner's ability to manage a DB project, the owner should check if they are qualified to use the DB delivery method. Similar to the case of the CMAR method, each state has an official procedure to allow the public agency to use the DB. For example, a public body in Washington State should apply for certification to utilize DB project delivery.[2] Once certified, a public body may use the contracting procedure for which it is approved on individual projects without seeking committee approval for three years.

In addition to legal qualification, the owner must decide whether or not they have the capability to craft solicitation documents internally or must hire a professional construction management firm or a DB consultant to take the lead in developing the needed documentation. If an external consultant firm is required, the owner typically solicits proposals from qualified consultants and selects the consultant with the expertise to guide the development of the owner's solicitation documents.

Project Requirements and the Owner's Need

The owner must determine the project requirements. This includes identifying all of the functions that the project must be capable of performing and specific performance requirements for each function. Operational and quality requirements must also be established for each function. The context for the project must also be established. What are the site characteristics, environmental restrictions, and aesthetic considerations that must be considered in designing and constructing the project? Site surveys and geological investigations of the project site are essential to establish the context in which the project will be executed. More importantly, the owner should decide how much control they wish to exercise over the design of the project. Is the owner willing to allow the contractor the freedom to develop the design as long as it meets the project program requirements?

The next issue to check is the size of the project. Is the project of sufficient size to be attractive to prospective design-builders? The size of the project must be sufficiently large to justify the proposer's cost and risk in developing a proposal. In other words, does the potential return justify the risk and effort to develop a proposal? Can the cost of developing a comprehensive proposal be justified by the potential profit from the contract for designing and constructing the project? If so, DB is an appropriate project delivery method. If the answer to any of these questions is no, the owner is advised to consider other project delivery methods.

Market Condition

The owner should be confident that there are multiple competitive design-builders, even though the project and owner are suited to DB. In addition, there should be competitive MEP design-builders available in the market who can work on your project.

Compensation Structure

The compensation structure often consists of two key factors: the level of design completion (or uncertainty) and the owner's risk tolerance at the time of contracting. The compensation structure will affect the commercial interests of both the owner and the DB team, influencing the contractor's incentives for cost savings or overruns. We will explore how different types of DB projects align with various compensation structures:

Progressive DB

The Progressive DB usually allows for design flexibility. A Guaranteed Maximum Price (GMP) structure is commonly used in this model, as it enables the owner to make design adjustments during the project. Typically, the GMP is established once the design reaches around 40%–60% completion, at which point much of the initial uncertainty has been resolved.

This compensation structure benefits both parties. The GMP structure enables the owner to make design changes till the GMP is fixed while protecting the owner from cost overruns once the GMP is in place. On the other hand, the design-builder is motivated to manage costs to stay within the agreed limit (i.e., the GMP).

Design Criteria DB

In the Design Criteria DB, the owner provides performance requirements and design criteria at the outset, guiding the design-builder in developing a design that meets these specifications. Since the design is partially defined in the RFP stage, the compensation structure often reflects the moderately low level of design uncertainty and the owner's desire for cost control.

A Lump Sum is a dominant contract type for the Design Criteria DB projects. Still, a GMP structure is sometimes used if the owner is not confident about the pricing at such a level of design uncertainty. Lump-sum contracts offer cost certainty for the owner and are appropriate when the scope is well-defined through the owner's initial design criteria.

On the other hand, GMP contracts are sometimes used, providing a cap on costs while allowing some flexibility for the design-builder to meet performance criteria. This structure protects the owner from cost overruns, while any savings can be shared between the owner and the design-builder.

Bridging (Preliminary Design DB and Design DB)

With a more defined project scope resulting from the bridging documents, a lump-sum contract is commonly used as the compensation structure. This approach provides cost predictability for the owner, as most design-related risks are transferred to the design-builder. The bridging documents help reduce uncertainty, enabling the design-builder to commit to a more accurate cost estimate at the outset. The owner has substantial control over design through detailed specifications in the bridging documents. Therefore, the owner's financial risk in pricing is low in Bridging DB.

Project Delivery Process

Selection Process

Owners often prefer a Best-Value method to balance price and qualifications in DB projects. However, in Progressive DB, owners prefer to use Qualification-Based Selection (QBS), focusing primarily on the qualifications of the design-builder.

Best-Value and QBS selection methods typically follow a two-step procurement process for public sector projects. Given the prevalence of this approach in the industry, this section will assume that the owner is using a two-step procurement process. The first step involves soliciting Statements of Qualification (SOQs) from prospective design-builders. Based on the evaluation of the qualification information, the top-ranked firms, usually 3–5, are short-listed and given the RFP. Proprietary meetings may be conducted between the project owner's staff and the short-listed design-builders to help the short-listed firms understand the project requirements and to obtain feedback from the owner on their preliminary design concepts.

Request for Qualification Process

The first step in the two-step procurement process for a DB contract typically is the issuance of the Request for Qualification (RFQ). The RFQ is a formal request for the necessary and desirable qualifications for design-builders who wish to be considered for the competitive proposal process. The RFQ usually includes the following information:

- Owner name and contact information
- Project description
- Estimated construction cost
- Competition time frame and key dates
- Information required to be submitted in the SOQ
- Evaluation criteria and process to be used for evaluating SOQ
- Format for SOQ, if the owner has a preference
- Location and date for submission of SOQ

In developing the RFQ, the project owner or a DB consultant must determine what information is needed to select the best-qualified design-builder for the project. Typical information requested in the SOQ includes:

- DB experience of the firm and the proposed project team
- Firm and project team experience with similar projects

- Design team members, including engineering consultants
- Proof of bonding capacity
- Financial strength
- Resumes of project team members
- Design and construction management approaches
- Organization charts
- Safety records

The information requested in the SOQ must be defined explicitly in the RFQ. The SOQ evaluation may weigh individual elements equally or allocate more weight to selected information.

Like other owner agencies, the California Department of Education[3] has detailed procedures for the pre-qualification and selection of DB contractors. The agency uses a standard questionnaire to obtain information on prospective design-builders. Information requested in the questionnaire includes:

- The contractor's license to offer construction services, the amount of liability insurance coverage purchased by the contractor, and audited company financial statements
- The organization's history, organizational performance, and compliance with civil and criminal law
- Ability to bond projects and the company's safety record
- Recent projects completed

RFP Process

Based on the evaluation of the individual SOQs, the short list of best-qualified design-builders is determined, and the shortlisted firms are issued the RFP. To ensure an adequate pool of qualified design-builders, the project owner may need to encourage qualified firms to submit their SOQs. One way to promote design-builder participation is to offer a stipend to unsuccessful proposers who are not selected for the award of the DB contract. If only one or two SOQs are received, the owner must review them to verify that the design-builders possess the requisite qualifications. If the answer is yes, then the RFP would be issued to both. If the answer is no, the project owner may terminate the solicitation and consider an alternative project delivery method.

The development of the RFP for a DB project requires considerable effort. Many project owners hire a DB consultant to assist in RFP development. The RFP needs to contain all of the information that a design-builder needs to be able to develop a proposal and specify what information is to

Table 6.1 Comparison of RFP Requirements

California Department of Education[a]	*United States Air Force*[b]	*United States Navy*[c]
Educational specifications Project description Site requirements Design standards and space requirements Geotechnical information Budget parameters Submission documentation requirements Selection process	Project functionality requirements Project description Site requirements Design standards and space requirements Contract management requirements Submission documentation requirements Selection process	General requirements Project description Site analysis Design standards and space requirements Performance specifications Prescriptive specifications Submission documentation requirements Selection process

[a] California Department of Education (2002). *AB 1402: Design-Build Projects Guidelines.* Sacramento, CA.
[b] U.S. Air Force (2012). *U.S. Air Force Design-Build Handbook.*
[c] Whole Building Design Guide (2003). *Naval Facilities Engineering Command Design-Build Request for Proposal.*

be submitted in the proposal. As a minimum, the RFP should contain the following information:

- Owner name and contact information
- Overall project description, budget, and schedule
- Proposal time frame, including any pre-proposal meetings and interviews
- Proposal submission requirements and honorarium or stipend (if any)
- Proposal selection procedure and weighted criteria
- Detailed site information, including a geotechnical survey report
- Detailed project program and performance requirements
- Aesthetic requirements
- Environmental restrictions
- Any constraints that must be considered in the design or construction of the project
- Contractual requirements
- Format for cost data (if required)

Table 6.1 presents three examples of typical RFP requirements by major public agencies in the United States. You may find similar types of information included in your RFP.

Proposal Evaluation

On the date stated in the RFP, written proposals are submitted to the owner and/or the owner's consultant. For a fair evaluation, the project owner would select a jury of five to seven individuals to review and score each proposal. All jury members should know the owner's needs and understand the project's context (e.g., uncertainties and complexity). Each jury member is responsible for evaluating the proposals using the evaluation criteria and selecting the proposal that best meets the owner's objectives.

Typically, the jury is composed of:

- one or two design professionals,
- one or two members who represent the users of the facility,
- one or two individuals representing the operations and maintenance staff, and
- at least one person who represents the project owner's construction management staff.

Generally, the owner or a senior staff member is the voting chair of the jury. In this approach, the jury members evaluate each proposal's technical merits and score using the criteria defined in the RFP. If interviews are to be held, the design-builders submitting the highest-ranked proposals are invited to participate. When you have the interview process, the proposal scores are added to the interview scores. If interviews are not conducted, the design-builder submitting the jury's highest-ranked proposal is selected for the contract award.

The two jury teams evaluate the technical and cost proposals separately, and their scores are combined at the end of the review process to select the design-builder for the project. Typically, technical proposals should score above 85% of the maximum allowable number of technical points. The lowest price proposal would receive 100% of the score allowed for the price, and the other proposals would receive lower price scores in proportion to the percentage of the proposed price exceeding the lowest price. For example, a proposal with a price that was 10% higher than the low price would receive a price score of 90% of the maximum score allowed for the price, and a proposal with a price 20% higher than the low price would receive a price score of 80% of the maximum score allowed for the price.

For example, the Washington Department of Transportation[4] used the following evaluation criteria for technical proposals on a DB highway project:

- Project management approach (100 points)
- Schedule (75 points)
- Quality management approach (125 points)

- Environmental compliance plan (100 points)
- Maintenance of traffic and staging plan[5] (175 points)
- Roadway (150 points)
- Structures (75 points)
- Drainage (50 points)
- Pavement design (50 points)
- Roadside restoration and aesthetic treatment (50 points)
- Utilities and third parties (50 points)

Project owners use several scoring techniques in the technical evaluation method of proposal evaluation. As you learned in Chapter 2, owners can use one of three scoring methods when evaluating the proposals: composite score system, best value score system, and adjusted score system. If you are not familiar with those scoring systems, please visit the scoring system in Chapter 2.

Design Phase

Once a contract is made between the design-builder and the owner, the design-builder leads the design process based on the design documents in the RFP. These initial documents (i.e., RFP) set the foundation for DD, aligning with the owner's requirements.

In the DB method, the owner has limited or no control over DD after contracting. The design-builder primarily manages the design. This limited owner's control over design can lead to concerns about the design quality. As discussed previously, the extent of design completed before issuing the RFP varies significantly, ranging from no design to a preliminary concept, based on the level of control the owner wants to retain. For owners seeking more control, the owner should use either a Bridging method, where a more substantial initial design is provided in the RFP, or a Progressive DB, where the owner is actively involved in DD, similar to the CMAR method.

The level of design in the accepted proposal is the starting point for developing the construction documents. The design-builder has a contract and must complete the project by the date prescribed in the contract. Thus, the design-builder must create a complete design that fully conforms to all of the performance requirements contained in the RFP and is executable within the cost and schedule constraints contained in the contract.

As the design approaches around 50% completion, the design-builder typically engages key trade partners (e.g., MEP contractors) who are to provide design-assist services. During this phase, trade partners offer feedback on design decisions, contribute to value engineering, and provide input on scheduling and interim cost estimates for their specific areas of expertise. During the construction document phase, the design-builder finalizes the design, produces construction drawings, and develops technical specifications.

Owners vary in their approach to reviewing design progress during the design phase. Some project owners wish to formally review the design at various points when developing the design documents. Other owners prefer an over-the-shoulder design review in which members of the owner's staff review design documents as the design-builder develops them without a formal design review process.

Each of these review approaches offers different benefits. Formal reviews provide a structured quality check, while over-the-shoulder reviews offer flexibility and quick turnaround. The choice of review method depends on the owner's preference and the need for control over the design. Because major subcontractors typically are involved in design decisions, fabrication/shop drawings are often developed early in the design process. DD in the DB process is a very collaborative process in that value engineering and constructability issues are addressed as design decisions are made.

The design-builder is not only responsible for DD, but equally important is the management of cost and schedule. How does the design-builder manage these aspects during the design phase? To develop a project design within the contract budget, the design-builder first creates a cost model for the entire project. This involves identifying the project's major components and assigning a target cost to each element. Then, the design-builder develops a design work breakdown structure, an organizational framework that depicts all project critical design tasks and their relationship to each other. In creating this design work breakdown structure, the design manager identifies all specific design tasks and their information dependencies and creates a schedule to manage the completion of the various design deliverables.

As the design evolves, cost estimates and budgets are updated to ensure that the final project can be constructed within the allowable budget. Completion of a cost-effective design depends on a collaborative process involving subcontractors, suppliers, and the design-builder's construction staff.

Construction Phase

Once construction begins, the design-builder manages the construction process, much as in other project delivery methods. However, the construction phase has some unique characteristics in DB projects. Most DB projects often progress more rapidly than in other delivery methods. The reduced project duration is primarily due to the significant overlap between design and construction (i.e., the fast-track approach).

In DB projects, construction can start on some project components (e.g., earthwork or civil works) while other design packages (e.g., MEP works) are still being finalized. For example, site preparation and foundation work can begin while the detailed designs for HVAC or interior finishes are still underway. This overlap reduces the project timeline by eliminating the need to wait for a fully completed design before starting construction.

By comparison, CMAR projects certain constraints for the fast-track: a GMP contract must be set before construction begins. In some states, such as Washington, public owners are restricted from entering into a GMP contract until design completion is at least 80%–85%. This requirement can limit the effectiveness of a fast-track approach in CMAR, whereas DB allows for flexibility in overlapping phases.

The construction process in DB projects usually generates fewer change orders. Because the contractor gets involved from the early design phase, constructability and value engineering reviews are easily performed in the design process. This collaborative approach helps reduce the likelihood of change orders during construction. Aside from these unique attributes, the construction management process in DB projects is similar to that in other delivery methods. The design-builder coordinates subcontractors while overseeing construction quality, adhering to the project schedule.

Risk Allocation in DB

In the DB delivery method, both design and construction services are provided under a single contract. The single contract generally places more responsibility on the design-builder, shifting a significant portion of project risk away from the owner. In many cases, the owner may limit the design-builder's discretion in specific areas when variables are predictable, thereby reducing contingencies and clarifying expectations. On the other hand, if the project involves uncertain elements, some owners may grant the design-builder more decision-making authority, transferring additional risks to the design-builder. This section examines the typical responsibilities and risk allocation assigned to each major stakeholder in DB projects.

Design Errors

As discussed in the "Spearin Doctrine" section, the design-builder is responsible for developing the design, which reduces the owner's risk of design errors. However, the owner retains some design responsibility, particularly regarding the initial project information (e.g., RFP documents) provided to the design-builder. For example, the RFP includes project information, such as site data or basic design requirements, that later proves inaccurate. In that case, they may be liable for costs associated with any necessary design adjustments. This can be more evident when an owner uses a "bridging" approach, where preliminary design or some level of detail design is already developed before the design-builder is selected. If those bridging documents contain inaccurate information, the design-builder may have grounds to seek compensation for the resulting impacts.

Schedule Delays

The design-builder in DB projects is also responsible for meeting the project schedule, including key milestones. Because design and construction are integrated, the design-builder has more flexibility to fast-track aspects of the project. However, this flexibility may lead to increased responsibility: any delays due to design or construction fall squarely on the design-builder.

In the DB method, the fast-track approach introduces specific risks when changes arise in design elements. For example, if the design-builder begins structural work based on a structural design at the DD stage, but the MEP design leads to structural design changes during the later stages, this can lead to rework on the structural work. To manage this, the design-builder must conduct periodic design reviews and coordinate with subcontractors. They help to identify potential issues early, minimize rework, and avoid delays.

Quality Risks

The design-builder is responsible for the design and construction quality. This includes a duty to meet the project's specified standards. Unlike in DBB projects, where the contractor may argue that any quality issues stem from flaws in the owner-provided design documents, the design-builder in DB is solely accountable for any design-related quality failures. The exception is that the error stems from an error in the RFP. This single-source responsibility makes quality control a critical risk area for the DB team.

The design-builder must develop a comprehensive quality control plan to manage quality risks. They coordinate all quality-related processes, conduct regular inspections, and warrant that materials and workmanship meet contractual standards. In a hospital construction project, for example, if the inspection reveals that the HVAC system does not meet the standards necessary for infection control, they would be held responsible for reworking the relevant system, including bearing the cost of any delays caused by this corrective work.

In DB projects, the quality risk is broader and comprehensive due to the design responsibility. Therefore, the design-builder must have rigorous quality control at every stage, from DD to final construction, to avoid rework and protect against potential claims from the owner. Getting involved in some quality-related claims from the owner would negatively impact your future project awarding.

The Level of Integration

One of the defining characteristics of the DB method is the integration between design and construction functions. Unlike in the CMAR approach, where the designer and contractor keep separate contractual relationships with the owner, DB has a contractual entity, a design-builder. This design-builder is leading the design and construction process throughout the project. This organizational structure inherently improves collaboration between the design and construction teams. In DB, the design and construction entities operate within the same organization or under a closely aligned partnership. This close bond allows interaction between the design and construction wings early in the project lifecycle. As a result, process design, such as constructability and logistics, is considered while developing the design. While CMAR allows the contractor to provide preconstruction services and advice, the contractual separation may inhibit organizational integration and shared accountability for the design outcomes.

However, it is essential to note that the level of owner involvement in DB projects can vary significantly depending on how the DB contract is structured. In DB projects, except for PDB, the owner's control or influence over the design almost diminishes once the contract is made. The design-builder then assumes responsibility for delivering both design and construction within the contractual parameters. In contrast, PDB allows the owner to keep control or influence design decisions, prior to the GMP being fixed.

Despite the integration between designer and contractor in DB, the integration often does not extend to the trade contractors. Specialty or trade contractors, such as MEP trades, are frequently brought into the project only for the design assist service. The design assist service is an independent contract that does not guarantee the construction contract. While the design-builder involves key trades through design-assist arrangements, the absence of a construction contract at this stage may discourage subcontractors from fully committing their expertise when there is no assurance that they will be awarded the construction scope. Suppose a mechanical subcontractor participates in the design coordination meeting to provide insights on mechanical system selection. However, this DB setting may limit their willingness to share their expertise. This presents a potential gap in the integration of the DB projects.

Case Study: Power Plant Facility[6]

Project: Power Plant Facility Upgrade
Location: Anchorage, AK
Budget: $8,500,000 (design and construction)
Project Delivery System: Design-build
Contract Type: Lump Sum

In a DB competitive selection process, owners or agencies have broad discretion to take corrective action where they determine that such action is necessary for fair competition. This case brings insight into the agency's discretion in the procurement stage. In this case, the RFP sought proposals to design and construct the central heat and power plant facility upgrades at the Clear Air Force Station in Alaska. The DB project included designing and building a facility to house three baghouse collection systems[7] designed to remove particulates from three existing coal-burning boilers. The award was to be made on a best-value basis, with the price and non-price factors considered equally important.

The RFP specified that the work must conform to detailed performance and prescriptive-based drawings and specifications, including certain baghouse specifications. The design criteria for the baghouse specifications required that: (a) the maximum net pressure differential between manifolds of the baghouse should be 6 inches; and (b) the minimum spacing between individual bags within the baghouse (i.e., bag-to-bag clearance) must be 2.5 inches. While the RFP did not specify a brand name, the baghouse specifications, including those set forth above, were written around a specific manufacturer.

Contractor S offered a baghouse system manufactured by a manufacturer different from that contemplated by the RFP specifications, and the system had a maximum net pressure differential of 7 inches and bag-to-bag clearance of 2 inches. While these variations were noted as problems by the Corps during technical discussions, Contractor S's proposal was ultimately accepted. Another bidder protested, arguing the Corp improperly accepted Contractor S's deviations from the RFP and gave Contractor S an advantage over the other bidders. The Corps agreed and admitted that it made a mistake in accepting Contractor S's nonconforming proposal. It is also believed that its original baghouse specification, which favored a single manufacturer, was proprietary and did not give potential proposers an accurate statement of the agency's minimum requirements. Consequently, the Corps decided to terminate S's contract, amend the RFP to permit using other baghouse manufacturers, and re-solicit proposals.

Case Study Questions

Question 1. Do you agree that the Owner's decision to use DB is the right decision? Defend your answer.

Question 2. Assuming that the Owner is the Army Corps of Engineers, what type of selection method do you think this project adopted? Why?

Question 3. Discuss the impact of prescribing the details of product requirements instead of functional requirements in the RFP.

Case Study: Wastewater Treatment Plant[8]

Project: Wastewater Treatment Plant
Location: Seattle, WA
Budget: $38,500,000–$44,000,000 (design and construction)
Project Delivery System: Design-build
Contract Type: Lump Sum

King County Washington Wastewater Treatment Division (WTD) is building a new wastewater treatment plant (36 MGD average wet weather flow, membrane bioreactor or MBR) to serve the northeast portion of its service territory around the Seattle, Washington metropolitan area. The project includes an associated 13-mile-long conveyance system (four tunnel segments) that will discharge effluent to Puget Sound, a marine estuary in western Washington. WTD's previous DBB experience for outfall construction suggested a different strategy was desirable for the newest outfall in the WTD system, and DB was selected.

Pre-Design Study

As part of the study for treatment plant siting and conveyance routing, WTD had a pre-design study prepared for the Brightwater Marine Outfall. While pre-design studies for other DB projects are generally limited to a conceptual or 10% level, this pre-design was carried to 30% or better. A complete alignment was provided for review by the Washington State Department of Natural Resources, which was the agency responsible for the underwater state lands on which the outfall was to be located. Local experienced firms also did pre-design geotechnical work, and WTD had confidence in the data developed from the soil investigations.

Selection Process

Using the pre-design studies and environmental monitoring, WTD was able to have the major permits in hand when the RFQ/RFP was advertised.

This relieved the design-builder from the uncertainty of not knowing what the regulatory agencies would approve and a lengthy permit process with labor effort and schedule that would have been difficult to estimate. Differences between the permitted WTD pre-design outfall and the final design facilities could be easily accommodated with a quicker submittal and review by the agencies as a revision instead of a new submittal.

WTD prepared the DB RFQ/RFP in-house with technical assistance from Vanir Construction Management and CDM. The owner used an "open & information-sharing" strategy. Sharing all geotechnical information and pre-design information allowed all the proposers to be on a level playing field for the proposal preparation and gave them all the same starting point. It eliminated the need for speculation by the contractor or designer, or worse, some costly pre-proposal soils investigations by some or all teams. There were no negative elements of sharing all the information, other than it made the RFQ/RFP a rather hefty document.

Request for Qualification

The SOQ was intended to be relatively quick and easy to prepare. SOQs were due about five weeks after advertising. The teams were asked to submit the expertise of the individuals involved in both design and construction, the project experience they had, and financial information on the DB team lead.

As seen in Table 6.2, the SOQ evaluation criteria were limited to three items: Specialized Experience and Technical Competence (75 points), Record of Past Performance and Project Examples (50 points), and Financial Resources (50 points). Five SOQs were received, and with the very close scoring by WTD, all five were asked to prepare proposals. No honorarium was paid for SOQ submittals. All five teams were contractor-led.

RFP and Proposal Evaluation

The proposal was intended to be a pre-design and construction planning effort. During proposal preparation, WTD held meetings individually with the teams to allow them to ask questions and provide some concept of their design. This was a face-to-face Q&A session that allowed WTD to

Table 6.2 Statement of Qualification Scoring Matrix

Specialized experience and technical competence	75
Record of past performance and project examples	75
Financial resources	50
Total	200

clarify the intent of proposal requirements and understand DB team comments or concerns.

Washington law requires payment of an honorarium to DB proposers who are not selected. Due to the effort and expense that this proposal needed, WTD paid a $115,000 honorarium to all proposers, which amounted to 0.26%–0.30% of the engineer's estimate for design and construction of $38,500,000–$44,000,000.

Table 6.3 presents the detailed scoring criteria for the Proposal. For most DB projects, such a scoring equation is published in their RFPs. A typical DB selection process may have eliminated the top three criteria from the proposal evaluation because they are a repeat of what was submitted for the SOQs. WTD expected new participants, both designers and subcontractors, to become part of the teams as the proposal developed, and the whole team should be re-evaluated. Price was evaluated based on the relation of each proposal to the lowest proposed price. The lowest proposed price received a maximum of 200 points. Any proposer within 3% of the lowest value received 192 points or 96% of the criteria maximum. A proposer within 6% was assigned 180 points or 90%. As the margin increased, the points decreased more rapidly until proposers whose price was greater than 18% more than the lowest proposed price received only 32 points.

The proposed price was submitted in a separate sealed envelope to the procurement staff. Their review of the price at the proposal evaluation was only to see if the design-builder was below the maximum allowable contract cost (design and construction) set forth in the RFP. The proposal evaluators were only told that all teams fell within the project budget. This allowed the evaluators to focus on the remaining criteria and find the best-qualified team with the best proposed outfall and not be swayed or appear to be swayed by price.

Table 6.3 Proposal Scoring Matrix

Hydraulic analysis	150
Proposer's approach to design (design concept)	100
Proposer's approach to construction	100
Geotechnical review and evaluation	75
Specialized experience and technical competence	75
Record of past performance and project examples	50
Updated financial resources	50
Approach to quality assurance and control (QA/QC)	50
Recent/current project workload for design-build team	25
Accident prevention program	25
Location and staging	25
Approach to environmental protection and mitigation	25
Project schedule	25
Permits and regulatory approvals	25
Sealed price proposal (separated sealed envelope)	200
Total	1000

Case Study Questions

Question 1. (Planning) Was the Owner's choice of project delivery method appropriate? Defend your answer.

Question 2. (Planning) Do you agree with the idea that the owner should spend more resources on preparing soil investigations? Why?

Question 3. Suppose that the owner finds out that several competent contractors are hesitant to submit a proposal because there might be unexpected soil conditions or discrepancies with the geotechnical report included in the RFP. How do you want to address them to attract more proposals?

Question 4. The evaluation factor did not include interview scores. Do you think that an interview with the proposers would help evaluate? Do you foresee any potential problems in the interview process as part of the evaluation?

Chapter Summary

1 There are four types of design-builders: (1) contractor-led design-builder, (2) designer-led design-builder, (3) integrated design-builder, and (4) joint venture.
2 DB can be classified according to the level of design completion at the time a design-builder is solicited: (1) PDB, (2) Design-Criteria DB, and (3) Bridging.
3 In PDB, the design-builder works directly with the owner's team from the early design phase, providing the owner with an opportunity for their input and feedback. The owners prefer the GMP contract for PDB projects, where the price is around 40%–60% of design completion. The owner retains control over DD with PDB until the GMP is locked in.
4 Bridging is a project delivery approach where the owner retains more control over the initial design stages by hiring a bridging architect before hiring a design-builder. There are two types of bridging: (1) Preliminary Design DB and (2) Design DB.
5 For concurrent engineering to be effective in the DB model, the design-builder needs to engage key subcontractors as early as possible. The level of concurrent engineering achievable depends on the type of DB project. Bridging methods may limit concurrent engineering opportunities.
6 In the DB projects, the owner interacts with one design-builder overseeing design and construction. This single contract structure places full accountability on the design-builder for managing project performance. In this regard, the Spearin Doctrine is not applied to the DB

projects. However, the Spearin Doctrine's applicability in DB projects can be complex in some contexts when the owner controls the design (e.g., PDB or Bridging).

7 The ability to fast-track is one of the key advantages of the DB delivery method. The contract with the design-builder, made early in the design phase, can move the project into the construction phase without waiting for a fully developed design or a separate construction contract.

8 Once the design-builder is selected, the owner has limited involvement in guiding the design, aside from specifying requirements in the RFP. However, the owner's control level varies depending on the type of DB project. Generally, the owner has more control over design in PDB or Bridging.

9 The owner should determine which delivery method is appropriate. Several factors should be considered to determine if the DB is the best project delivery method for your project. (1) The owner should be qualified to manage the DB projects. The qualification includes not only legal qualifications but also experience/expertise. (2) The project requirements should be clearly defined. (3) The owner should allow the contractor the freedom to develop the design as long as it meets the project program requirements.

10 A Lump Sum is a dominant contract type for the DB projects. However, a GMP is commonly used in this model, as it enables the owner to make design adjustments during the project.

11 In DB projects, owners often prefer to use a Best-Value method to balance price and qualifications. However, in Progressive DB, owners prefer to use QBS, focusing primarily on the qualifications of the design-builder. Both Best-Value and QBS selection methods generally follow a two-step procurement process for public sector projects.

12 Once a contract is made between the design-builder and the owner, the design-builder leads the design process. The level of design contained in the accepted proposal is the starting point for developing the construction documents. To develop a project design within the contract budget, the design-builder typically develops a cost model for the entire project. As the design approaches around 50% completion, the design-builder typically engages key trade partners (e.g., MEP contractors) who are to provide design-assist services.

13 Once construction begins, the design-builder manages the construction process, much as in other project delivery methods. Most DB projects often progress more rapidly than in other delivery methods due to the significant overlap between design and construction. It is noted that the construction process in DB projects usually generates a reduced number of change orders.

14 The single contract generally places more responsibility on the design-builder and shifts a significant portion of project risk away from the

owner. (1) Design Error: The design-builder is responsible for developing the design. However, the owner retains some design responsibility, particularly regarding the initial project information (e.g., RFP documents) provided to the design-builder. In the Bridging method, if the bridging documents contain inaccurate information, the design-builder may hold the owner accountable for the resulting impacts. (2) Schedule Delays: The design-builder in DB projects is also responsible for meeting the project schedule, including meeting key milestones. (3) Quality Risks: The design-builder is responsible for the design and construction quality. This includes a duty to meet the project's specified standards.

Review Questions

1 Please respond to the following questions.

1.1 Which of the following describes the owner's role in DB? Choose only one answer choice.

- a To coordinate design and construction in the construction phase
- b To establish its project requirements, review and approve the design.
- c No different than in any other project delivery system
- d Ask a contractor to choose a designer with whom the owner has a good relationship

1.2 Identify one advantage the DB method offers owners compared to CMAR. How does this advantage benefit the owner?

2 Discuss the level of organizational integration you (as an owner) expect from DB compared to CMAR.

3 Below is a short description of a new bridge construction project. Please respond to the questions at the end of the scenario.

The Washington State Department of Transportation decided to construct a new bridge on State Route 78 in Western Washington. Because of past success with DB projects, the department used a Preliminary Design DB method for this new bridge. You are tasked with developing an acquisition plan for this project.

3.1 On what basis do you recommend the successful contractor be selected?

3.2 What type of pricing method do you recommend the department adopt for this DB contract – Lump Sum, Unit Price, or Cost-Plus? Why did you select this pricing method?

3.3 What information should the department provide in the solicitation package for this contract (i.e., RFP)?

Notes

1 48 CFR Parts 1, 5, 14, and 36. Federal Acquisition Regulation; Two-Phase Design Build Selection Procedures, *Federal Register*, 62(1), 271–273.
2 RCW.39.10.260, https://app.leg.wa.gov/rcw/default.aspx?cite=39.10.260.
3 California Department of Education (2002). AB 1402: Design-Build Projects Guidelines, the School Facilities Planning Division, Sacramento, CA.
4 Washington State Department of Transportation (2022). *Design-Build Manual*. Construction Division. Olympia, WA.
5 Strand Hunt Construction, Inc. (2003). *Comptroller General Decision, B-292415*. U.S. Government Accountability Office. https://www.gao.gov/products/b-292415.
6 Maintenance of traffic and staging plan addresses separation of work zones and traffic, temporary lanes and shoulder closures, and temporary lanes delineation.
7 Lundt, J. (2011). Using Design-Build Procurement for a Deep Water Marine Outfall Pipeline Project. *Conference Proceedings of Pipelines 201*, ASCE, Seattle, WA.

Bibliography

Al Khalil, M. I. (2002). Selecting the Appropriate Project Delivery Method Using AHP. *International Journal of Project Management*, 20(6), 469–474.

Anderson, S. D. and Oyetunji, A. (2003). Selection Procedure for Project Delivery and Contract Strategy, ASCE Construction Institute. *Construction Research Congress 2003*, Oahu, HI, March 19–21.

Associated General Contractors of America (AGC) (2004). *Project Delivery Systems for Construction*. AGC, Arlington, VA.

Beard, J. L., Loulakis, M. C., and E. C. Wundram. (2001). *Design-Build: Planning through Development*. McGraw-Hill Professional, New York, NY.

Cushman, R. F., Loulakis, M. C. (2001). Design-Build Contracting Handbook. Aspen Publishers, New York, NY.

Ghavamifar, K. and Touran, A. (2008). Alternative Project Delivery Systems: Applications and Legal Limits in Transportation Projects. *Journal of Professional Issues in Engineering Education and Practice*, *134*(1), 106.

Love, P. E. D., Gunasekaran, A., and Li, H. (1998). Concurrent Engineering: A Strategy for Procuring Construction Projects. *International Journal of Project Management*, 16(6), 375–383.

Mearig, T. (2004). *Project Delivery Method Handbook*., State of Alaska Department of Education & Early Development, Juneau, AK.

Nguyen, P., Tran, D., Scott, S., and Konrath, L. (2024). Implementation of Pro gressive Design-Build in Airport Capital Construction Projects. *Transportation Research Record*, 2678(12), 1098–1109.

Rahman, M. M., and Kumaraswamy, M. M. (2004). Contracting Relationship Trends and Transitions. *Journal of Management in Engineering*, 20(4), 147–161.

Chapter 7

Integrated Project Delivery (IPD)

In the last three chapters, we examined the evolution of project delivery methods, observing a shift from fragmented approaches toward more collaborative systems. This chapter will explore Integrated Project Delivery (IPD), a newer delivery method that has gained popularity. IPD enhances the level of collaboration by integrating organizational structure and aligning the interests of all team members. At this point, you might wonder how the IPD differs from Construction Manager at Risk (CMAR) or Design-Build (DB)?

Over the last couple of decades, construction professionals have increasingly recognized the benefits of collaboration among all project participants, starting from the design phase. However, a key limitation in CMAR and DB lies in the traditional project organization structure, especially regarding the involvement of specialty contractors. In both DB and CMAR, mechanical, electrical, and plumbing (MEP) subcontractors, for instance, are often only brought into the process for the design-assist service once the design is 50% or more complete.[1] This late involvement is often intended to establish competitive pricing but can limit subcontractors' ability to influence the design. While these subcontractors may provide feedback at this stage, they typically reserve their most innovative ideas to gain a competitive edge during the bidding process. Consequently, valuable insights that could reduce costs are sometimes withheld.

Let me take an example where a mechanical contractor is invited to advise on the design without a commitment to hire them for the construction phase. Even if the contractor identifies cost-saving opportunities – such as adopting a new technology or altering the design to reduce costs – they may be reluctant to share these ideas, preferring to use them later in their bid to secure the contract. This example highlights a key barrier to integration in either CMAR or DB methods, where contractual structures can discourage integration.

IPD addresses these integration challenges by fostering a collaborative environment. One significant difference between IPD and other methods is

DOI: 10.1201/9781003628842-10

that it brings all primary stakeholders, often called Primary Team Members (PTMs), into the project early and binds them together under a single contract. In IPD, unlike previous delivery methods, each team member shares responsibility for both risks and rewards. In other words, if the project performs well, all participants benefit; if challenges arise, the impact is managed collectively.

This IPD method can be seen as an extension of the DB method due to the single-contract structure with the owner. However, IPD differentiates itself by sharing "pains and gains" among all PTMs, including key subcontractors. This shared responsibility incentivizes each party to prioritize the project's overall success. By aligning the interests of team members, IPD promotes a shared accountability, reducing adversarial relationships.

In IPD, project participants contribute their knowledge and expertise from the early design. The collaborative nature of IPD allows for creative problem-solving, leading to faster project delivery, reduced costs, and fewer disputes. This feature contrasts with traditional delivery methods, where contractual limitations can silo each stakeholder. As we delve further into IPD in this chapter, we'll explore primary attributes and main project delivery processes and discuss how the risks are allocated and shared among team members.

Major Attributes of the IPD

More Inclusive and Integrated Team

While CMAR and DB methods foster collaboration between design and construction teams, the IPD method takes a further step. One of the significant attributes of IPD is its inclusiveness: having downstream stakeholders (i.e., major specialty contractors) early in the design phase. In traditional delivery methods, specialty contractors like MEP contractors often join the project after the design has progressed substantially, limiting their ability to contribute to design decisions. In IPD, on the other hand, these key trade contractors are considered part of the core project team, giving them a voice when making decisions. This inclusive approach enables the project to take advantage of their expertise for design alternatives or new technology.

Let's continue using the case of a mechanical trade contractor on a hospital project. In an IPD framework, this contractor can collaborate with the design team from the early design to optimize the HVAC (Heating, Ventilation, and Air Conditioning) system layout for energy efficiency and maintenance. Their input might lead to decisions about equipment placement, duct routing, or alternative materials that could reduce both upfront costs and operating expenses. Without early involvement, these insights might only surface during construction, leading to design changes and delays. With such inclusiveness, IPD not only improves communication but also brings the expertise of downstream stakeholders upfront in developing the design.

Concurrent Engineering

Concurrent engineering is an approach in which product design and process design progress simultaneously. As discussed in previous chapters, CMAR and DB methods facilitate concurrent engineering by involving contractors during the design phase. However, IPD is better positioned to fully implement concurrent engineering because of its inclusiveness: it integrates the general contractor (GC) and key specialty contractors into the design process from the very beginning.

In CMAR and DB methods, specialty trades typically join the project in the middle of design development. While their input is valuable, it occurs relatively late. In contrast, IPD involves specialty trades much earlier, often at the schematic design phase. These specialty contractors are no longer merely reviewers but are instead integral members of the design development team, working collaboratively with architects, GCs, and other stakeholders. Therefore, process design (e.g., construction methods or logistics) can be more effectively developed concurrently with product design development in the IPD framework. For example, the curtain wall supply chain, including its logistics, can be considered when developing a façade design with concurrent engineering.

Fast Track

As with CMAR and DB methods, IPD is well-suited to using a fast-track approach, which allows construction to begin before the construction documents are finalized. IPD is widely recognized for reducing total project duration. This reduced duration results from its ability to overlap between design and construction and the early involvement of key specialty contractors, which reduces the number of Requests for Information (RFIs) and change orders. For example, the HVAC contractor participates in the early design phases, helping to refine the layout and systems collaboratively with the other PTMs. This early engagement eliminates the need for late-stage corrections.

However, it is noted that it usually requires a more extended design phase, while IPD shortens the total project duration. This is due to the active design coordination of multiple stakeholders, such as specialty contractors, consultants, and the GC, who contribute to design development. Let's consider two scenarios. In the first scenario, common in traditional methods, the architect/engineer (A/E) develops the design independently, and other stakeholders provide feedback only after the design is effectively complete. In the second scenario, typical of IPD, the A/E collaborates with the GC and specialty contractors throughout the design phase from the early design stages. The second scenario may seem slower because of the collaboration and coordination required, though it results in a higher-quality design with fewer RFIs or design changes during construction.

Owner's Control over Design

While the DB method often limits the owner's control over design, the IPD method enables the owner to participate in the design development from the early stages. The owner's design control level in IPD is similar to that in CMAR or Progressive DB projects, but IPD promotes a more integrated decision-making process. In IPD projects, the owner typically sets the target price early in the design phase and provides performance parameters. Throughout the design process, the owner attends regular design coordination meetings with the PTMs, which include the designer, GC, and key specialty contractors. These meetings allow the owner to review the design and offer feedback.

When faced with multiple design alternatives, the team often uses structured decision-making tools like Choose-By-Advantages. While the owner does not govern the entire process, their input is critical in selecting solutions. For example, in deciding on a chiller system for a building, the design team might present three options: a water-based system, an air-based system, and a hybrid system. Each option has distinct trade-offs regarding construction cost, energy efficiency, and maintenance requirements. Though all team members (i.e., PTMs) participate in such a decision-making process, the owner contributes to evaluating these options by sharing their operational priorities, such as long-term maintenance costs. This collaborative approach helps the owner's objectives be integrated into the design while leveraging the technical expertise of the PTMs.

Pain-Gain Sharing

As previously discussed, IPD is distinguished by its integrated structure. A critical component of team integration is aligning stakeholders' interests by sharing opportunities and risks. This approach incentivizes collaboration, as all PTMs are responsible for managing challenges.

The current IPD framework has two primary pain-gain sharing mechanisms. The first one is the Estimated Maximum Price (EMP) model, which shifts some financial risk to the owner compared to the traditional Guaranteed Maximum Price (GMP) model. With the EMP, the owner shares in both the financial risks and the benefits, leading to a better partnership.

The second mechanism is a shared pool. This shared pool aggregates a portion of the fees and contingency from each stakeholder and ties their compensation to the project outcomes. The final balance of the shared pool will be distributed to every PTM according to the pre-determined ratio, regardless of who is at fault. The specifics of these mechanisms are further explored in this chapter's section on compensation structure.

Does IPD Fit Your Needs?

Though IPD is an excellent project delivery method for promoting organizational integration and interest alignment, it is not for every project. In choosing a project delivery method, there are three issues that you should take into account: (1) project characteristics and requirements, (2) the owner's ability, and (3) market conditions.

Project Characteristics and Requirements

The rule of thumb for determining if your project is right for IPD is that the benefit of organizational integration, such as early involvement of the PTMs, should be greater than the cost of IPD. When your project is complex and uncertain, you would expect greater benefits of early involvement than in simple projects. In addition, the portion of electrical and mechanical work should be critical and substantial in that MEP trade contractors get involved in the early design phase.

As far as project size is concerned, small projects are not suitable for IPD. If a team has not undertaken an IPD project, it must invest resources in learning. Small projects may not be able to absorb such costs. Moreover, the IPD method is appropriate when the project involves complex and significant MEP systems. One of the core benefits of IPD is the early integration of MEP trade contractors into the design process. Therefore, for this early involvement to be cost-justifiable from the owner's perspective, the project must be of sufficient scale and complexity to leverage the expertise of the MEP specialists. In addition, schedule tightness would be a component that prompts the owners to lean toward IPD since IPD allows a significant level of fast track, an overlap between design and construction.

The Owner's Capability

The owner's capability to lead an IPD process is critical in an IPD project. The owner defines its goals, provides direction, and sets the tone of collaboration. Therefore, not every owner is qualified to lead the IPD method. Generally speaking, the successful owner has the following characteristics:

First, the owner should engage in a project management team, which differs from the DB method. The owner must participate in every project phase, including the design process. The owner should evaluate his organization. If you are an owner, you should ask yourself: Do you want to get deeply involved in project implementation? Do you have the capacity to engage in the project? Suppose the owner is willing to participate in project implementation but does not have the capacity. In that case, he may need professional services, such as professional project management or agent construction management.

Second, the owner should be collaborative and transparent in project implementation. If the project team feels the owner is authoritarian and prioritizes himself first, the project team tends to place their company's needs above the project's. It does conflict with "project-first" and "project-level optimization." Traditionally, many owners are accustomed to using command-and-control approaches to management rather than the collaborative and transparent approach. The owner should evaluate their organization to see if it is ready to adopt a new collaboration model.

In addition, the owner should check to see if he is eligible to employ the multi-party agreement and get trade contractors involved in the early design phase without price competition. If the owner is a public agency, it must consider how to comply with the state procurement law.

Market Conditions

The successful IPD process requires a knowledgeable and experienced project team that includes designers/engineers, a construction manager/general contractor (CM/GC), and MEP trade contractors. If the owner cannot find an eligible project team in his market, IPD may not be suitable for your project.

Designer: In IPD, the design process has a continuous workflow with all PTMs developing and providing commentary and suggestions. This is contrary to the traditional design process, in which only designers work directly with the owner. In one of the IPD projects the author was involved as a consultant, the designer asked for additional compensation because it took more effort and time to coordinate all members throughout the design phase. However, many designers report that it is more invigorating than traditional practice and that the relationship and trust built between designers and contractor teams allows the designers to focus on the creative aspects of their practice. In addition, the design discussions encompass not only design issues but also construction issues, requiring the designer to have a knowledge of project management and leadership, as well as design capacity.

Construction Manager/General Contractor

The CM/GC is central to the PTMs. The CM/GC should be a collaborator and facilitator, leading to the overarching project implementation, as opposed to focusing on cost and schedule. From the view of communication, the CM/GC should be transparent in handling project information, including the decision-making process, and view designers and trade contractors as partners throughout the process. From the view of technical qualification, the CM/GC should have skills and experiences in preconstruction activities (especially Target Value Design [TVD] and conceptual cost development); reliable planning (as known as the Last Planner System [LPS]); and Building Information Modeling (BIM).

Trade Contractors

Trade contractors are expected to do more than they usually do in traditional projects, including GC/CM (or CM at Risk). They are expected to get involved in design development, cost analysis, design review, and constructability analysis in addition to traditional construction tasks. They are jointly responsible for developing means (i.e., developing construction methods, schedule, supply chain management plan) with CM/GC. If they have experience in design-assist service, which includes preconstruction services, IPD requires more than design-assist service.

Compensation Structure for Collaboration: EMP and Shared Pool in Integrated Form of Agreement

Traditional project deliveries have a contractual structure that sets up a conflict where project participants should look out for their own commercial interests as well as those of the project. Moreover, they organize different project parties separately in a way that their diverse interests sometimes converge and sometimes oppose.[2] To avoid conflicts of interest in construction projects, project participants' commercial interests and objectives should be reconciled in the early stage of a project.[3] In other words, sub-goals of project team members should be aligned with the project's goals in a way that all those involved in the project have a sense of winning.

The construction industry has historically responded to these conflicts with various project delivery systems, such as DB and CMAR methods. Among them, the DB method shares some IPD principles in that constructors' expertise can be leveraged in the design phase. As a result, the project quality and speed can be improved compared to DBB because of the involvement of the contractor in the design stage and the active coordination between the contractor and the designer. However, there is a difference in DB and IPD in that:

- IPD uses a relational contract[4] and
- Trade contractors are contractually involved as stakeholders in the design phase instead of just "being invited" as a design reviewer.

The IPD uses a unique Integrated Form of Agreement (IFOA), which includes two contractual vehicles to align the interests of the various project stakeholders: EMP (IFOA) and a shared pool.

Integrated Form of Agreement

IFOA is a form of agreement that can facilitate the IPD project. The purpose of IFOA is to define a legal relationship among the owner, design team, and construction team that aligns them on what is best for the project as a whole, by promoting five big ideas:

- Increase relatedness of the project team;
- Collaborate, really collaborate;
- Manage the Project as a network of commitments;
- Optimize the whole Project, not the pieces; and
- Tightly couple learning with action

The owner, designer, and CM/GC sign one agreement (i.e., the IFOA) while key trade contractors and engineers would sign on the same contract. In IFOA, the owner, designer, CM/GC, and trade contractors form an "IPD team," which is the PTMs, subject to the terms of IFOA. In some IPD projects, only three members (i.e., the owner, designer, and CM/GC) are the PTMs, excluding trade contractors. The IFOA includes two pain-gain sharing mechanisms as follows:

EMP as a Mechanism for Pain-Gain Sharing

The EMP is the sum of the design costs, construction costs, and a negotiated team contingency and profit. The team agrees to put 100% of their profit at risk if the estimated costs and contingency are exceeded. In exchange, the owner will pay the cost of the work, including project direct costs and project overhead costs, regardless of the final amount. In the case of actual costs exceeding EMP, the project team would get no profits at all. The details of the EMP calculation can be found in Chapter 3.

Shared Pools as a Mechanism for Pain-Gain Sharing

The IFOA, with a shared pool, has been known to change the mindset of project members.[5] The IFOA has two shared pools: (1) a Contingency Pool and (2) an At-Risk Pool. The PTMs place their contingency in a "Contingency Pool" and expected profit (a certain percentage of their COW) in an "At-Risk Pool." Each IPD team member has a pre-agreed share of those pools.

A Contingency Pool

This pool is sometimes referred to as a project reserve to absorb unforeseen conditions. While each stakeholder is accountable for its own contingency in other project delivery methods, all PTMs are responsible for managing this contingency pool with transparency. When an unforeseen event leads to excessive costs, the PTMs must document the cost event, log it in the contingency log, and periodically report it to the owner.

Can you guess who owns the balance when the project is complete? Typically, if any balance remains at the end of the project, it reverts to the owner. However, the remaining contingency may be rolled into the At-Risk Pool in some cases. It is noted that moving into an at-risk pool is less common and should be defined in the contract.

An At-Risk Pool

This pool is a collective fee of all PTMs to cover their general overhead and profits. This "At-Risk Pool" is also managed collaboratively by the PTMs and used to handle project risks and reward success. The combined approach strengthens the alignment of interests by ensuring that all stakeholders are jointly responsible for managing risks, avoiding unnecessary costs, and improving project outcomes.

For example, if a project uses more contingency than anticipated, the PTMs do not finger-point at who is responsible. When actual costs exceed the contracted cost of work (COW), the At-Risk Pool gets smaller because the difference should be funded from the At-Risk Pool. These fees are at-risk because the team earns them only if the project completes less than the EMP.

The pain-gain sharing mechanism works if the actual project cost exceeds the EMP. The owner should reimburse 100% of the actual project costs, including general conditions or project overhead costs, even beyond the EMP. This represents the owner's "pain," as they pay more than the contracted price. Then, what is the pain of the PTMs? The PTMs do not receive any portion of their fee beyond the EMP. This results in the PTM's "pain," as they forgo overhead recovery and profit.

Importantly, before the owner's additional reimbursement begins, the following must occur in sequence: (1) the shared contingency pool must be depleted first and (2) the at-risk pool must also be completely depleted. Only after these two shared pools are consumed does the owner begin reimbursing the excess costs.

Since we have covered how pain is shared when costs exceed the EMP, how is the gain distributed when actual costs fall below the EMP? A certain percentage of project savings is allocated to a shared pool. Most owners are willing to split cost savings in a 50:50 or 60:40 ratio (owner to PTMs) when there is a surplus. Consequently, the shared pool increases when actual costs come in under budget (compared to the EMP). The total cost matters, but not individual performance. For example, a mechanical trade may be under budget and lose profit if the project is over budget due to another PTM's fault.

Table 7.1 presents the allocation of a $500 million COW among key project stakeholders – Designer, CM/GC, Mechanical, and Electrical – and illustrates their respective shares in both the Contingency Pool ($25 million) and the At-Risk Pool ($50 million). Each PTM's contribution to the pool is proportional to their COW in the EMP. However, the fee ratio of the at-risk pool is subject to negotiation. For instance, the GC may receive a higher fee percentage due to their role in project coordination.

Assuming the project is complete with a total incurred cost of $530 million, the $25 million Contingency Pool was fully depleted, and an additional $5 million was drawn from the At-Risk Pool to cover the overrun. As a result, the remaining balance in the At-Risk Pool is $45 million.

Table 7.1 Example of Shared Pools (Unit = $million)

COW	500.00	Contingency Pool	25.00	At-Risk Pool	50.00
Designer	50.00	Designer	2.50	Designer	5.00
CM/GC	250.00	CM/GC	12.50	CM/GC	25.00
Mechanical	100.00	Mechanical	5.00	Mechanical	10.00
Electrical	100.00	Electrical	5.00	Electrical	10.00

Table 7.2 Example of At-Risk Pool Leftover (Unit = $million)

Assumption: Actual Costs at Completion = $530 million	
Actual Costs	530.00
Contingency Pool	–
At-Risk Pool	45.00
Designer	4.50
CM/GC	22.50
Mechanical	9.00
Electrical	9.00

This balance will be distributed to the PTMs in accordance with their original allocation ratios as presented in Table 7.2.

Delivery Process of IPD Projects

Planning and Team Selection Phase

Once the owner decides to pursue the IPD method, selecting a team to deliver the project is one of the first steps. The objective of team selection is to bring together stakeholders who are competent and aligned with the project context. If the owner already has an established relationship with trusted designers and contractors, they can help form the core of the IPD team. Together, the owner and these trusted partners can develop strategies to secure the remaining stakeholders. This process can be informal, such as conducting joint interviews with potential team members, to make sure they fit the project.

If an IPD team should be selected in an objective way, the owner can use a structured procurement process. IPD emphasizes collaboration and trust among stakeholders over cost competitiveness. This philosophy aligns with the QBS method, which focuses on selecting teams based on their qualifications. This approach allows the owner to evaluate a broad pool of candidates and shortlist the most qualified ones for further consideration. Shortlisted firms – usually two to four – are then invited to proprietary meetings, where they engage with the owner to discuss project requirements and receive feedback on their preliminary concepts.

While the selection process for IPD may be similar, the evaluation criteria might differ from those of other project delivery methods. In IPD, cultural alignment is just as important as technical expertise because the IPD framework emphasizes collaboration, trust, and innovation – qualities that depend on the team's ability to work together effectively. Selecting team members who have a commitment to collective problem-solving is critical for the project's success. For example, team members should be willing to work transparently, contribute to shared decision-making, and embrace the project's financial model of shared risks and rewards. A cultural fit among the stakeholders is needed for communication, collaboration, and innovative solutions throughout the project lifecycle.

In the selection process, it is also important to have potential team members understand the contractual relationship as well as the project's financial structure during the selection phase. In this regard, many owners include detailed information about the IPD financial model in the Request for Proposal (RFP). For instance, the RFP should clearly explain how the team will be compensated as well as how risks and rewards will be distributed (e.g., the split ratio when the actual costs are less than the contracted amount). This process helps prospective PTMs understand the gain-pain-sharing mechanism.

Validation Phase

The validation phase is a component of IPD projects, a critical gateway early in the process. This phase tests the feasibility for both the IPD team and the owner by evaluating whether the project can meet the owner's Conditions of Satisfaction (CoS) within the allowable cost and schedule constraints. The validation is more than a mere schematic design process. The IPD team (i.e., the PTMs) prepares and presents a comprehensive validation study to the owner. This study typically includes a report on whether the project can meet the required CoS, along with a conceptual schedule and allowable costs.

During this phase, the IPD team organizes available information, explores design and construction solutions, and evaluates project feasibility. The team's efforts are organized to meet the owner's requirements while minimizing unnecessary investments. Though needed for this validation process, the design evolves only to the degree necessary to achieve certainty about the project's viability. The team takes the owner's performance criteria and budget as a framework and evaluates various alternatives. The focus is on aligning the schematic design with the owner's performance requirements while the conceptual cost and schedule remain achievable.

Once the validation study is completed, the IPD team presents its findings to the owner. If the expected cost meets or is close to the owner's allowable cost, the owner provides written notice directing the team to proceed with TVD. However, if the expected cost exceeds the allowable cost, the owner has three options:

- Increase the Allowable Cost: Adjust the budget to accommodate the expected cost.
- Reduce the Project Scope: Collaborate with the IPD team to identify scope reductions and revise the validation study accordingly.
- Terminate the Project: Close the project for the owner's convenience if neither cost adjustments nor scope reductions are feasible.

With a validation phase, the owner can reduce risks associated with unrealistic budgets or misaligned expectations. The validation phase itself needs the collaborative culture of IPD to pave the way for successful project delivery.

Design Phase (TVD)

The design phase in IPD projects is a collaborative process. A key process in the design phase of IPD is TVD, which challenges the traditional paradigm of the design development process. After the validation phase, the owner establishes the target cost, which represents the financial framework for the project. Once the target cost is agreed upon, the IPD team allocates it into smaller, manageable budgets for each building system (e.g., mechanical, electrical, structural). The allocated budget for each system is managed by a cluster. Each cluster, consisting of contractors and designers, needs to develop designs for each building system. These sub-targets serve as the foundation for the design and construction processes, guiding each cluster to achieve the design solutions within their budget constraints (i.e., sub-targets).

While developing the design, each cluster uses cost modeling to align its design with the target budget. Most clusters use the BIM not only as a design tool but also as a communication tool. The design coordination meetings are periodically held, where the IPD team (all clusters) reviews the updated design, discusses the expected project costs and the assumptions, and addresses inquiries. This collaborative approach helps all parties understand the financial implications of design decisions. For example, suppose the HVAC trade contractor identifies an opportunity to replace a traditional air-based chiller system with an energy-efficient hybrid system. In this case, the proposal should be discussed in the TVD meeting. Before the design development phase concludes, the IPD team proposes an EMP and negotiates its price and terms with the owner.

Construction Phase

A defining characteristic of IPD projects is that the contract sets the transactional terms and addresses how the project is managed. Beyond standard project scheduling requirements, IPD projects frequently mandate the use of Lean Construction planning and control methods, the so-called LPS, to improve workflow reliability. This involves several key processes in the construction phase:

- Pull Planning: Collaborative scheduling where tasks are planned in reverse order, focusing on what needs to be completed to meet milestones.
- Look-Ahead Planning: Identifying constraints for assignments with a three to six-week span.
- Weekly Work Plans: Finalizing short-term schedule commitments
- Learning: Measuring how effectively the team adheres to the plan and regularly reporting these metrics to the owner, alongside traditional performance measures such as schedule and budget updates.

The early involvement of all PTMs, including key trade contractors, is a critical feature of this delivery method. This collaborative approach during the design phase enables the IPD process to reduce RFIs and change orders during the construction phase.

Another attribute of IPD during the construction phase is the extensive use of BIM. Most IPD contracts require the PTMs to develop a BIM Protocol, which establishes guidelines for the BIM applications throughout the project. One example would be the interference detection (or clash detection). BIM platforms such as Navisworks are commonly used by the MEP trades for clash detection. Though those tools can be used in any project delivery method, the contract specifies those guidelines (i.e., the use of BIM) in the IPD. As you can see, the construction phase in IPD projects integrates Lean Construction principles and BIM to improve workflow reliability and project performance.

Risk Allocation in IPD

IPD is built upon the principle of shared risk and reward, promoting collaboration among all stakeholders. While the PTMs bear most of the construction risks, the owner also shares these risks. In IPD, risk allocation is structured to align incentives to reduce costs, improve workflow reliability, and foster innovation. Below, we explore how risks are distributed among the project stakeholders.

Owner's Risk

In IPD projects, the owner actively shares risks with the PTMs, mainly through the compensation structure. Unlike GMP contracts, where the owner is protected from costs exceeding the GMP, IPD projects promote shared financial accountability. The EMP serves as a baseline, with any cost savings allocated according to the agreed-upon risk/reward mechanism as discussed in the compensation structure section. Specifically, the owner is responsible for reimbursing all COWs if actual costs exceed the EMP baseline. This approach fosters collaboration between the owner and the PTMs to achieve project goals.

The PTM's Responsibility

The IPD team or PTM assumes primary responsibility for most construction-related risks. These include:

Design Error

Since the PTMs lead the design process, the PTMs are responsible for any design error unless the information the owner provides includes critical errors.

Quality Control

The PTMs make sure that the construction meets specified quality standards.

Schedule Progress

The PTMs should control progress to meet the project milestones and mitigate delays.

Cost Management

The PTMs should control expenses, including contingency, to stay within the EMP.

One of the critical issues is to make sure that the roles and responsibilities are clearly defined. The IPD team develops a Responsibility Allocation Matrix (RAX) to allocate and manage these risks effectively. This RAX clarifies the roles and responsibilities in managing the project. The matrix reduces ambiguities in accountability. For instance, in a hospital construction project, the matrix might assign the HVAC contractor responsibility for coordinating with the design team to finalize MEP system layouts. If an unforeseen event occurs, the matrix clarifies which party is accountable for addressing the issue.

Risk allocation in IPD projects shows a fundamental shift from adversarial to collaborative relationships. The owner and the IPD team promote an environment where accountability and collaboration thrive by clearly sharing risks and accountability. Tools like the RAX and the shared risk mechanisms help manage project risks effectively.

The Level of Integration

The fragmented nature of traditional project delivery methods has historically hindered collaboration among stakeholders. Although methods such as CMAR and DB have introduced improvements to foster collaboration, they still fall short of achieving the integration for project success. In this chapter, we explored the defining characteristics and processes of the IPD method as well as its approach to risk allocation among stakeholders. As you've learned, IPD is not merely a delivery method; it establishes a framework that governs the responsibilities and how the project is managed. This transformative approach enables IPD to elevate the level of integration.

Several characteristics of IPD play a direct or indirect role in enhancing integration among project stakeholders:

- Shared Risk/Reward Mechanism: The IFOA contract improves collaboration and integration through (1) the EMP and (2) a Shared Pool. This financial alignment incentivizes all parties to work toward shared project goals.
- Selection Method Considering Cultural Alignment: By emphasizing qualifications and shared values through the selection process, the project will likely have team members culturally aligned. This alignment is critical for fostering a collaborative atmosphere throughout the project.
- Lean Construction Management Tools: TVD and the LPS promote reliable planning and collaboration among team members.
- BIM: With a BIM protocol, BIM facilitates communication and coordination throughout the design and construction processes.

These elements do not work in isolation. Instead, they work with each other for an integrated project context. For example, the shared risk/reward mechanism with cultural alignment fosters stakeholder trust and collaboration. Trust and collaboration enable the team to facilitate lean implementation (e.g., TVD and LPS) and use BIM for collaborative design and construction. By combining these elements, IPD can achieve an unparalleled integration with other delivery methods. It enhances the relationships among stakeholders while aligning their interests and objectives.

Case Study: Children's Hospital[6]

Project: Seattle Children's Hospital
Location: Bellevue, WA
Budget: $5,500,000 (design and construction)
Project Delivery System: Integrated Project Delivery
Contract Type: Guaranteed Maximum Price

Planning

Children's Hospital and Regional Medical Center (Children's Hospital), the project's owner, initiated the first IPD in the northwestern area. The Children's Hospital project used an IFOA to align risks and rewards to optimize project success, ensuring the implementation of lean principles at the project level. Though the project was not a megaproject in its size, it was complex, uncertain, and fast-tracked: the physical site and building type were complex, the delivery method was new to stakeholders, and the schedule and budget were tight. Initiated by the owner, there was a strong focus on implementing lean construction practices throughout the project, particularly in the design phase. To that end, a lean specialist was hired to facilitate lean activities. Given the affordances of the IFOA contract, the project participants also tried to take advantage of BIM techniques.

Selection Process

The Owner went through the qualification-based selection (QBS) process to select a GC and a designer. In the case of the Children's Hospital project, they used an IFOA to establish the framework for the IPD, which included a shared pool. The IFOA bound three parties – the owner, designers, and CM/GC into a single agreement that required them to share risks and rewards, and the project team had a statement of intent and guide to standard work on which all parties agreed. This encouraged everyone on the team to think of the project first, as their commercial interests were contingent upon the project-level performance. As one of the owner's representatives stated, "Everyone's role changed....it was not really Seneca[7] leading it, it was a team effort."

Design Process

In the schematic design phase, an IPD team selected the MEP contractors who participated in the design phase. Although the IFOA did not include the trade contractors, the mechanical and electrical contractors were involved throughout the design phase. As one designer put it, "good ideas came from everywhere." Consequently, the project team incorporated the contractors' expertise into the design and rallied around the design intent and program. This was illustrated by a late change in the budget. Right before the start of construction, due to the economic downturn, the owner had to reduce the budget by 10%. The project team had already worked hard to meet the existing budget and was now asked to make a big cut. This time, the teamwork really paid off. They could sit down together and pull off the seemingly impossible: provide the owner's original program, keep the architect's design intent intact, and achieve a 10% reduction in construction costs. The CM/GC commented that this would not have happened on a regular project.

Case Study Questions

Question 1. Given the Children's Hospital case information, discuss the context of the project that suits the IDP.
Question 2. Discuss the pain-gain-sharing mechanism used in the Children's Hospital project.
Question 3. The project used GMP instead of EMP. Discuss how using GMP would impact the interest alignments compared to the EMP contract.
Question 4. The MEP trade contractors participated as subcontractors instead of part of the IPD team. How does it impact the project performance compared to MEP contractors as IPD team members?

Chapter Summary

1 IPD is a relatively new delivery method that has gained popularity. IPD enhances the level of collaboration by integrating organizational structure and aligning the interests of all team members. In IPD, unlike other delivery methods, each team member shares responsibility for both risks and rewards.
2 One of the significant attributes of IPD is its inclusiveness: having downstream stakeholders (i.e., major specialty contractors) early in the design phase. Even key trade contractors are considered part of the core project team, giving them a voice when making decisions.
3 Both CMAR and DB methods facilitate concurrent engineering by involving contractors during the design phase. However, IPD is better positioned to fully implement concurrent engineering because of its inclusiveness: it integrates not only the GC but also key specialty contractors into the design process from the very beginning.
4 IPD is widely recognized for reducing total project duration. This reduced duration results from its ability to overlap between design and construction and the early involvement of key specialty contractors, which reduces the number of RFIs and change orders.
5 While the DB method often limits the owner's control over design, the IPD method enables the owner to participate in the design development from the early stages. The owner's design control level in IPD is similar to that in CMAR or Progressive DB projects, but IPD promotes a more integrated decision-making process.

6 The IPD contract has two primary pain-gain sharing mechanisms. The first one is the EMP model. With the EMP, the owner shares in both the financial risks and the benefits, leading to a better partnership. The team agrees to put 100% of their profit at risk if the estimated costs and contingency are exceeded. In exchange, the owner will pay the cost of the work, including project direct costs and project overhead costs, regardless of the final amount. The second mechanism is a shared pool. This shared pool aggregates a portion of the fees and contingency from each stakeholder and ties their compensation to the project outcomes. The final balance of the shared pool will be distributed to every PTM according to the pre-determined ratio, regardless of who is at fault.
7 Consider the three issues to determine if the IPD is the best for your project. (1) Project Characteristics and Requirements: When your project is complex and uncertain, you expect greater benefits of early involvement than in simple projects. In addition, the portion of electrical and mechanical work should be critical and substantial in that MEP trade contractors get involved in the early design phase. (2) The Owner's Capability: The owner should engage in a project management team and be transparent and collaborative. The owner should employ the multi-party agreement and involve trade contractors in the early design phase without price competition. (3) Market Conditions: The owner should be able to find an eligible IPD project team in his market.
8 Once the owner decides to pursue the IPD method, selecting a team to deliver the project is one of the first steps. The objective of team selection is to bring together stakeholders who are competent and aligned with the project context. IPD emphasizes collaboration and trust among stakeholders over cost competitiveness. The QBS method is the most commonly used selection method, though Best Value Fee may be used in some cases.
9 The validation phase tests the feasibility for both the IPD team and the owner by evaluating whether the project can meet the owner's CoS within the allowable cost and schedule constraints. During this phase, the IPD team organizes available information, explores design and construction solutions, and evaluates project feasibility. If the expected cost exceeds the allowable cost, the owner has three options: (1) increase the allowable cost, (2) reduce the project scope, or (3) terminate the project.
10 A key process in the design phase of IPD is TVD. After the validation phase, the owner establishes the target cost, representing the project's financial framework. Once the target cost is agreed upon, the IPD team

allocates it into smaller, manageable budgets for each building system. While developing the design, each cluster uses cost modeling to align its design with the target budget. Before the design development phase concludes, the IPD team proposes an EMP and negotiates its price and terms with the owner.

11 In the construction phase, IPD projects frequently mandate using Lean Construction planning and control methods, the so-called LPS, to improve workflow reliability. The early involvement of all PTMs, including key trade contractors, is a critical feature of this delivery method. This collaborative approach during the design phase enables the IPD process to reduce RFIs and change orders during construction. Another attribute of IPD during the construction phase is the extensive use of BIM. Most IPD contracts require the PTMs to develop a BIM Protocol, which establishes guidelines for the BIM applications throughout the project.
12 In IPD projects, the owner actively shares risks with the PTMs, mainly through the compensation structure. The owner is responsible for reimbursing all COWs if actual costs exceed the EMP baseline. On the other hand, the IPD team or PTM assumes primary responsibility for most construction-related risks: design error, quality control, schedule progress, and cost management.
13 IPD is not merely a delivery method. This transformative approach elevates the level of integration in several ways: through a shared risk/reward mechanism, the selection method considers cultural alignment, lean construction, and BIM.

Review Questions

1 Please respond to the following questions.

  1.1 Which of the following is not considered a principle for IPD

    a shared risks
    b shared reward
    c early involvement of major specialty contractors
    d competitive bidding

  1.2 List at least one difference between CMAR and IPD.

2 Many owners try to integrate their projects. You learned how the project is integrated using IPD in this lesson. It is noted that there are more ways to integrate the project. What do you think makes the project integrated?

3 A nationwide technology company is considering building a new data center in the Midwest. The company decided to use a fast-track since they needed the data center urgently. The tech company determined to use an IPD method because IPD could save costs while accelerating progress. The owner decided to use QBS to select an IPD team.

The owner has internal project management staff who have experience managing the design and construction of similar projects. Using the QBS, the owner chose an IPD team with data center construction experience. The team made an IFOA contract and had a three-day workshop in which the team determined the specific project goals. The owner set a target price of $680 million, including all fees. From the schematic design, the owner participated in the design coordination meeting. At the end of the schematic design, the IPD team's estimate was around $700 million. When the team moved into design development, the team used a co-location in which the team of diverse disciplines worked together in one office. At the end of design development, the team submitted an EMP proposal of $726, including all fees. They negotiated and finalized the EMP of $690 million ($640 m COW).

3.1 Why did the owner use the IPD method? Do you agree with this decision?
3.2 Discuss the difference between DB and IPD. Discuss the project context in which IPD is a better option than DB.
3.3 At the end of the project, the contractor team incurred $740 m COW. What is the total amount the contract team expects to receive from this project?

Notes

1 Most cases of design-assist service do not give a commitment to hire them for the construction phase.
2 Construction Management Association of America (CMAA) (2009). *Managing Integrated Project Delivery.*
3 Olander, S. and Landin, A. (2005). Evaluation of stakeholder influence in the implementation of construction projects. *International Journal of Project Management*, 23, 321–328.
4 Contract that defines the relationship that takes on the properties of a mini society with a vast array of norms beyond those centered on the exchange of goods and services.
5 Matthews, O. and Howell, A. (2005). Integrated Project Delivery: An Example of Relational Contracting. *Lean Construction Journal*, 2(1), 46–61.
6 Kim, Y. and Dossick, C. (2011). What Makes a Project Integrated? A Case Study of Children's Hospital, Bellevue, WA. *Lean Construction Journal*, 8(1), 53–56.
7 The owner's representative in the project.

Bibliography

AIA (2012). IPD Case Studies. https://rp.design.umn.edu/resources/documents/IPD-Case-Study-Matrix-2012_corrected02.pdf.

AIA California Council (2007). Integrated Project Delivery. https://assets.aiacontracts.com/ctrzdweb02/zdpdfs/ipd_guide.pdf

Azari, R., Kim, Y., Ballard, G., and Cho. S. (2014). Starting from Scratch: A New Project Delivery Paradigm. *Proc., Construction Research Congress*, 140–149. ASCE, Reston, VA.

Baiden, B. K., Price, A. D. F., and Dainty, A. R. J. (2006). The Extent of Team Integration within Construction Projects. *International Journal of Project Management*, 24(1), 13–23.

Cho, S., Ballard, G., Azari, R., and Kim, Y. W. (2010). Structuring an Ideal Project Delivery System. *International Public Procurement Conference 2010 Proceedings*, Seoul, Korea.

Choi, J., Yun, S., Leite, F., and Mulva, S. (2018). Team integration and owner satisfaction: Comparing Integrated Project Delivery with Construction Management at Risk in Health Care Projects. *Journal of Management in Engineering*, 35(1), https://doi.org/10.1061/(ASCE)ME.1943-5479.0000654

Construction Management Association of America (CMAA). (2009). *Managing Integrated Project Delivery*, Construction Management Association of America, McLean, VA.

Kim, Y. and Dossick, C. (2011). What Makes a Project Integrated? A Case Study of the Children's Hospital, Bellevue, WA. *Lean Construction Journal*, 8(1), 53–66.

Matthews, O. and Howell, A. (2005). Integrated Project Delivery: An Example of Relational Contracting. *Lean Construction Journal*, 2(1), 46–61.

Part 4

Moving Forward

Chapter 8

New Trends in the Project Delivery System

In the previous chapters, we explored project delivery systems, discussing how the project delivery systems have moved from fragmented to more integrated and collaborative approaches. Specifically, we highlighted how collaborative delivery methods such as Construction Manager at Risk (CMAR), Design-Build (DB), and Integrated Project Delivery have improved collaborative processes among stakeholders. However, these systems are primarily focused on the delivery system of a single project. In this chapter, we shift our attention to an emerging trend in project delivery systems that concerns integration not just at the project level but at the program level. This shift is driven by the construction industry's growing need for productivity improvements and sustained stakeholder relationships.

While those collaborative project delivery methods have proven effective for collaboration among project stakeholders, their scope remains confined to individual projects. In such systems, even with collaborative practices on a specific project, owners often restart procurement processes from scratch when initiating a new project. However, how do you guarantee the relationships beyond a single project if you restart the procurement process? These limitations have necessitated some kind of supplementary systems that address these gaps.

This chapter addresses these needs and introduces two emerging approaches: framework agreements and program-based procurement. They are not designed to replace existing project delivery systems, such as DB, but to extend their value for long-term relationships across multiple projects. Framework agreements allow owners to pre-establish contractual relationships with a few service providers, without the need for repetitive administrative processes on each job or project. Similarly, program-based procurement focuses on procurement at the program level for a series of related projects.

In this chapter, we will examine the defining attributes of these methods, the major processes involved, and the context in which each method is applicable. By exploring these new trends, we want to expose you (i.e.,

DOI: 10.1201/9781003628842-12

owners and construction students) to those emerging procurement vehicles. Note that these approaches represent the next evolution in project delivery systems, addressing the industry's need for integration on a broader scale.

Framework

A framework is a procurement method centered on long-term collaboration agreements. It allows an owner to work with a pre-approved single or group of suppliers/contractors under predefined terms and conditions. Unlike traditional selection methods that require tendering and negotiations for each project, frameworks expedite the procurement process with less administrative effort. The owner can reduce the procurement time significantly with a framework. The framework often leads to trust and performance improvements by establishing a long-term relationship between the owner and the suppliers/contractors.

The application of frameworks remains relatively limited in the construction industry. Still, some examples include the UK's Procure 21/22 program and the Job Order Contracting (JOC) model used in the public sector. In addition to public sector initiatives, private sector clients have adopted similar procurement methods, reflecting the growing recognition of the benefits of long-term collaboration between owners and construction service providers. Frameworks shorten project setup time, reduce administration costs, and encourage suppliers to invest in innovation and customer satisfaction with a stable relationship with the owner.

This section will explore framework-based procurement's attributes and processes using two representative examples: the Procurement 21/22 program in the UK and the JOC model in the US. These case studies will show you how frameworks can be implemented in practice. We hope you will grasp how frameworks represent a paradigm shift in construction project delivery systems.

Case 1. Procurement 21/22 in the UK[1]

The Department of Health (DOH) in the UK launched the Procure 21 program in the mid-2000s to reform procurement and delivery systems. The objective of the innovative framework is to improve the delivery process of healthcare facilities in the UK. Over a decade later, the program continues under its updated version, Procure 22, signaling the ongoing movement to innovate procurement practices in the public sector. Central to this initiative are the long-term relationships and integrated supply chains.

One of the defining features of Procure 21/22 is the terminology shift from traditional labels like "contractor" or "supplier" to Principal Supply Chain Partner (PSCP). This change shows the program's focus on supply chain integration and partnership rather than conventional transactional relationships. PSCPs are selected through a framework process, making them eligible to compete for healthcare projects under the program for a fixed five-year period. Importantly, this framework contract does not guarantee automatic project awards; PSCPs must still participate in competitive selection processes on individual projects. In this regard, PSCPs can be considered pre-qualified and shortlisted contractors or suppliers under the framework.

The Procure 21/22 program offers benefits for both owners and supply partners. The framework gives owners access to competitive contractors and suppliers with reduced procurement costs and time. For supply partners, it provides opportunities for long-term project pipelines. Here, we will explore the key processes and attributes of the Procure 21/22 framework.

Selection Process

The selection process for the PSCPs in the Procure 21/22 framework is structured; it is designed so that only the most qualified and capable partners are chosen to deliver healthcare construction projects. This process evaluates not only the primary contractor but also their key collaborators. The selection process of the Procure 21/22 has some distinctive characteristics as follows:

Evaluation of PSCPs and Their Collaborators

During the selection phase, the DOH assesses the PSCP candidates and their principal subcontractors/consultants. Key collaborators are evaluated alongside the main construction contractor to verify their competence and alignment with the project's objectives.

Commitment to Approved Collaborators

Once selected, PSCPs are expected to maintain their key collaborators throughout the framework agreement period. Any proposed changes to these collaborators require prior approval from the owner. Such a process regarding the variation of their collaborators is intended to have a reliable supply chain, fostering relationships and consistent performance over the five-year period.

Streamlined Selection Timeline

One of the major attributes of the Procure 21/22 framework is the expedited selection process for individual projects. On average, clients can select a PSCP and their key collaborators within one month. This rapid procurement process significantly reduces the time and costs related to bidding processes, enabling projects to commence in a timely manner. Such a reduction in the procurement timeline is significant given the UK's increasing demands on public healthcare.

Major Attributes of the Procure 21/22 Projects

The Procure 22 program, through the framework approach, has three key features. These attributes have promoted participation from public owners and construction service providers.

Guaranteed Pipeline for the PSCPs: Streamlining Procurement

One of the key attributes of the Procure 21/22 program is to guarantee a project pipeline exclusively for PSCPs during the framework's duration. The program provides these suppliers with a predictable pipeline by restricting the pool of project bidders to pre-selected PSCPs. It is particularly significant for projects like Off-Site Construction (OSC), which often require significant upfront investment. A reliable pipeline mitigates the financial risks while motivating service providers or suppliers to invest in innovation and relationships with the project owner.

For owners, a framework agreement with PSCPs works as a pre-established shortlist. Instead of issuing a Request for Qualifications (RFQ) and evaluating Statements of Qualification to create a shortlist for every individual project, owners can directly invite the PSCPs to submit expressions of interest for a particular project. For each project, those showing interest from a pre-selected group (i.e., PSCPs) go through an interview and evaluation process, expediting the procurement process, as addressed in the selection process. The Procure 22 framework significantly reduces this administrative burden.

Guaranteed Maximum Price for Transparency

The second key feature of the Procure 21/22 program is using a Guaranteed Maximum Price (GMP) contract based on open-book accounting. This approach improves cost transparency between the owner and the PSCPs (Table 8.1). The GMP contracts in the Procure 21/22 program, similar to those in US commercial construction projects, include two notable differences.[2]

First, fees are determined during the selection process in typical GMP projects. This project-by-project fee requires significant time and effort for proposal evaluation or contract negotiation. In contrast, the Procure 21/22 framework determines pre-agreed rates for items such as fees and training

Table 8.1 Comparison of Conventional GMP and Procure 22 GMP

	Conventional GMP	*Procure 22 GMP*
Fee and saving-sharing	Project-by-project negotiation	Pre-agreed rates during the framework period
If (COW + Fee) exceeds or equals GMP	GMP = total transfer price	GMP = total transfer price
If (COW + Fee) becomes less than GMP	COW + Fee + Pre-agreed ratio of the savings = total transfer price	If (COW + Fee) becomes less than 95% of GMP, then (COW + Fee) + 2.5% GMP = total transfer price If 95% GMP < (COW+Fee) < GMP, then (COW+Fee) + 50% of the savings = total transfer price

costs at the beginning of the framework. These predetermined rates are applied across all projects in the Procure 21/22 program during the framework period. This preemptive agreement guarantees a baseline profit for PSCPs and eliminates the need for project-based negotiations, reducing project durations.

The second key distinction lies in how the total payment to contractors (i.e., total transfer price) is calculated relative to GMP. Under the Procure 21/22 program, GMP is negotiated between the PSCP and the owner during the detailed design phase. While the actual costs are not necessarily equal to GMP, several potential outcomes occur. If the actual costs exceed the GMP, the owner is only responsible for paying up to the GMP. If there are savings, how can they handle them?

In conventional GMP contracts,[3] the cost savings are shared between the contractor and owner according to pre-agreed ratios if actual costs fall below the GMP. However, in many cases, the owner retains 80%–100% of the savings, disincentivizing contractors to reduce construction costs below GMP. However, the Procure 21/22 program introduces a more balanced approach. The first 5% of savings below GMP are shared equally between the owner client and the PSCP. Any savings exceeding 5% of the GMP are retained entirely by the owner. In implementing GMP contracts, the Procure 21/22 framework requires open-book accounting, which is also true for conventional GMP contracts. This system gives the owner visibility of all project costs, fostering transparency and trust between parties.

Standardization and Knowledge Sharing

The Procure 21/22 program focuses on standardization and systematically sharing knowledge across projects. You may wonder why such attributes are related to the procurement system. Though they are not the attributes of the procurement system, they are feasible since the Procure 21/22 program promotes sustainable relationships and a reliable pipeline. The Procure 21/22 framework agreement requires Building Information Modeling (BIM) and promotes collaborative platforms like Standard Share and Project Share. These tools enable project participants to share standardized components and design information, benefiting all stakeholders.

As PSCP, project teams gain access to a suite of standardization tools for their design process. For example, the Standard Share platform provides pre-designed room templates for various healthcare facility types, such as single-patient rooms.[4] These standardized templates help the PSCPs improve the design productivity with minor modifications to meet specific requirements. By leveraging pre-designed templates with associated BIM data, project teams can also improve cost estimation accuracy. As a result, using standardized room designs has contributed to an 11% reduction in overall project costs.[5] This cost efficiency is critical for public healthcare projects.

Beyond room design standardization, the Standard Share program extends to other building components. In November 2017, an agreement was established to standardize critical components used in healthcare facilities. This effort, which began during the Procure 21+ program, has been expanded under Procure 22 to include a broader range of products and manufacturers. Through these agreements, high-quality standardized components are made available to project participants at fixed prices for the duration of the framework. This not only improves productivity but also quality outcomes across projects.

In addition, the Standard Share platform improves knowledge sharing among stakeholders. The lessons learned from completed projects are systematically incorporated into future work. This cyclical process of knowledge transfer fosters continuous improvement.

Case 2. JOC in the US

JOC is a project delivery method designed for multiple construction and maintenance projects using competitively bid and unit-price contracts. It is effective for public agencies and organizations that regularly undertake a number of small to medium-sized projects. Large corporations can easily use framework agreements or JOC methods for the private market for repetitive maintenance jobs. These approaches involve pre-selecting a limited number of contractors and defining predefined unit costs across projects. Unlike traditional procurement methods, JOC enables owners to establish a long-term partnership with a group of contractors for an indefinite delivery–indefinite quantity (IDIQ) of work, typically spanning three to five years.

Developed by the United States military in the 1980s to expedite smaller projects, JOC has since been adopted by many public entities across the U.S. JOC is considered a type of framework contract. However, it has specific characteristics that differentiate it from other framework agreements.

Selection Process

The selection process for JOC consists of two primary phases: selecting JOC contractors for a multi-year contract (e.g., five years) and selecting a contractor from the pool of JOC contractors for each job order. The JOC procurement process usually follows a two-step procedure. The process begins with an RFQ to solicit interest from contractors. Contractors are evaluated based on criteria such as past performance, JOC experience, financial stability, safety record, and quality control capability. An essential factor in this evaluation is the contractor's capacity to efficiently handle IDIQ contracts.

Shortlisted contractors receive a Request for Proposals (RFP), which includes contract terms, requirements, and expectations. At this stage, contractors submit their price coefficient – a numerical multiplier applied to the Unit Price Book (UPB) to determine project costs. The UPB is the pre-priced catalog of construction services for standardized pricing. The owner compares the proposed price coefficients against representative completed projects. The contract is either negotiated with the top-ranked firm or awarded based on predefined selection criteria. Importantly, each JOC contract establishes minimum and maximum guaranteed job order amounts for each contractor.

Once a pool of JOC contractors is selected, the owner should choose a contractor from a pool of JOC contractors for each specific job order unless there is only one JOC contractor in the pool. The owner may directly assign job orders based on contractor availability or solicit competitive proposals from multiple JOC contractors. Most JOC policies require that at least 80% of cost items in each job order proposal must come from the UPB, multiplied by the predetermined cost coefficient. This policy prevents cost inflation by excessively using non-pre-priced items.

Major Attributes of JOC Projects

While JOC aligns with the principles of framework contracts, it is more specialized, focusing on recurring construction or repair tasks rather than traditional construction projects. JOC projects usually have the following attributes:

Cost Predictability through Predefined Unit Pricing

One of the key advantages of JOC is cost predictability, which is achieved through a well-structured UPB, which serves as the foundation for pricing job orders. Though some public owners use an RS Means for the UPB, owners should have a UPB that is locally adjusted, reflecting regional labor rates, material costs, and other relevant pricing factors. During the contractor selection process, prospective JOC contractors submit a cost coefficient, which remains unchanged throughout the framework period.

Efficiency Through Standardized Product Use

Though not required for JOC, the use of UPB promotes the use of standardized products by reducing the variability of products. The standardization of products and materials improves cost efficiency and predictability in JOC projects. Owners who use standardized products, such as fixed window sizes/types or uniform flooring materials, reduce the number of unit-price items in the UPB. This practice reduces the administrative costs, such as purchasing costs. Additionally, standardized products allow for better pricing with suppliers.

In addition to procurement benefits, product standardization contributes to long-term maintenance efficiency. When owners consistently use the same materials and components across multiple projects, JOC contractor(s) can manage repairs and replacements more effectively, reducing inventory complexity. In this regard, this approach contributes to facility management.

Rapid Procurement Process for Repetitive and Small Projects

One of the key advantages of JOC is that it expedites the procurement process for repetitive and small projects. Unlike traditional procurement methods, JOC employs a framework under which the owner develops a long-term agreement with a qualified contractor or multiple contractors, allowing individual job orders to be issued as needed. Since each job order (i.e., each small project) operates under the overarching JOC contract, procurement steps for each job order are significantly reduced. This efficiency makes JOC an ideal delivery method for organizations managing ongoing facility maintenance, renovations, and infrastructure repairs.

According to the University of Washington Facilities Department,[6] the typical job order procurement process can take six to eight weeks for a framework contract. Once a framework agreement is made, individual work orders can move faster: it usually takes less than two weeks till the individual work order agreement is made. This is particularly beneficial for public agencies, educational institutions, and healthcare facilities with constant maintenance needs.

Long-Term Relationship with Small Contractors

One of the defining attributes of JOC is to foster long-term relationships with small contractors. Unlike traditional procurement methods, JOC operates as a multi-year program over a typical period of three to five years. This extended timeframe allows owners to build strong working relationships with their selected JOC contractors. Working consistently with those small contractors benefits both parties in the long run.

Many public sector owners actively support women- and minority-owned enterprises by giving them credits in the evaluation process. JOC serves as a vehicle for fostering small business growth by offering a steady pipeline of work under JOC. This model reduces the business uncertainty that small contractors face in the competitive bidding environment. With predictable workloads, these businesses can focus on their capabilities and competitiveness without the constant pressure of securing new contracts.

Program-Based Procurement

Most traditional project delivery methods, such as Design-Bid-Build (DBB), CMAR, and DB, are structured for a single project. However, certain types of projects, particularly those involving OSC, demand significant upfront investment. For instance, OSC projects usually require contractors or manufacturers to set up production lines – investments much larger than in traditional construction. One of the major barriers to market expansion for OSC suppliers is the substantial capital investment burden. From an owner's perspective, a critical question arises: how can procurement strategies be structured to mitigate these initial investment risks for contractors? One of the solutions is program-based procurement, which grants a pipeline of projects. With a steady demand over multiple projects, suppliers can scale operations and reduce unit costs.

This approach has been successfully implemented in initiatives such as the public housing program in Vancouver, Canada, where multiple housing developments were procured under a single program. By grouping similar projects together, owners can streamline procurement processes and attract more capable contractors. This approach is particularly valuable for sectors that require standardized, repeatable unit design, such as public housing. Here, we will explore the selection process and the major attributes, followed by a case study of from a case of a housing development program in Vancouver, BC, Canada.

Selection Process

Each project can be procured using any project delivery method. However, the owners prefer a DB approach, where both design and construction are contracted in a single contract. For public housing programs, some owners include operation in the scope of the work. In most cases, a two-phased procurement is widely considered a best practice for public projects.

Phase 1: RFQ: In this stage, interested teams are invited to submit their qualifications. A shortlist of qualified teams was developed based on their experience, technical capability, and financial stability.

Phase 2: RFP: The shortlisted teams are then invited to submit detailed proposals, including technical solutions, project approach, and pricing. The final selection is based on a Best Value-Total Cost evaluation, which combines qualitative (technical and design approach) and quantitative (cost) criteria.

A key feature of the procurement process is the inclusion of preliminary design work in the RFP. The modular unit design (e.g., modular unit dimensions, unit types, and system performance specifications) should be clearly defined if the program involves modular construction. With these prescriptive guidelines, the service providers or contractors can develop reliable pricing for each unit type.

Major Attributes of the Program-based Procurement

Program-based procurement bundles multiple, similar projects into a single procurement, creating advantages for both owners and service providers. This is particularly useful for projects such as OSC, with significant early-stage capital investment. Below are the key attributes that define this procurement strategy:

Procurement Cost Efficiency Through Bundled Procurement

By grouping similar projects into a single program, public or private owners can reduce repetitive tendering time and costs. For example, Canada's Vancouver Modular Housing Program delivered residential buildings under a single procurement umbrella, grouping multiple modular projects.

Economies of Scale

Programmatic procurement supports the repeated use of standard components and assemblies. A similar design with standard components across multiple projects enables the service provider (or contractors) to achieve economies of scale. This reduces not only construction costs but also maintenance and operation costs.

Hedging the Financial Risk of Upfront Investments

Some projects, like OSC, require substantial upfront capital, such as setting up prefabrication lines. Program-based procurement offers suppliers a predictable pipeline of projects, justifying such investments and reducing financial risk.

Case Study: The City of Vancouver's Modular Project

Project: The City of Vancouver's First Modular Project
Location: Vancouver, BC, Canada
Budget: 66,000,000 CAD
Project Delivery System: Design-Build
Contract Type: Unit Price

The City of Vancouver launched an affordable housing initiative[7] to develop ten volumetric modular apartment buildings across ten city-owned sites, delivering a total of 611 housing units. This project was procured as a program-based initiative. Instead of awarding contracts on a per-project basis, the entire program – ten apartment building projects, including design, manufacturing, transportation, and on-site construction – was procured as a single package. However, the operation of each housing complex was outsourced separately to public or non-profit entities. Therefore, facility operation and maintenance were not included within the procurement scope.

The total budget for the program was 66 million CAD, covering all phases of the modular housing development except facility maintenance and operations. The owner's agency was the BC Housing Management Commission (BC Housing), a public agency established by the Province of British Columbia in 1967 to oversee social housing initiatives. BC Housing manages various affordable housing programs and collaborates with both public and private sector partners to deliver housing solutions. Additionally, BC Housing operates the Provincial Rental Housing Corporation (PRHC), a subsidiary responsible for land acquisition and management. For this particular program, PRHC provided the land required for the modular housing developments, ensuring a streamlined and cost-effective approach to site selection and development.

Table 8.2 provides details on the ten individual projects within the program, including their respective locations and the organizations

Table 8.2 Operation Companies for First Temporary Modular Program

Address	*Units*	*Operation*
Reiderman Residence 7430 & 7460 Heather Street	77	Community Builders Group
Chartrand Place 1131 Franklin Street	39	PHS Community Services
Aneki Housing for Women 525 Powell Street	39	Atira
Sarah Ross House 4480 Kaslo Street	52	Atira
Margaret Mitchell Place 2132 Ash Street	52	PHS Community Services
Little Mountain 137 East 37th Avenue	46	Coast Mental Health
Larwill Place 610 & 620 Cambie Street	98	MPA Society
New Beginnings 5077 and 5095 Heather Street	104	Lu'ma
Hummingbird Place 265 West 1st Avenue	52	PHS Community Services
Nora Hendrix Place 258 Union Street	52	PHS Community Services
Sum	611	

responsible for their operation and management. These operating units, primarily public or non-profit organizations, were separately contracted to ensure effective long-term management of the affordable housing units.

Selection Process

The Vancouver Affordable Housing Program was procured using a DB approach, where both design and construction were contracted in a single contract. A two-phased procurement method was employed to ensure a competitive and transparent selection process: RFQ and RFP.

In the second phase, pricing in the proposal was structured into three distinct cost components:

- Modular manufacturing cost
- Transportation cost
- Installation and on-site construction cost

For modular manufacturing, bidders were required to submit pricing on a unit-price basis for each modular type as presented in Table 8.3. Once the final design was confirmed, the contract would be converted into a Lump Sum agreement.

Table 8.3 Unit Pricing Form Example

Modular Unit #	*Modular Unit Drawing*	*Type of Unit*	*Information for Pricing*	*Price Per Unit*
Number	Modular unit layout with dimensions		Any details for pricing	

The RFP provides detailed guidelines to assist the design-builders in preparing the proposal, including the guidelines for estimating costs. The RFP addresses information on performance criteria and design variables, helping bidders develop proposals.

For example, the design guidelines for key building elements such as external walls and flooring were as follows:

- External Walls: Required to meet one-hour fire-rated standards with an STC 50 rating for soundproofing; additional design considerations were specified for compatibility with the Air Handling Unit system.
- Flooring: Required the use of resilient sheet flooring, with anti-slip flooring for bathrooms.

Among the competing firms, Horizon North, a modular manufacturer with extensive experience and in-house design capabilities, was selected as the design-builder.

Question 1. There are several selection methods when evaluating the proposals: Qualification-Based Selection, Low-Bid, Best Value-Fee, and Best Value-Total Cost. Which method do you think the owner should use?

Question 2. What would be the impact on the selection process and cost performance if the owner uses a project-based procurement (i.e., ten RFPs instead of a single RFP)?

Question 3. Why do you think the owner contracted separately with the public units for the operation and maintenance?

Chapter Summary

1. A framework is a procurement method for long-term collaboration, which allows an owner to work with a pre-approved single or group of suppliers/contractors. With predefined terms and conditions, the owner can reduce the procurement time significantly with a framework.
2. The Procure 21/22 Program uses a framework to improve the delivery process of healthcare facilities. The Procure 21/22 program offers benefits for both owners and supply partners. For owners, the framework

gives them access to a pool of competitive contractors and suppliers with reduced procurement costs and time. For supply partners, it provides opportunities for long-term project pipelines. Here, we will explore the key processes and attributes of the Procure 21/22 framework. With the long-term relationship, the Procure 21/22 Program promotes standardization and knowledge sharing across the projects.

3 JOC is another example of a framework designed for multiple construction and maintenance projects using competitively bid and unit-price contracts. Unlike traditional procurement methods, JOC enables owners to establish a long-term partnership with a group of contractors for an IDIQ of work, typically spanning three to five years.

4 A program-based procurement, which grants a pipeline of projects, has been used to mitigate these initial investment risks for contractors. With a steady demand over multiple projects, suppliers can scale operations and reduce unit costs. This approach is particularly valuable for sectors that require standardized, repeatable unit design, such as public housing.

Review Questions

1 Please respond to the following questions.

1.1 Which of the following is NOT a characteristic of a framework agreement?

a It promotes a long-term relationship.
b It provides a pipeline of future projects for contractors.
c Contractors in the framework spend less on the bidding process and more on improving their capabilities, leading to added value for the owner.
d Construction costs in a framework agreement are generally lower than those in a DB contract.

1.2 What is the main reason why some owners/developers prefer program-based procurement for modular housing projects?

1.3 Which of the following is NOT a typical application of JOC?

a School maintenance projects
b Maintenance projects in a pharmaceutical production facility
c Building maintenance projects for Microsoft
d A new campus construction project for Amazon

2 The City of Seattle recently made a JOC (job order contract) with ABC Construction as allowed under RCW 39.10.420–460. JOC is an alternative to the traditional DBB public works contracting process, where the City designs the project and posts plans and specifications for public

bid. In the JOC, the City of Seattle does not solicit bids and selects contractors for every project. Instead, the City may issue work orders directly to a JOC prime contractor for construction projects not exceeding $500,000. The City can award up to $6 million yearly to each JOC prime contractor, chosen through a competitive selection process.

2.1 How can the City of Seattle ensure that project costs remain within budget in this scenario?

2.2 JOC is similar to program-based procurement in that both cover multiple jobs (or small projects). What are the major differences between JOC and program-based procurement?

Notes

1 https://www.procure21plus.nhs.uk/.
2 Department of Health. (n.d.). *The ProCure21+ Guide: Achieving Excellence in NHS Construction.*
3 See Chapter 4 if you are not sure about the conventional GMP contract.
4 Department of Health. (n.d.). *Repeatable Rooms.* https://www.ihprojects.co.uk/documents/P22_RepeatableRooms_CaseStudy.pdf.
5 Department of Health. (n.d.). *The ProCure21+ Guide: Achieving Excellence in NHS Construction.* Department of Health, Leeds, UK
6 Capital Projects Review Board (2021). *Job Order Contracting (JOC) Best Practice Guideline.* JOC Evaluation Committee, Washington State Department of Enterprise Services, Olympia, WA.
7 Journal of Commerce News Service (2018). Vancouver Unveils Latest Modular Housing Proposal. https://canada.constructconnect.com/joc.

Bibliography

Center for Job Order Contracting Excellence (2019). *Introduction to Job Order Contracting.* https://jocexcellence.org/

Department of Health. (n.d.). *The ProCure21+ Guide: Achieving Excellence in NHS Construction.* Department of Health, Leeds, UK

Francom, T., Ariaratnam, S. T., & El Asmar, M. (2016). Cost Performance of Job Order Contracting (JOC) Applied to Wastewater Pipeline Engineering and Construction Projects. *Proceedings of Pipelines 2016*, Reston, VA.

Lam, T. and Gale, K. (2014). Highway Maintenance: Impact of Framework Agreements Upon Project Financial Performance. *Construction Management and Economics*, 32(5), 460–472.

Moore D. R. and Dainty, A. R. J. (2001). Intra-Team Boundaries as Inhibitors of Performance Improvement in UK Design and Build Projects: A Call for Change. *Construction Management and Economics*, 19(6), 559–562.

Stanford, M. S. and Molenaar, K. R. (2016). Indefinite Delivery–Indefinite Quantity Construction Contracts: Procurement Practices and Owner Priorities. *Journal of Management in Engineering*, 32(5), 04016015.

Chapter 9

Shift of Paradigm and Future

Now we are in the last chapter. In this chapter, you will identify the trends of the project delivery systems by comparing different delivery systems and exploring the intricate relationship between societal & industry culture and the construction of project delivery mechanisms. The author has observed that there have been cases where the projects struggle when a new delivery method or a new form of contract is introduced. There are also many cases where the industry or society resists accepting a new project delivery system because some of the features in the system are incompatible with the values shared in the community. While some project delivery systems prioritize efficiency and innovation, they may overlook fairness and objectivity. Therefore, the choice of a project delivery system cannot be divorced from the cultural context in which it operates, as societal acceptance of fairness plays a crucial role in its implementation. This chapter further discusses the dynamics of the project delivery systems in the future.

Shift of the Project Delivery System Paradigm

Throughout the chapters, you may have observed emerging trends and shifts in project delivery systems. In this section, we will revisit the various types of project delivery systems from a comprehensive perspective, with a particular focus on four key criteria: (1) objectivity in the selection process, (2) organizational integration, (3) risk allocation in the compensation structure, and (4) the owner's level of control over the design process.

Selection Criteria: Objectiveness vs. Subjectiveness

When an owner selects a contractor, on what basis is that decision made? The selection process can vary in objective or subjective, and this spectrum can be used to categorize different approaches. As we discussed in Chapter 2, the owner typically selects one of four methods: Low Bid, Best Value – Fee, Best Value – Total Cost, and Qualification-Based Selection (QBS).

DOI: 10.1201/9781003628842-13

Table 9.1 categorizes the four selection methods based on how much weight is placed on price in the decision-making process. Traditionally, Low-Bid has been the preferred method among most public agencies. However, as the construction industry increasingly values collaboration and integrated approaches, subjective evaluation criteria, such as qualifications and experience, have gained weight in the evaluation factors.

Level of Integration: Fragmented vs. Integrated Approach

Another important criterion is the level of organizational integration. This refers to how effectively all project participants are brought together to collaborate throughout the project lifecycle, especially the design phase. Generally speaking, integration in the project delivery systems involves collaboration between the contractor and the designer (i.e., the architect and engineer) in the design process. The main criteria for the integration lie in how they work together beyond the organizational boundary. The concept of integration can also be extended to include integration among the contractor, designer, and owner, promoting more collaborative planning and decision-making. In the Integrated Project Delivery (IPD) method, integration even includes specialty trades to leverage their expertise early and work as a team across all project phases.

At the far left of Table 9.2, Design-Bid-Build (DBB) is the most typical example of a fragmented approach. In this method, the design and construction functions are completely separated, with no collaboration during the design phase. There is no way for the construction contractor to provide their construction (i.e., construction methods and materials) expertise during the design phase. In contrast, Construction Manager at Risk (CMAR) introduces a collaborative approach. The contractor becomes involved during the design phase, working alongside the designer to provide preconstruction service during the design phase, including constructability reviews. However, because there is no contractual relationship

Table 9.1 Contractor Selection Criteria

More Objective			*More Subjective*
Low-Bid	Best Value – Total Cost	Best Value – Fee	Qualification-Based Selection (QBS)

Table 9.2 Project Delivery Methods by the Level of Integration

Fragmented					*Integrated*
DBB	CMAR	Bridge DB	DB	Progressive DB	IPD

between the contractor and designer, it is challenging to consider CMAR an integrated approach.

Design-Build (DB) creates a legally bound relationship between the designer and contractor. They operate as a single legal entity, making it a more integrated method than CMAR. Bridging DB allows the owner to retain design control in DB by hiring a bridging designer. This allows the owner to retain design control up to approximately 30% design completion before handing it over to the design-builder. On the other hand, Progressive DB allows the owner and the DB team to collaborate from the beginning of the design phase, resulting in a more integrated delivery process compared to other types of DB.

IPD represents the most collaborative and integrated model at the far right of the table. In IPD, all core team members (i.e., the owner, designer, contractor, and key trade contractors) enter into a single multi-party agreement, sharing their risks and rewards from the early design.

Financial Risk Approach: Shifted vs. Shared

One of the critical factors in designing a project delivery system is how risks are allocated among stakeholders. In particular, financial risk influences how each stakeholder behaves. Throughout this book, we have discussed each compensation structure in depth, as it governs the commercial interests of each party. Table 9.3 presents a classification of various compensation structures based on how financial risks are transferred or shared among project participants.

On the left side of Table 9.3 are the Cost-Plus and Guaranteed Maximum Price (GMP) contracts. The owner takes on most of the financial risk in a Cost-Plus contract. The contractor is reimbursed for all actual construction costs, plus an additional fee for overhead and profit. When the scope of work is not clearly defined, the project's final cost can be uncertain until the very end, making budgeting challenging for the owner.

A GMP contract is a variation of the Cost-Plus model but includes a built-in safeguard for the owner: a maximum price cap. This protects the owner from cost overruns beyond the agreed-upon limit. While it still reimburses actual costs like a cost-plus contract, the GMP arrangement shifts more financial risk to the contractor, especially if costs exceed

Table 9.3 Compensation Structures by Financial Risk Allocation

Financial Risk Transferred			*Financial Risk Transferred*
Cost-Plu (owner) GMP (contractor)	Lump sum	Unit price	EMP

expectations. Even if the actual costs fall under the agreed-upon price, the owner pays less than the agreed-upon price (actual costs + split ratio of the difference). For the contractor, a GMP contract carries significant risk if actual costs exceed the agreed maximum – the contractor is responsible for covering the difference. However, there is room to improve the GMP mechanism to support collaboration better.

On the other hand, a Lump Sum contract motivates the contractor to reduce construction costs, as their revenue is fixed. Therefore, the contractor tends to use the most economical methods to increase their profits as long as they meet the design requirements. While this model shifts the risk of cost overruns to the contractor, it provides the owner with price certainty, locking the total project cost before construction begins. Of course, the owner should keep an eye on the potential change order issue because the change order easily increases the contract price.

A Unit Price contract shares risk between the owner and the contractor. The owner takes on the risk related to quantity variations since the final price depends on the actual quantities. The owner pays more if the quantity exceeds what was estimated. Meanwhile, the contractor takes on the risk of unit price, which reflects their productivity and cost per unit of work.

The risk-shared type of contract that is in line with an integrated approach is the Estimated Maximum Price (EMP) contract. In this model, the owner and contractor share the financial outcomes through a pain-gain mechanism. Unlike GMP contracts, where the contractor carries more risk for overruns, EMP spreads that risk more equally between both parties, in that the owner is to reimburse up to 100% project costs when the actual costs exceed the EMP. This pain-gain sharing mechanism helps align the interests, fostering collaboration and joint responsibility.

Owner's Risks in Design Issues

One of the key themes discussed throughout this book is the owner's level of control over the design process. In many ways, this control can be seen as an indicator of the owner's exposure to design-related risks. The more the owner controls the design process, the more risks the owner bears for the design errors. Table 9.4 categorizes various project delivery methods based on the degree of design control retained by the owner.

Table 9.4 Project Delivery Methods by the Owner's Design Control

Less Owner's Design Control				*More Owner's Design Control*
DB	Bridge DB	Progressive DB IPD	CMAR	DBB

In Table 9.4, DBB, located at the far right, is the project delivery method where the owner has the greatest control over the design process by directly hiring and communicating with the designer throughout the entire design phase. The next delivery method in terms of the owner's design control is CMAR. In CMAR, the owner still hires the designer independently, with a contractor (CM/GC) engaged during the design phase. In the design phase, the contractor works as a construction manager, providing preconstruction services. The Spearin Doctrine typically applies because the owner keeps control over the design in both DBB and CMAR, holding the owner responsible for the design issues.

IPD and Progressive DB are even further along the spectrum, where the designer is contractually bound to the contractor. While the owner remains engaged in the entire design process and can influence design decisions, control power is reduced compared to DBB or CMAR.

To the left on the spectrum, indicating reduced owner control over design, is Bridging DB. In this method, the owner hires a bridging designer and retains design control up to approximately 30% completion, after which the design responsibilities are handed off to the design-builder. Usually, the owner is responsible for the bridge document, not the entire design.

At the far left in Table 9.4 is DB, where the owner's design influence is the most limited. The owner may define the design criteria in the Request for Proposals, but after the design builder is selected, the owner's role in design decisions diminishes significantly. This limited involvement means the owner assumes minimal design risk, as the design responsibility largely shifts to the design-builder.

Paradigm Shift: Fragmented to Integrated

As briefly revisited above, the project delivery system has undergone notable changes over the last few decades. These include a change from price-focused selection criteria toward those that incorporate qualifications and other non-price factors, a movement from fragmented approaches (where design and construction are separated) toward integrated models, and a rethinking of risk allocation – from transferring risks solely to one party to adopting more shared risk structures. What, then, is driving such a shift in project delivery systems?

A key driver is the misalignment between traditional project delivery systems and the evolving demands of construction projects. Construction projects in today's market differ from those of the past regarding speed, complexity, and uncertainty, as presented in Table 9.5. Project owners are demanding faster delivery; construction structures are becoming more complex; and the level of uncertainty in projects has increased. Notably,

Table 9.5 Evolving Project Contexts and Project Delivery Systems

Project Context			
Speed Complexity Uncertainty	Slow Simple Certain	→	Fast Complex Uncertain
Project Delivery System			
Selection criteria Management approach Risk allocation	Objective Fragmented Shifted to one party	→	Subjective Integrated Shared

efforts to meet accelerated schedules through fast-track delivery, such as a fast-track approach, only increase the uncertainty.

For example, the owner would have contractors or specialty trades involved in the design process to address this rising uncertainty. However, engaging contractors and specialty trades early in the design phase poses significant challenges under traditional systems like DBB, with Low Bid selection. Moreover, contractors would not accept the conventional way of risk allocation in a more uncertain project context. These traditional models do not support the collaboration and integration that are critical in the evolving context. As a result, new delivery methods, selection strategies, and risk allocation models have emerged, and many of these have evolved into various project delivery systems, as we explored in the previous chapters.

Integrated Approach vs. Project Performance

In our book, one of the assumptions is that the integrated approach is effective in challenging projects (i.e., complex and uncertain context). Do we have any work to support this assumption? Yes. Here is the one. The author led Construction Industry Institute (CII)'s research team (RT) 271 on the ideal project delivery system.[1] The RT used case studies to explore how project delivery processes are designed and managed for various projects, ranging from less to more challenging. The case studies investigated (1) project description in terms of purpose, scope, and performance as conditions of satisfaction for value delivery, (2) characterization of each project in terms of challenges (e.g., complexity, uncertainty, and speed), and (3) characterization of the project delivery processes, which encompasses alignment of interests and organizational integration.

Two indices were generated to analyze the cases: the challenge index (CI) and the performance index (PI). The CI was created from the average

Low CI, High PI	Mean Value
AI	5.1
OI	5.9

High CI, High PI	Mean Value
AI	6.5
OI	4.4

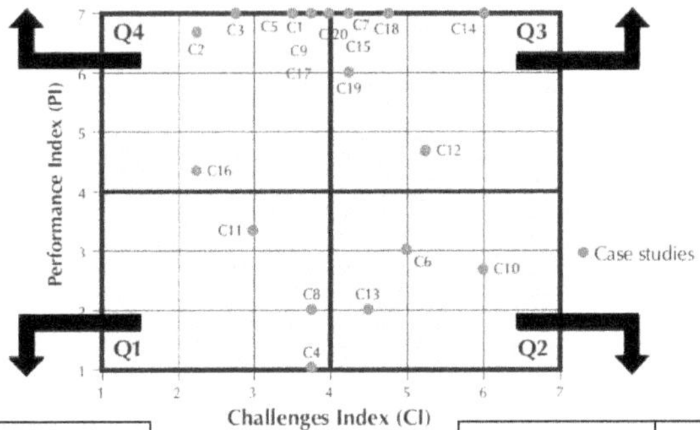

Low CI, Low PI	Mean Value
AI	3
OI	3.5

High CI, Low PI	Mean Value
AI	4
OI	3.7

Figure 9.1 CI-PI Diagram. Construction Industry Institute (2011). Starting from scratch: a new project delivery paradigm. Research Summary of Research Team 234, Austin, TX.

of complexity, uncertainty, and speed ratings in each case study. The PI was created from the average of the indicators measuring project success and quality of relationships. Illustrating the relationship between CI and PI, Figure 9.1 also includes mean scores (1–7) for alignment of integration (AI) and organizational integration (OI) for 20 completed construction projects by quadrant. To move from quadrants 1 and 2 to quadrants 3 and 4, a shift that represents an increase in project performance, there must be a corresponding increase in levels of AI and OI. Though statistically insignificant, the integrated approach is more efficient in challenging projects.

Some of the findings include:

- The high levels of organizational integration were achieved through the involvement of all parties in decision-making processes, starting from the early phases of project development.
- The alignment of interest is more effective when the project is more challenging.
- When asked what they would do differently if given the chance, the respondents from the unsuccessful projects agreed on the need for more integration to achieve the project goals.

Compatibility and Resistance: Culture Issue

In addition to CII's RT 271 research, there is a plethora of studies attesting to the effectiveness of integrated approaches compared to the traditional fragmented approach. The integrated approach is more important and effective when the project is uncertain and complex. However, we have observed industry or organization-level resistance to adopting the collaborative and integrated approach in managing projects, including the project delivery systems. Not everyone "gets" collaboration. Many organizations, and the people within them, still hold a "win–lose" view, believing that they can only "win" if someone else "loses."

Even if business leaders have committed to delivering the project more collaboratively, there can be pockets of resistance at the project level. You may observe that some parts of construction organizations (either contractors or owners), particularly larger ones, can be risk-averse. How can we overcome such obstacles? How can we reduce change resistance? The first thing we need to do before finding a solution is to understand the organizational culture.

Organizational Culture

In 2011, when I led the RT of 18 industry professionals on a new project delivery system, the team conducted interviews with construction organizations, including the owner's agency and construction contractors, to understand our industry culture. Most of the interviewees had a hard time representing their organizational culture. It might be a better approach to define the word "culture" in the first place. We should pause and reflect on the word "culture." When the project went through a hard time in its delivery, we often hear contract managers say things like "It's all about culture" and "It's a people matter." On the positive side, such statements suggest our industry appreciates the importance of culture in the project delivery system. The downside is that most construction professionals do not have a clear definition of the word "culture."

One prevailing definition of culture that fits with the project delivery system is the pattern of basic assumptions that a given group has developed in learning to cope with its problems.[2] From this definition, culture shapes how members of the organization perceive, think, and even behave. Schein's model shows three levels of culture, which are:

Artifacts and Behavior

Artifacts and behavior represent a tangible aspect of the organizational culture based on what is written or seen at the surface level. At this level, we see documents or objects such as procurement manuals or boiler templates of construction contracts. For example, suppose a project delivery manual mandates that procurement requires at least three contractors to submit low bids. This requirement suggests that the organization prioritizes price as the most critical factor in project procurement.

Norm and Value

Norms are generally accepted rules of behavior. They are not necessarily written down in the documents, but most people in the organization or community know the rules of behavior. Values are the principles people live by and consider important in their decision-making processes.

Norms in an organization's project delivery system influence decisions, including the project delivery system owners choose to implement and the project delivery system contractors are comfortable with. While they may not be explicitly stated in manuals or policies, they have an impact on how they behave or their daily practices. For example, in an organization where collaboration is a norm, project teams might be comfortable with an integrated or collaborative approach such as IPD or DB. Similarly, decision-making may be centralized if a hierarchical structure or command-and-control is the norm. Even when unwritten, these norms impact how projects are executed.

Values, on the other hand, reflect the principles that guide an organization's approach to project delivery. When asked what their organization stands for, employees may respond with words like "trust," "efficiency," or "fairness." These values influence contractor selection methods and contract structures. For instance, a public-sector agency that values "fair competition" may prefer a Low Bid, while a company that prioritizes "relationship" might favor Best Value or QBS. Values are embedded in how project teams set priorities, shaping the procedures of the project delivery system.

Underlying Assumptions

What do we assume when we communicate our thoughts regarding decisions? When we make decisions or communicate our thoughts, we rarely articulate the assumptions that guide our reasoning. In many cases, we are not even consciously aware of these assumptions. It is no wonder that people often share similar underlying beliefs or assumptions that shape their reasoning. If you are a public owner's agency contracting officer, these assumptions may influence procurement and contract strategies in a project delivery system.

For instance, in many construction organizations (either contractors or owners), there is an implicit assumption that shorter project timelines inevitably increase expenses. These assumptions drive decision-making, even if they are not explicitly stated. Consider the case of selection methods – many public sector owners require at least three proposals before evaluating a bid. If asked why, employees may simply shrug and say, "Isn't it too obvious?" This reaction suggests a deeply held assumption: competition leads to fairness and cost-effectiveness. Similarly, in selecting a compensation structure or contract type, some owners who favor Lump Sum contracts often operate under the assumption that cost certainty outweighs cost flexibility, such as a Cost-Plus contract. Meanwhile, those who favor GMP or EMP contracts may assume that uncertainty is inevitable and collaboration is essential to manage it. These underlying assumptions, though unspoken, shape the project delivery landscape and define how organizations approach procurement and project risks.

Dimensions of Culture and Project Delivery System Compatibility

Now that we have explored what culture is and how it manifests in project delivery systems, the next challenge is to understand how organizational culture is related to compatible project delivery methods. Even with the help of the previous section, culture feels intangible, making it challenging to assess systematically. However, before analyzing its impact on project delivery systems, it is helpful to explore dimensions of the culture that provide a structured way to compare different cultural environments.

One of the most well-known quantitative studies of culture is the work of Geert Hofstede and his team,[3] who identified primary dimensions that differentiate cultures across countries. These dimensions provide a framework for understanding cultural attributes. The cultural dimensions relevant to the project delivery system include Power Distance (degree of hierarchy acceptance), Individualism, Uncertainty Avoidance (tolerance for ambiguity), and Long-Term Orientation. The following will explore these dimensions in detail.

Power Distance

Power distance refers to the extent to which a community accepts an unequal power distribution. It reflects the degree to which people at the lower hierarchy levels accept this power imbalance. In general, communities with high power distance tend to have strong hierarchical structures. Decision-making is often centralized, with senior leaders holding significant control. This cultural characteristic influences the selection of project delivery systems in several ways.

First, organizations with a high-power distance culture typically have well-defined hierarchies and a chain of command. In such environments, owner organizations prefer to retain control over the entire project delivery process, including design decisions. Likewise, general contractors tend to view trade contractors as subcontractors in a hierarchical relationship rather than equal collaborators. Given these attributes, organizations with high power distance are more likely to prefer the DBB delivery method. They prefer a well-defined hierarchy over more integrated approaches like DB, in which the owner does not retain control over the design process. Similarly, contractors with a high power distance culture are not likely to accept the IPD, where the mechanical and electrical contractors are their partners and share the risks and rewards.

Second, rule-based decision-making and chain-of-command are favored over negotiation or collaborative processes. Low Bid procurement tends to be preferred over QBS to have a transparent, rule-based decision process in selecting the contractors. Similarly, in terms of compensation structures, lump-sum contracts are favored over cost-plus or EMP contracts, as the latter requires negotiation and flexibility – elements that may be less compatible with high-power distance cultures.

Individualism

Individualism represents how much they prioritize personal interests over group interests. In individualistic societies, individuals tend to prioritize personal interests and self-reliance. Each individual tends to take care of themselves, and professional relationships are often transactional. In contrast, collectivistic societies accept group cohesion and mutual support. The success of the group is often seen as more important than individual achievements.

This cultural dimension can impact the way they manage the project and interact with other stakeholders. In communities with high collectivism, project teams may prefer a collaborative or integrated approach, such as IPD, where stakeholders need to work closely. The project goals should be prioritized over silo minded. In contrast, organizations with individualistic cultures may be comfortable with silo minded. They are not likely to accept the shared risks/reward model. Instead, they tend to prefer transactional project delivery methods, where contracts focus on competition rather than long-term collaboration. In such environments, stakeholders operate independently, and selection methods often prioritize the lowest.

Uncertainty Avoidance

Uncertainty avoidance addresses the extent to which a community or organization feels comfortable with risk and unpredictability. Communities with high uncertainty avoidance prefer project environments with defined rules and formalized processes. In contrast, communities with low uncertainty avoidance are more open to flexible arrangements and uncertainties. This cultural attribute influences how organizations approach compensation structures and management strategies.

With high uncertainty avoidance cultures, owner organizations tend to minimize financial and contractual risks by not favoring contingencies or any mechanism that can change the final compensation at the end of the project. These organizations often prefer lump-sum contracts, where the project's cost is predetermined before construction begins. The preference for fixed-price arrangements reflects a need for cost certainty.

This risk-averse mindset can also create resistance to adopting a collaborative or integrated approach, where the organization should accept some uncertainties or a risk-sharing compensation structure. For instance, IPD and CMAR usually involve contract types that introduce financial uncertainty, such as GMP or Cost-Plus contracts. They may be viewed as too uncertain, as the final project cost remains unknown until completion.

Though they adopt CMAR with GMP, the owners with a risk-averse mindset tend to finalize the GMP as late as possible to minimize the design contingency in the contract. IPD can be viewed as risky, especially as it involves aggressive risk-sharing mechanisms that carry uncertainty and financial risks. Organizations with high uncertainty avoidance may perceive these models as lacking predictability. As a result, despite the potential benefits of efficiency, such organizations may be reluctant to move away from innovative project delivery methods such as IPD.

Long-Term Orientation

The community with a long-term orientation values future-focused thinking, sustained growth, and long-term relationships rather than short-term gains. These societies value collaboration and trust rather than competition and transactional gain. This perspective influences how project delivery systems are structured, particularly in the contractor relationships and procurement strategies.

The long-term oriented organizations tend to prefer developing long-term partnerships between owners and contractors or between contractors and subcontractors. As a result, IPD or framework agreements can be favored, as they focus on collaborative and sustainable relationships over multiple projects. On the other hand, short-term-oriented organizations are likely to rely on transactional delivery models, such as DBB with Low Bid selection, where each project is treated as an independent contract with little regard for future partnerships.

For example, the Procure21/22 program uses a framework contract model, where a primary supply chain partner (usually a general contractor) develops a dedicated supply chain team. The supply chain team remains unchanged during the framework contract period. The owner then enters into a framework contract for an extended period, fostering a long-term relationship. On the other hand, organizations or governments that focus on immediate cost savings may be more resistant to adopting such relationship-based models.

The Next Decades of the Project Delivery Systems

Construction Project Context: Evolving Demands

Over the past decades, we have observed changes in the context of construction projects. We expect this shift to continue and even intensify in the years ahead. These changes are not random; they follow recognizable patterns, and understanding these patterns is key to getting you better prepared for the future of project delivery systems.

One of the most notable trends is that project owners place a premium on speedy delivery and relationship-based collaboration. More projects are being delivered under fast-track approaches, where construction begins even before the design is fully complete. At the same time, owners are more interested in long-term, trust-based relationships with contractors and suppliers rather than transactional ones. The need for speed inevitably increases the level of project uncertainty. As this pattern continues, uncertainty and complexity will likely become even more defining characteristics of construction projects in the coming decades.

Another noticeable trend is the use of prefabrication or modular construction in construction projects.[4] The prefabrication shifts work from construction sites to fabrication shops, reducing site congestion and risks of construction projects. As prefabrication/modular construction has increased in the construction industry, the supply chains are a key to the success of construction projects. In this regard, long-term relationships and a collaborative/integrated approach should be core values in the project delivery system.

To properly respond to these evolving demands, project management must adapt on two fronts. First, it requires project management tools and techniques that support speed, flexibility, and early decision-making under uncertainty. Second, and just as importantly, the industry needs organizational structures – namely, project delivery systems – that enable these processes to be implemented effectively.

Project Management Tools and Technologies

The future of project delivery systems should be compatible with how we manage future construction projects. How we manage projects (i.e., design and construction) must also evolve to better respond to the construction context and needs. Collaborative/integrated approaches have proven more effective and efficient in this changing project context and needs. In addition, more construction organizations are likely to move toward a reliable supply chain.

Today's integrated approach in project management should go beyond simply involving downstream stakeholders (e.g., specialty trades) early in the design phase. To mature and advance the project delivery system, two components must support one another: (1) information technology and (2) project management processes.

Information Technology for Integration

Information technologies have developed rapidly to support an integrated approach in project management. Building Information Modeling (BIM) is critical in most contemporary construction projects. BIM allows project stakeholders to share a digital model of the building throughout the project lifecycle, improving communication, coordination, and decision-making. Using BIM, project stakeholders (e.g., designers, contractors, and owners) can identify design conflicts early, simulate construction sequences, and better predict time and cost impacts. These capabilities help reduce uncertainty and make the project more predictable from an early stage. It is no wonder that you find the VDC (virtual design and construction) department in most construction companies.

In addition to BIM, other technologies, such as cloud-based collaboration platforms and digital project management tools, also contribute to more transparent and timely communication. These tools allow real-time updates and shared access to documents, making it easier to coordinate across teams. Such technological advancements have been and will continue to work with project delivery systems.

First, some delivery methods, such as IPD, require the team to work collaboratively. There are cases where specific technology or software (e.g., BIM) is required in the contract document. As you've seen in Chapter 9, Procure 21/22 enforces the team to use a collaboration platform. The technology fosters and enhances collaboration efficiency, such as data sharing.

Second, emerging technologies have the potential to significantly improve project delivery systems, particularly in the areas of compensation structures and contractual practices. For example, the blockchain or distributed ledger technology is a protocol that functions as a decentralized digital database where transactions are recorded in a distributed ledger. The smart contracts, built on blockchain platforms, have added a powerful tool to automate the execution of contractual terms and conditions. Such technology can potentially improve transparency and streamline compensation workflow in construction projects.

Project Management Processes

Alongside technology, project management methods have also evolved to support integrated approaches in project management. One of the leading movements is Lean Construction, which emphasizes a pull strategy that involves the downstream stakeholders in the upstream decision-making process. The general goal of Lean Construction is to improve workflow reliability through better coordination in planning and scheduling. Tools like the Target Value Design (TVD) process help teams align design decisions with budget constraints and project goals. These processes assume that downstream stakeholders, such as MEP (Mechanical, Electrical, and Plumbing) trade contractors, get involved in early design, encouraging collaboration during design.

The project delivery systems have also evolved to support the project management methods. A project delivery system provides the organizational and contractual framework within which a project is executed, and it must align with the project management tools and strategies. Historically, this relationship became evident in the early 21st century, when the industry tried to change traditional delivery systems to better accommodate Lean Construction and BIM. Recent models like IPD and EMP were developed as a result. Project delivery systems will continue to adapt emerging project management tools and methods.

Future and Dynamics of the Project Delivery Systems

As we have explored throughout this book, project delivery systems are not static – they evolve in response to the changing context of construction projects. How a project (i.e., design and construction) is managed is influenced by the project context: the level of uncertainty, complexity, and the need to fast-track. The project delivery system should be designed to support how the design and construction are managed.

For example, if an owner wants to implement TVD or an Integrated Design process, the delivery system should be designed to support them. This means that not only the general contractor but also specialty trade contractors need to be engaged as early as the Design Development phase. To support this early involvement, selection should no longer be based on the lowest price, but rather on qualifications and relationship. In other words, the project delivery system must be equipped to support the way the project will be managed in three terms: (1) project delivery method, (2) selection method, and (3) compensation structure.

As we've seen in previous chapters of this book, the last few decades have seen a shift toward a more integrated approach in the project delivery systems. In this sense, the future of project delivery systems is not a far stretch. We can expect several trends in the project delivery system:

- Project delivery systems will increasingly allow and encourage the participation of specialty contractors and key suppliers in the design phase.
- Risk will be shared among stakeholders, rather than transferred onto a single party.
- The industry will move from one-time transactional relationships to trust-based, long-term partnerships.
- Selection criteria will further shift from being cost-focused to qualifications and relationships.
- Compensation structures will evolve to foster innovation and improved productivity better.
- Emerging technologies (e.g., blockchain) will be actively employed for speedy and transparent data sharing and transactions.

However, designing a well-structured project delivery system does not guarantee project success. Even the project management methods and tools do not guarantee project success. One of the most critical factors is organizational culture. As discussed in the previous section, even a well-structured delivery system can lead to friction, resistance, or worse outcomes than traditional methods if the organizational or project culture is incompatible with integrated approaches.

The author has a key question: Is organizational culture fixed, or can it evolve over time? The answer is: Yes, it can change – but slowly. Culture tends to lag behind changes in management practices and technology, but history shows that it does evolve. Consider how many traditionally Confucian Asian societies have embraced modernization and economic development, shifting significantly toward more Westernized values over time. Similarly, as project management environments change, through changes in technology and management practices, so will organizational culture.

Changes in construction management curricula at the Universities will also play an important role in this transformation. It is easily observed that many construction programs at universities have adopted an interdisciplinary curriculum where construction students work closely with architecture students on the projects. Students learn how to collaborate and coordinate with each other. Though gradual, these shifts will eventually support the organizational or industry-level adoption of integrated approaches in project delivery systems.

This dynamic nature of project delivery – where culture, process, technology, and education evolve together – suggests that the future of the project delivery method is not just about how we design the project delivery system. People and organizational culture will shape the future of the project delivery system. If we can align organizational culture with emerging project delivery strategies, we are not just building better projects but a better industry.

Chapter Summary

1. Low Bid has been the preferred method among most public agencies. However, as the construction industry increasingly values collaboration and integrated approaches, subjective evaluation criteria, such as qualifications and experience, have gained weight in the evaluation factors.
2. DBB is the most typical example of a fragmented approach. In contrast, CMAR introduces a collaborative approach. DB creates a legally bound relationship between the designer and contractor. They operate as a single legal entity, making it a more integrated method than CMAR. IPD represents the most collaborative and integrated model.
3. Regarding risk allocation, the owner takes on most of the financial risk in a Cost-Plus contract. The GMP arrangement shifts more financial risk to the contractor. In contrast, a Lump-Sum contract motivates the contractor to reduce construction costs, as their revenue is fixed. A Unit Price contract shares risk between the owner and the contractor. The owner takes on the risk related to quantity variations, while the contractor takes on the unit price risk, reflecting their productivity. The risk-shared type of contract that is in line with an integrated approach is the EMP contract.
4. The project delivery system has undergone notable changes over the last few decades. There is a movement from fragmented approaches (where design and construction are separated) toward integrated models. A key driver for this movement is the misalignment between traditional project delivery systems and the evolving demands of construction projects.
5. The high levels of organizational integration were achieved through the involvement of all parties in decision-making processes, starting from the early phases of project development. The alignment of interest is more effective when the project is more challenging.
6. Even if business leaders have committed to delivering the project more collaboratively, there can be pockets of resistance at the project level. Organizations with high power distance are more likely to prefer the DBB delivery method. They prefer a well-defined hierarchy in terms of the roles. Low Bid procurement tends to be preferred over QBS because it has a transparent, rule-based decision process when selecting contractors.

7 In highly collectivist communities, project teams may prefer a collaborative or integrated approach, such as IPD, where stakeholders must work closely. In contrast, organizations with individualistic cultures may be comfortable with silo minded. They are not likely to accept the shared risks/reward model.
8 The risk-averse mindset can also resist adopting a collaborative or integrated approach, where the organization should accept some uncertainties or a risk-sharing compensation structure. Though they adopt CMAR with GMP, the owners with a risk-averse mindset tend to finalize the GMP as late as possible to minimize the design contingency in the contract.
9 The long-term oriented organizations tend to prefer developing long-term partnerships between owners and contractors or between contractors and subcontractors. As a result, IPD or framework agreements can be favored.
10 Project owners place a premium on speedy delivery and relationship-based collaboration. The need for speed inevitably increases the level of project uncertainty. As this pattern continues, uncertainty and complexity will likely become even more defining characteristics of construction projects. To appropriately respond to these evolving demands, project management must adapt on two fronts. First, it requires new processes and tools supporting speed, flexibility, and uncertainty-free decision-making. Second, and just as importantly, the industry needs project delivery systems that enable these processes to be implemented effectively.
11 As we have explored throughout this book, project delivery systems are not static – they evolve in response to the changing context of construction projects. The project delivery system should be designed to support how the design and construction are managed. However, designing a well-structured project delivery system does not guarantee project success. The dynamic nature of project delivery – where culture, process, technology, and education evolve together – suggests that the future of the project delivery method is not just about how we design the project delivery system.

Review Questions

1 What are the three components of the organizational culture?
2 What is the impact of the owner's weight on delivery speed on the project delivery system?
3 What types of project delivery systems are more compatible with an organization with a culture of uncertainty avoidance?

Notes

1 Construction Industry Institute (CII) (2011). *Starting from Scratch: A New Project Delivery Paradigm*. Research Summary of Research Team 234, Austin, TX.
2 Schein, E. (1984). Coming to a New Awareness of Organizational Culture. *Sloan Management Review*, 25(2), 3–16.
3 Hofstede, G. (2010). *Cultures and Organizations: Software of the Mind* (3rd Edition). McGraw-Hill, New York.
4 McGraw Hill Construction (2011). *Prefabrication and Modularization: Increasing productivity in the construction industry*. Smart Market Report.

Bibliography

Bygballe, L. and Swärd, A. (2019). Collaborative Project Delivery Models and the Role of Routines in Institutionalizing Partnering. *Project Management Journal*, 50, 161–176.
Construction Industry Institute (CII) (2011). *Starting from Scratch: A New Project Delivery Paradigm*. Research Summary of Research Team 234, Austin, TX.
Hofstede, G. (2010). *Cultures and Organizations: Software of the Mind* (3rd Edition). McGraw-Hill, New York.
Kelly, D. and Ilozor, B. D. (2022). Performance Outcome Assessment of the Integrated Project Delivery (IPD) Method for Commercial Construction Projects in the USA. *International Journal of Construction Management*, 22(14), 2808–2816.
Mitropoulos, P. and Taum, C. B. (2000). Management-Driven Integration. *Journal of Management in Engineering*, 16(1), 48–58.
Schein, E. (1984). Coming to a New Awareness of Organizational Culture. *Sloan Management Review*, 25(2), 3–16.

Index

For Product Safety Concerns and Information please contact our EU representative GPSR@taylorandfrancis.com
Taylor & Francis Verlag GmbH, Kaufingerstraße 24, 80331 München, Germany

www.ingramcontent.com/pod-product-compliance
Ingram Content Group UK Ltd.
Pitfield, Milton Keynes, MK11 3LW, UK
UKHW021502240926
475702UK00035B/166
* 9 7 8 1 0 4 1 0 4 1 8 4 9 *

Waste Valorisation to Nanomaterials

In recent years, the focus has shifted towards the development of nanomaterials using green synthetic methods, and the concept of *waste-treat-waste* has led to the valorisation of waste materials into nanomaterials and related products. This book reflects on the status of the synthesis of nanomaterials from waste materials using green synthetic approaches. Fundamental aspects involving the synthesis, characterisation, and applications of waste-derived nanomaterials are explored.

Features:

- Covers the mechanisms involved in the synthesis of nanomaterials from economically and environmentally feasible sources
- Explains the processes involved in waste valorisation
- Discusses the impact of relevant processes on the circular economy
- Reviews the use of waste-derived nanomaterials as alternative sorbents to commercial and expensive solid-phase extraction columns
- Explores inexpensive and readily available precursors for the development of functional nanomaterials

This book is aimed at researchers and graduate students in environmental and chemical engineering, waste management, remediation, and environmental chemistry.

Emerging Materials and Technologies

Series Editor: Boris I. Kharissov

The *Emerging Materials and Technologies* series is devoted to highlighting publications centered on emerging advanced materials and novel technologies. Attention is paid to those newly discovered or applied materials with potential to solve pressing societal problems and improve quality of life, corresponding to environmental protection, medicine, communications, energy, transportation, advanced manufacturing, and related areas.

The series takes into account that, under present strong demands for energy, material, and cost savings, as well as heavy contamination problems and worldwide pandemic conditions, the area of emerging materials and related scalable technologies is a highly interdisciplinary field, with the need for researchers, professionals, and academics across the spectrum of engineering and technological disciplines. The main objective of this book series is to attract more attention to these materials and technologies and invite conversation among the international R&D community.

Lightweight Materials for Electric Vehicles: Sustainable Materials, Production Process, and Modeling Techniques
Edited by R. Suresh, C. Durga Prasad, Satish Kumar, Bharath K.N. and Ajith G. Joshi

Graphitic Carbon Nitrides
Michio Inagaki, Masahiro Toyoda and Yasushi Soneda

The Science and Applications of 3D Bioprinting
Edited by Thiago Domingues Stocco and Anderson de Oliveira Lobo

Metal-Organic Frameworks for Anticorrosive Coatings
Edited by Ashish Kumar, Ambrish Singh and Humira Assad

Evaluation of Electrochemical Corrosion: Experimental Procedures and Characterization
Nam Nguyen-Dang

Smart Functional Wearables: From Nanofibers to Textiles
Majid Montazer and Ali Mianehro

Waste Valorisation to Nanomaterials: Classification, Synthesis Methods, and Applications
Anele Mpupa, Shirley Kholofelo Selahle, Philiswa Nosizo Nomngongo, Mthokozisi Mnguni, Ngwako Joseas Waleng and Siyabonga Shoba

For more information about this series, please visit: www.routledge.com/Emerging-Materials-and-Technologies/book-series/CRCEMT

Waste Valorisation to Nanomaterials

Classification, Synthesis Methods, and Applications

Anele Mpupa, Shirley Kholofelo Selahle,
Philiswa Nosizo Nomngongo, Mthokozisi Mnguni,
Ngwako Joseas Waleng, and Siyabonga Shoba

CRC Press
Taylor & Francis Group
Boca Raton London New York

CRC Press is an imprint of the
Taylor & Francis Group, an **informa** business

Designed cover image: Getty Images

First edition published 2027
by CRC Press
2385 NW Executive Center Drive, Suite 320, Boca Raton FL 33431

and by CRC Press
4 Park Square, Milton Park, Abingdon, Oxon, OX14 4RN

CRC Press is an imprint of Taylor & Francis Group, LLC

ISBN: 978-1-032-64544-5 (hbk)
ISBN: 978-1-041-25257-3 (pbk)
ISBN: 978-1-003-74584-6 (ebk)

DOI: 10.1201/9781003745846

Typeset in Times
by Apex CoVantage, LLC

Contents

About the Authors

Anele Mpupa obtained his Bachelor of Science (Chemistry and Microbiology) and BSc (Hons) in Chemistry from Rhodes University. Later, he completed his MSc and PhD in chemistry from the University of Johannesburg. He has received funding from the National Research Foundation (NRF) from 2017 to 2020 for his MSc and PhD, respectively; he was also a recipient of the highly competitive Deutscher Akademischer Austauschdienst (DAAD) German academic exchange service for a 4-month short-stay research grant at Ulm University. He then served as a postdoctoral research fellow at the University of Johannesburg under the Department of Science, Technology and Innovation-National Research Foundation South African Research Chair Initiative (DSTI-NRF SARChI) in nanotechnology for water. His research is centred around water research, which includes water quality monitoring and the analysis of chemical and biological pollutants in water. Furthermore, Dr Mpupa focuses on the development of methodologies based on chromatography and spectroscopic detection of water pollutants. Additionally, Dr Mpupa's research also adds an element of material science by preparing different materials for use as sorbents; these include metal oxides, nanomaterial composites, and functional materials, such as molecularly imprinted polymers, for the extraction and removal of pollutants. Therefore, the research focuses on the development of adsorbent materials for the determination of emerging pollutants in water matrices. Furthermore, some of the developed materials are tested for their potential for the mitigation of pollutants in water, with special attention being paid to efforts on using recyclable feedstocks.

Shirley Kholofelo Selahle is an emerging researcher who graduated with a Doctor of Philosophy degree in chemistry from the University of Johannesburg. After obtaining her PhD degree, she started her postdoctoral research fellowship at the University of Johannesburg. She joined the University of Venda later in 2023 as a lecturer in analytical chemistry. Dr Selahle's arrival at the University of Venda saw her establish an Analytical Environmental Chemistry Research Laboratory/Group that is growing yearly. This group is currently composed of four MSc students, one PhD student, and three Honors students under her supervision. This research group focuses on the development of sample preparation methods for the analysis and removal of emerging pollutants in water matrices using nanomaterials. Emerging pollutants are chemical substances that fall out of standard monitoring and regulatory programmes. These chemicals occur at low concentration levels, which makes their monitoring a challenge. Therefore, her research focus also expands to the development of adsorbent materials for the determination and quantification of emerging pollutants in environmental water matrices. Taking into consideration green chemistry, her research also involves the use of waste materials to make composites that are friendly to the environment and to utilise them for water treatment. The materials are designed in such a way that they result in a high percentage of recoveries of the targeted pollutants and improve the detection limits of the analytical instruments used for analysis. During her academic journey, she has received several funds, including the Global Excellence Stature (GES) fellowship and the National Research Foundation in support of her postdoctoral fellowship. She is currently an emerging researcher who is funded by the THUTHUKA grant from the National Research Foundation. She has contributed to the body of knowledge through research publications in reputable journals and presentations at local and international conferences. She also serves as a reviewer in international journals such as *MDPI* and *Frontiers in Chemistry*, among others. In 2024, she was recognised as one of the 200 young South Africans making a positive impact on the country by the *Mail* and *Guardian*.

Philiswa Nosizo Nomngongo is a full professor of environmental analytical chemistry at the University of Johannesburg, Faculty of Science (Department of Chemical Sciences). She holds a DSI-NRF SARChI in Nanotechnology for Water. Her scientific career has been dedicated to environmental analytical chemistry and nanotechnology, aiming to solve environmental problems in water quality and environmental protection. Her most significant contributions have been in developing sample preparation methodologies using nanostructured materials as adsorbents for monitoring inorganic (toxic trace metals and metalloids) and organic pollutants (more specifically, emerging contaminants, which include pesticides, endocrine disruptors, and personal care and pharmaceutical products) in water systems. Her current research includes the application of various nanomaterials in fields that are attracting increasing interest, such as wastewater treatment, groundwater remediation, water quality monitoring, development of water quality sensors, and treatment and recovery of precious metals, among others. Her research achievements have been recognised through numerous awards such as the L'Oréal-UNESCO Sub-Saharan Women in Science Regional Fellowship in 2014, the South African Women in Science Award in the Distinguished Young Woman Research in the Natural and Engineering Sciences category in 2017, and the Vice-Chancellor's Distinguished Award: Most Promising Young Researcher of the Year in 2017. Her other recognition includes African Academy of Science Affiliate, Southern African Women in Leadership Trailblazers Award winner (Top 10), 2020/2021 NSTF-South23 Award winner of the Engineering Research Capacity Development category; NSTF-South23 Award Finalist (Categories: 2019/2020 TW Kambule: Emerging Researcher & Engineering Research Capacity Development; 2020/2021 TW Kambule: Researcher), 2025 James J. Morgan Early Career Award (American Chemical Society), among others. During her academic years at the University of Johannesburg (2015–present), Prof Nomngongo has supervised/co-supervised 48 MSc and 32 PhD students. She is currently the supervisor/co-supervisor to several MSc and PhD students. She is the author/co-author of over 170 peer-reviewed publications, 28 book chapters, and two peer-reviewed conference proceedings. She has presented 50+ keynotes and invited contributions at scientific conferences. Her Google Scholar, Scopus, and Web of Science H indices are 43, 37, and 34, respectively. Her work has been cited 5,781 (Google Scholar), 4,521 (Scopus), and 3,832 (Web of Science) times. She is the associate editor of leading and prestigious international journals and publishing houses. She is also the topic editor for *ACS Environmental Science & Technology Letters*. She serves as an editorial board member of *Discover Applied Sciences*, *Energy & Environmental Materials*, *ACS Environmental Science & Technology* and *Environmental Toxicology and Chemistry*. She served as the guest editor for some international journals that are DHET accredited, including *Frontiers in Chemistry*, *Frontiers in Environmental Science*, *Molecules*, and *Materials*.

Mthokozisi Mnguni completed his Matric in 2010 at Adams College in Durban. He completed his BSc in chemistry and chemical technology (2014), Honors in chemistry (2015), MSc in chemistry (2017), and PhD in chemistry (2022) at the University of KwaZulu-Natal. The National Research Foundation (NRF) funded all the degrees besides the BSc. Throughout his MSc research project, he acquired expertise in organic synthesis and studied the mechanisms and kinetics of biological reactions. In his PhD project, Mthokozisi Mnguni obtained expertise in designing nanomaterials with specialised morphologies for sample preparation of emerging contaminants from water. The knowledge obtained here enabled the appointment as a postdoctoral research fellow in September 2023 at the University of Johannesburg under Prof Nomngongo, a DSI-NRF SARChI in Nanotechnology for Water. The following year (2024), Mthokozisi Mnguni obtained NRF postdoctoral research funding for two years to research the fabrication of nanomaterials for extraction and removal of organic and inorganic contaminants from water. Mthokozisi Mnguni has co-supervised four MSc one Honors, students to completion and is currently co-supervising one MSc, and five PhD students. Mthokozisi Mnguni has presented his research at six local and international conferences.

Ngwako Joseas Waleng is an assistant lecturer and postdoctoral researcher in the Department of Chemical Sciences at the University of Johannesburg, South Africa. He specialises in analytical chemistry, with expertise in developing and applying advanced nanomaterials for wastewater remediation. He earned his PhD in analytical chemistry from the University of Johannesburg, where his research focused on developing sample preparation methods for the preconcentration and removal of pharmaceuticals from the environment. His research interests include environmental monitoring, wastewater remediation, nanomaterials for pollutant removal, chromatography, and chemometrics. His work integrates green-chemistry principles, functional nanomaterials, and data-driven approaches to address pressing challenges in pollution control, water quality management, and sustainable development. He has authored and co-authored 10 peer-reviewed publications, contributing to the advancement of knowledge in analytical and environmental chemistry. He has also presented his research at both local and international platforms, engaging multidisciplinary audiences on innovative approaches to environmental remediation and analytical method development. Beyond research, he is actively involved in teaching and mentoring undergraduate and postgraduate students. He has experience in teaching courses such as general chemistry, instrumental analysis, and research methodology. His long-term goal is to bridge fundamental research in analytical chemistry with practical solutions for environmental and industrial challenges, thereby advancing both scientific innovation and societal impact.

Siyabonga Shoba is a chemist and researcher with a passion for discovery and storytelling. He earned his master's degree from Nelson Mandela University and holds a PhD in inorganic and analytical electrochemistry from the University of KwaZulu-Natal. He did a postdoctoral fellowship at Stellenbosch University under the mentorship of Prof Wessel du Toit and Prof Robert Pott, where he investigated oxygen mass transport in wine during nitrogen sparging and applied computational fluid dynamics (CFD). While also mentoring emerging researchers, Siyabonga is currently a postdoctoral fellow at the University of Johannesburg under the guidance of Prof Philiswa Nosizo Nomngongo. His research focuses on converting waste-derived materials into nanomaterials for the chemical modification of electrodes to detect and monitor heavy metals and emerging organic contaminants in water. He is also continuing to mentor the next generation of scientists.

Preface

Global population growth, urbanisation, rising living standards, and economic growth have led to unprecedented levels of waste generation, exacerbating environmental degradation and negative health impacts, thereby putting pressure on existing waste management systems and natural ecosystems. Electronic, agricultural, municipal, and industrial activities contribute to the global solid waste stream, which often has limited options for safe disposal or reuse. Against this background, waste valorisation has emerged as a compelling strategy for transforming various waste materials into valuable resources or value-added products. Among the most promising outcomes of the waste valorisation approach is the conversion of waste materials into advanced functional nanomaterials, with applications spanning environmental monitoring and remediation, sensing technology, biomedical science, and clean energy production.

This book, *Waste Valorisation to Nanomaterials: Classification, Synthesis Methods, and Applications*, is a response to the rapid and extensive research on waste management, nanotechnology, material science, and the sustainability nexus. It brings together fundamental concepts and advances in methodologies for preparing waste-derived nanomaterials, thereby providing a comprehensive and logical account of how various waste streams are transformed into functional waste-derived nanomaterials. These materials proved promising candidates for applications in sensing technology, sample preparation, wastewater and water remediation, and the pharmaceutical, cosmetic, and food industries. Therefore, the book contributes to the broader objectives of the circular economy, resource efficiency, and sustainable development.

The book is structured to guide readers logically from foundational principles to advanced applications. It begins with an overview and classification of waste-derived nanomaterials, emphasising the physicochemical properties that make them well-suited to specific applications. Successive chapters detail synthesis ways, including physical, chemical, and green methodologies. In addition to synthesis and classification of waste-derived materials, the book critically examines real-world applications of these nanomaterials. This book is intended for a broad readership, including postgraduate students, researchers, academic staff, and professionals working in various fields such as waste management, chemistry, materials science and environmental engineering, among others.

The authors declare that during the preparation of this work, Grammarly (v1.2.279.1925) and ChatGPT (v5.5) were solely used to improve and refine their English language and writing, while the originality and scientific content remain entirely the work of the authors, and to enhance readability and clarity. The authors have reviewed and edited the AI-generated content as necessary and take full responsibility for the contents of this publication.

Anele Mpupa
Shirley Kholofelo Selahle
Philiswa Nosizo Nomngongo
Mthokozisi Mnguni
Ngwako Joseas Waleng
Siyabonga Shoba

Acknowledgements

The authors gratefully acknowledge the contributions of the book proposal reviewers, whose suggestions and expertise have shaped the structure of this volume, as well as the wider scientific community, whose published articles and book chapters continue to push the boundaries of waste valorisation for sustainable nanotechnology and environmental research. The authors wish to thank the University of Johannesburg and the University of Venda, South Africa, for continued support and encouragement. They also wish to extend their appreciation to Routledge, Taylor & Francis Group, for the opportunity to write and publish this book.

1 General introduction

Waste-derived nanomaterials are emerging as a sustainable solution that integrates nanotechnology with waste management to address environmental challenges. These materials offer unique properties that make them suitable for applications in analytical chemistry, environmental remediation, sensing technologies, and industrial sectors such as food, pharmaceuticals, and cosmetics. Despite their advantages, concerns about toxicity and safety persist, underscoring the need for continued research and regulatory development. Overall, waste-derived nanomaterials offer a promising pathway to advance the circular economy and sustainable innovation.

The emergence of nanotechnology as a field of science and technology has led to significant efforts to investigate the remarkable properties of nanomaterials and their potential applications (Bhattacharjee et al., 2024). The conversion of waste materials into nanomaterials using various synthesis methods, as discussed in Chapter 5, is a fast-growing field with great potential across various applications and has introduced a new era in nanotechnology and environmental research (Samaddar et al., 2018; Dutta et al., 2018; Amdeha, 2021). The connection between global waste management crises and the urgent need for green, sustainable material design is particularly evident in the role of waste-derived nanomaterials as an innovative pathway (Venkataraman & Sundramoorthy, 2026). This book emphasises the potential of recycling and upcycling various waste streams (Figure 1.1; see also Chapter 5) into high-value nanomaterials, offering the opportunity to redefine waste as an indispensable, untapped reservoir of advanced material precursors. As a result, recent studies have shown that nanotechnology plays a significant role in waste minimisation and in converting waste into value-added products used for environmental monitoring and remediation, sensing technology, biomedical science, and clean energy production, among others (Rao et al., 2018; Abd Elkodous et al., 2021; Bhattacharya et al., 2021; Mubarik et al., 2021; Olaiya et al., 2025).

The direct impact of waste-derived materials is evident in the development of analytical sample preparation methods, wastewater/water decontamination, and electrochemical and chemical sensing technologies, among others. In environmental analytical chemistry, waste-derived nanomaterials are used in three ways: sample preparation for analytical purposes, sensor development for environmental monitoring, and environmental remediation (Antonelli et al., 2026; Bagheri et al., 2025; Olaiya et al., 2025). Additionally, waste-derived nanomaterials are attractive in analytical chemistry due to their remarkable properties, including facile functionalisation, greenness, biocompatibility, low cost, good chemical stability, and abundance (Antonelli et al., 2026; Bagheri et al., 2025). They are used in sorbent-based extraction techniques, such as solid-phase extraction (SPE) and solid phase microextraction techniques, as well as their derivatives (Figure 1.1). Figure 1.1 also outlines possible extraction mechanisms of diverse analytes from different sample matrices using waste-derived nanomaterials. Chapter 7 discusses the application of various waste-derived nanomaterials and the development of sample preparation methods for the preconcentration of organic and inorganic pollutants.

DOI: 10.1201/9781003745846-1

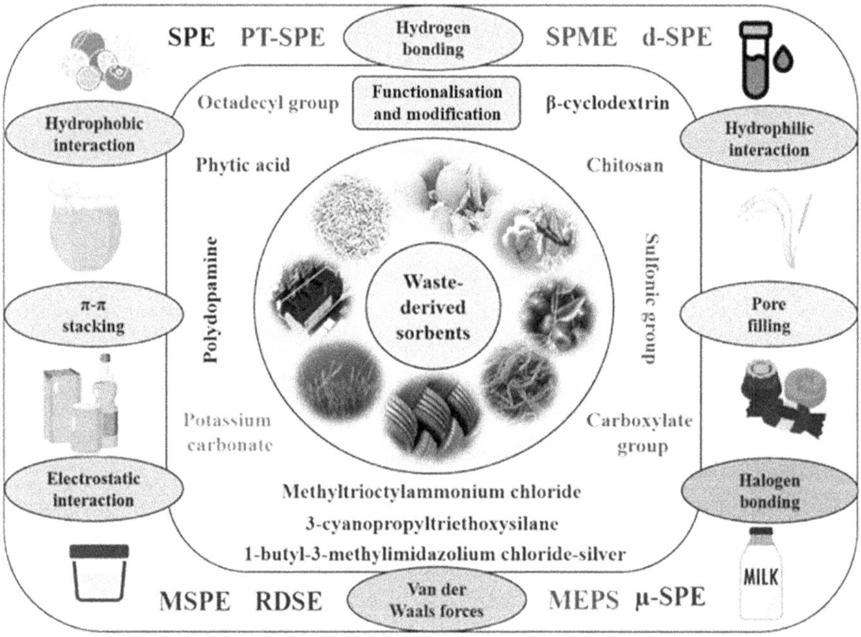

FIGURE 1.1 Various waste materials used for the preparation of waste-derived functional nanomaterials in their application as adsorbents in various sorbent-based extraction techniques. SPE: solid-phase extraction; RDSE: rotating-disk sorptive extraction; d-SPE: dispersive solid-phase extraction; MSPE: magnetic solid-phase extraction; μ-SPE: micro-solid-phase extraction; PT-SPE: pipette-tip solid-phase extraction; SPME: solid-phase microextraction; MEPS: microextraction by packed sorbent (Bagheri et al., 2025).

The integration of waste management and nanotechnology to produce advanced functional nanomaterials from waste streams enables the development of next-generation sorbents for sample preparation for analytical purposes, aligning with circular-economy and green analytical chemistry principles (Antonelli et al., 2026).

Previous studies have reported that waste-derived nanomaterials are used for water/wastewater treatment in two main operational modes: batch systems (waste-derived adsorbent is in contact with contaminated water for a fixed contact time), and flow-through systems (contaminated water is passed through a column or membrane containing the waste-derived adsorbent material) (Mukhopadhyay et al., 2022). As a proof of concept, batch systems are widely used because they are adaptable to small-laboratory-scale waste-derived adsorbents, whereas flow-through systems require structurally robust, uniform nanomaterials, which in turn require additional steps and equipment (Candido et al., 2021). Additionally, flow-through systems require large amounts of adsorbent materials, which encourages the scaling up of the waste-derived nanomaterials. The possibility of producing industrial-scale waste-derived nanomaterials has been discussed in detail in Chapter 6. Additionally, the application of waste-derived nanomaterials for water and wastewater treatment is discussed in detail in Chapter 10.

Compared with sample preparation for environmental monitoring and water treatment, the application of waste-derived nanomaterials in sensing technology (e.g., the development of chemical and electrochemical sensors (see Chapters 8 and 9)) is relatively emergent, and the volume of research work is lower (Bhattacharya et al., 2021). Studies have shown that waste-derived nanomaterials have enabled the development of sensitive, selective, and robust sensor interfaces for the detection of targeted analytes in environmental, clinical, and food matrices (Onfray & Thiam, 2023; Bressi et al., 2024; Venkataraman & Sundramoorthy, 2026).

In addition to their application in environmental analytical chemistry and sensing technology, waste-derived nanomaterials are increasingly being explored for use in the food, pharmaceutical, and cosmetic industries due to their unique properties and sustainable production methodologies (Latini et al., 2025; Bharti et al., 2024, see Chapter 11). In the food industry, waste-derived nanostructured materials are used as ingredients and packaging materials because of their remarkable properties that enhance solubility and bioavailability and improve the shelf life of food products (Yesilsu et al., 2024; Srivastava et al., 2024; Rao & Poonia, 2024). The pharmaceutical application of waste-derived nanomaterials for drug delivery and wound healing is an emerging field that leverages sustainability and circular economy principles (Abd Kadir et al., 2022; Ilangovan et al., 2021). Studies have reported that the unique properties and environmental benefits of waste-derived nanomaterials offer a promising avenue for the sustainable, effective development of cosmetic products (Saba et al., 2025; Latini et al., 2025; Piccinino et al., 2021). Waste-derived nanomaterials such as lignin nanoparticles, nanostructured silica, and biogenic nanoparticles have been reported to have good antioxidant, anti-ageing, and UV-shielding properties, making them suitable for sunscreen, anti-ageing formulations and delivery of bioactive agents to improve skin hydration and appearance (Maghraby et al., 2024; Javaid et al., 2022; Sharma & Garg, 2023).

Despite the above benefits, concerns about the toxicity and safety of nanomaterials continue to dominate the discussion. Studies reporting the potential adverse effects of nanomaterials on human health and the environment necessitate thorough risk assessments and regulatory measures (Mogharabi et al., 2014; Srivastava et al., 2024). Furthermore, ongoing research is being conducted to address the safety, regulatory, characterisation, and risk assessment of nanomaterials to ensure their safe use in the food, cosmetics, and pharmaceutical industries, as well as in environmental monitoring and remediation. Current regulations and safety assessments are evolving to address these concerns. Chapter 12 provides an overview of the toxic effects of nanomaterials on ecosystems and public health.

REFERENCES

Abd Elkodous, M., El-Husseiny, H. M., El-Sayyad, G. S., Hashem, A. H., Doghish, A. S., Elfadil, D., . . . & Matsuda, A. (2021). Recent advances in waste-recycled nanomaterials for biomedical applications: Waste-to-wealth. *Nanotechnology Reviews*, *10*(1), 1662–1739.

Abd Kadir, N. H., Mig, T. C., Ramli, A., Yashim, M. M., & Mohammad, M. (2022). The prospect of microcrystalline cellulose and nanocrystalline cellulose composites from agricultural residues in biomedical/pharmaceutical. In *Applications of nanocomposites* (pp. 211–235). CRC Press.

Amdeha, E. (2021). Recovery of nanomaterials from agricultural and industrial wastes for water treatment applications. In *Waste recycling technologies for nanomaterials manufacturing* (pp. 385–417). Cham: Springer International Publishing.

Antonelli, L., Felli, N., De Cesaris, M. G., Cinti, S., & Gentili, A. (2026). From waste to sorbent: Recycling polymers and natural materials for sample preparation and water remediation. *Talanta, 301*, 129263.

Bagheri, A. R., Shokrollahi, A., & Madani-Nejad, E. (2025). Waste-derived sorbents in analytical extraction processes: A review. *TrAC Trends in Analytical Chemistry, 193*, 118429.

Bharti, A. S., Baran, C., Bhardwaj, A. K., Tripathi, S., Pandey, R., & Uttam, K. N. (2024). Domestic waste utilization in the synthesis of functional nanomaterial. In *Green and sustainable approaches using wastes for the production of multifunctional nanomaterials* (pp. 61–76). Elsevier.

Bhattacharjee, S., Girigoswami, A., Nag, M., Rubab, T., & Girigoswami, K. (2024). Role of nanotechnology as a zero waste tool. *Environmental Quality Management, 33*(4), 193–209.

Bhattacharya, G., Fishlock, S. J., McLaughlin, J. A., & Roy, S. S. (2021). Metal-oxide nanomaterials recycled from E-waste and metal industries: A concise review of applications in energy storage, catalysis, and sensing. *International Journal of Energy Research, 45*(5), 8091–8102.

Bressi, V., Celesti, C., Ferlazzo, A., Len, T., Moulaee, K., Neri, G., Luque, R., & Espro, C. (2024). Waste-derived carbon nanodots for fluorimetric and simultaneous electrochemical detection of heavy metals in water. *Environmental Science: Nano, 11*(3), 1245–1258.

Candido, I. C. M., Pires, I. C. B., & de Oliveira, H. P. (2021). Natural and synthetic fiber-based adsorbents for water remediation. *CLEAN—Soil, Air, Water, 49*(6), 2000189.

Dutta, T., Kim, K. H., Deep, A., Szulejko, J. E., Vellingiri, K., Kumar, S., Kwon, E.E., & Yun, S. T. (2018). Recovery of nanomaterials from battery and electronic wastes: A new paradigm of environmental waste management. *Renewable and Sustainable Energy Reviews, 82*, 3694–3704.

Ilangovan, R., Subha, V., Ravindran, R. E., Kirubanandan, S., & Renganathan, S. (2021). Nanomaterials: Synthesis, physicochemical characterization, and biopharmaceutical applications. In *Nanoscale processing* (pp. 33–70). Elsevier.

Javaid, A., Imran, M., Latif, S., Hussain, N., Iqbal, H. M., & Bilal, M. (2022). Multifunctional attributes of nanostructured materials, toxicology, safety considerations, and regulations. *Journal of Materials Science, 57*(36), 17021–17051.

Latini, V., Feliczak-Guzik, A., & Wawrzyńczak, A. (2025). Application possibilities of sustainable nanostructured silica-based materials in cosmetics. *Cosmetics, 12*(4), 134.

Maghraby, Y. R., Ibrahim, A. H., El-Shabasy, R. M., & Azzazy, H. M. E. S. (2024). Overview of nanocosmetics with emphasis on those incorporating natural extracts. *ACS Omega, 9*(34), 36001–36022.

Mogharabi, M., Abdollahi, M., & Faramarzi, M. A. (2014). Toxicity of nanomaterials; an undermined issue. *DARU Journal of Pharmaceutical Sciences, 22*(1), 59.

Mubarik, S., Qureshi, N., Sattar, Z., Shaheen, A., Kalsoom, A., Imran, M., & Hanif, F. (2021). Synthetic approach to rice waste-derived carbon-based nanomaterials and their applications. *Nanomanufacturing, 1*(3), 109–159.

Mukhopadhyay, A., Duttagupta, S., & Mukherjee, A. (2022). Emerging organic contaminants in global community drinking water sources and supply: A review of occurrence, processes and remediation. *Journal of Environmental Chemical Engineering, 10*(3), 107560.

Olaiya, N. G., Al-Amin, M., Rashed, K., & Maraveas, C. (2025). Nanomaterials from textile waste for purification and environmental applications. *Polymers*, *17*(23), 3098.

Onfray, C., & Thiam, A. (2023). Biomass-derived carbon-based electrodes for electrochemical sensing: A review. *Micromachines*, *14*(9), 1688.

Piccinino, D., Capecchi, E., Tomaino, E., Gabellone, S., Gigli, V., Avitabile, D., & Saladino, R. (2021). Nano-structured lignin as green antioxidant and UV shielding ingredient for sunscreen applications. *Antioxidants*, *10*(2), 274.

Rao, M., Jha, A. K., & Prasad, K. (2018). Nanomaterials: An upcoming fortune to waste recycling. In *Exploring the realms of nature for nanosynthesis* (pp. 241–271). Cham: Springer International Publishing.

Rao, V., & Poonia, A. (2024). Role and application of nanostructures in food preservation and it's use in active food packaging. In *Nanotechnology and nanomaterials in the agri-food industries* (pp. 205–234). Elsevier.

Saba, S., Aslam, L., Hussain, A., & Arooj, I. (2025). A comprehensive introduction on biogenic nanoparticles: Fabrication, implementation, and environmental concern. In *Biogenic nanoparticles* (pp. 3–23). Elsevier.

Samaddar, P., Ok, Y. S., Kim, K. H., Kwon, E. E., & Tsang, D. C. (2018). Synthesis of nanomaterials from various wastes and their new age applications. *Journal of Cleaner Production*, *197*, 1190–1209.

Sharma, R., & Garg, A. (2023, October). Nanomaterials: Applications and challenges in the cosmetics sector. *International Journal of Nano Dimension*, *14*(4).

Srivastava, K., Srivastava, L., Singh, S., Desai, H. S., & Jain, T. (2024). Nanotechnology-based approaches applied to nutraceuticals in connection with its legal aspects. In *Nutraceuticals inspiring the contemporary therapy for lifestyle diseases* (pp. 158–168). CRC Press.

Venkataraman, S., & Sundramoorthy, A. K. (2026). Waste-derived carbon nanomaterials for electrochemical applications: Toward a circular and sustainable future. *RSC Advances*, *16*(1), 733–747.

Yesilsu, A. F., Alp-Erbay, E., Kahyaoğlu, L. N., & Aydin, İ. (2024). Fish waste to valuable products: Nanotechnological approach. *BioNanoScience*, *14*(5), 4733–4751.

2 Impact of waste valorisation on the circular economy

2.1 INTRODUCTION

The foreseeable consequences of rapid global urbanisation, population, and economic growth are characterised by the generation of large amounts of waste (Wang et al., 2022). Various waste materials are generated in abundance from major sectors, including agriculture, municipality, food, textiles, plastics, glass, construction, mining, metal processing, power generation, and electronics, among others (Wang et al., 2022; Hajam et al., 2023). The waste generated from these streams comprises both biodegradable and non-biodegradable inorganic and organic materials, which cause environmental pollution and potentially hazardous impacts that degrade the ecosystem, thereby adversely affecting public health (Wang et al., 2022; Mustafa et al., 2017; Xiong et al., 2019). Waste management strategies worldwide have been developed to mitigate the hazardous impacts of waste on the environment and human health (Huang et al., 2022; Zuberi & Ali, 2015; Hajam et al., 2023). According to previous studies, landfilling and incineration are the most widely used waste management strategies globally (Huang et al., 2022; Zuberi & Ali, 2015; Hajam et al., 2023). However, these strategies have been reported to be major contributors to surface water and groundwater pollution, as well as to greenhouse gas emissions (Zuberi & Ali, 2015; Hajam et al., 2023; Nanda & Berruti, 2021). Therefore, research should focus on developing alternative technologies for sustainable waste management by converting or recycling waste into functional products using suitable scientific methods (Wang et al., 2022; Mong et al., 2024).

Waste valorisation and the circular economy are gradually acknowledged as key strategies for achieving sustainability, resource efficiency, and economic development (Ferronato et al., 2019; Elroi et al., 2023). Waste valorisation, also known as resource recovery or waste utilisation, has emerged as a promising strategy for converting waste materials from various waste streams into value-added products such as energy, fuels, and materials (Priyadarshini & Abhilash, 2020; Gueboudji, 2024). The waste valorisation concept aligns closely with the principles of the circular economy, where waste is reimagined as a hidden resource rather than a liability (Gueboudji, 2024). The circular economy, on the other hand, is a systemic approach that aims to effectively utilise resources through balancing environmental protection and socio-economic benefits (Morseletto, 2020; Velenturf & Purnell, 2021). These are achieved by embracing long-term value retention, minimising primary resource exploitation and waste reduction, promoting closed loops of

DOI: 10.1201/9781003745846-2

products and materials, and minimising environmental impact (Morseletto, 2020; Velenturf & Purnell, 2021; Gueboudji, 2024; Kokkinos et al., 2025; Kulkarni & Singh, 2025).

The objective of waste valorisation and the circular economy is to deviate from the traditional linear waste management model, characterised by "take-make-dispose," and embrace closed-loop systems that promote a more environmentally and economically sound, resource-efficient framework (Priyadarshini & Abhilash, 2020; Gueboudji, 2024). In the circular economy, waste management practices employ the 3R approach, which involves "Reduce, Reuse, and Recycle" (Middha et al., 2024; Thakker & Narayanamoorthi, 2015). In the 3R approach, waste materials are not disposed of in the environment; instead, they are recycled and repurposed for reuse, thereby enhancing their value, generating revenue, and creating employment opportunities. However, the circular economy concept requires the efficient use of waste materials as a resource by improving design and production practices (Middha et al., 2024; Thakker & Narayanamoorthi, 2015). This chapter provides a brief overview of the circular economy model and its relationship with the United Nations Sustainable Development Goals (UNSDGs). Additionally, the waste valorisation and circular economy phases, along with their benefits and challenges, are discussed, as are the benefits and challenges of adopting these practices in developing countries such as South Africa.

2.2 CIRCULAR ECONOMY MODEL

The circular economy is a systemic and innovative approach to economic development, where production and consumption are designed to reduce waste, extend product lifecycles, and promote the efficient use of resources to benefit businesses, society, and the environment (Kandpal et al., 2024a; Reddy et al., 2023; Aborujilah et al., 2023; Hysa et al., 2020). The circular economy operates through both the technical and biological cycles (Jansen et al., 2022; Stephan, 2022; Kusumo et al., 2022; Guo et al., 2023; Finamore & Oltean-Dumbrava, 2024).

In technical cycles, non-renewable resources that cannot be replenished by the biosphere are adequately managed (Navare et al., 2021). These resources are kept in use through various processes, such as maintenance, recycling, reuse, redistribution, sharing, refurbishment, and remanufacturing, to maximise product value and minimise waste (O'Connor, 2021; Navare et al., 2021; Stephan, 2022; Jansen et al., 2022; Guo et al., 2023). Although technical cycles of the circular economy have well-established recycling technologies and are more mature, the main challenges include energy inefficiencies and the need for continuous innovation to improve efficiency and reduce waste generation (Gnekpe & Plantec, 2023). On the other hand, in the circular economy, biological cycles play a fundamental role in ensuring sustainability by enabling the flow and regeneration of biological materials (O'Connor, 2021; Navare et al., 2021; Stephan, 2022). The processes involved in the biological cycles of the circular economy include sustainable harvesting of biotic resources, anaerobic digestion, composting, and natural degradation to ensure that these resources return to the natural environment to support ecosystem regeneration without causing harm or waste (O'Connor, 2021; Navare et al., 2021; Stephan, 2022; Guo et al.,

2023). However, biological cycles suffer from a significant gap in assessment and monitoring compared to well-established technical cycles, which could lead to over-exploitation and ecosystem degradation (Rocchi et al., 2021; Aragonés & Torralvo, 2024). Therefore, for the successful implementation of biological cycles, issues such as overexploitation and ecosystem degradation must be assessed, monitored, and addressed.

2.3 THE RELATIONSHIP BETWEEN THE CIRCULAR ECONOMY AND SUSTAINABLE DEVELOPMENT GOALS (SDGS)

Sustainable development is a major global priority due to climate change, rising resource consumption and global resource scarcity (Knäble et al., 2022). As a result, the circular economy has garnered significant attention because it promotes regenerative production and consumption approaches, which directly align with the United Nations SDGs (Knäble et al., 2022; Milhem et al., 2024). Therefore, the relationship between the circular economy and the SDGs provides a critical framework for tackling global challenges, such as waste generation, resource depletion, and environmental degradation caused by pollution (Knäble et al., 2022; Provenzano & Seminara, 2024). This is because the circular economy emphasises waste reduction while maximising resource use through practices such as upcycling, recycling, reusing, recovering, and remanufacturing (Alnafrah et al., 2025; Milhem et al., 2024). According to Provenzano and Seminara (2024), the circular economy has emerged as a powerful catalyst for addressing the complexities of sustainable development. The principles of the circular economy resonate with the 17 SDGs (Provenzano & Seminara, 2024; Ghisellini et al., 2016). According to previous studies, circular economy practices have direct contributions to the advancement and achievement of SDGs 6, 7, 8, 9, 11, 12, 13, 14, and 15. Also, studies have reported a strong relationship between circular economy practices and these SDGs (Saidani et al., 2019; Schroeder et al., 2019; Panchal et al., 2021; Fatimah et al., 2020; Rodríguez-Antón et al., 2022; de Oliveira & Oliveira, 2023). Table 2.1 summarises the contribution and role of the circular economy in advancing and achieving the above-listed SDGs.

2.4 ROLE OF WASTE VALORISATION IN THE CIRCULAR ECONOMY

Waste valorisation plays a crucial role in implementing the circular economy. This is achieved by transforming waste materials into valuable products, thereby offering remarkable benefits such as resource efficiency, economic development, sustainability, and technological innovation (Karungamye, 2024; Alaghemandi, 2024). By embracing reuse, recycling, and converting waste materials into energy and value-added products, waste valorisation enables the transition to a circular economy by reducing reliance on non-renewable resources and minimising waste generation (Hajam et al., 2024; Kokkinos et al., 2025; Kulkarni & Singh, 2025). Waste valorisation presents significant environmental benefits while driving innovation and economic growth. For instance, life cycle assessments

TABLE 2.1
The contribution of the circular economy to achieving United Nations Sustainable Development Goals (SDGs).

SDG	Goal Description	Contribution of the Circular Economy
SDG 6	Clean water and sanitation	In water management, the principles of the circular economy promote the reuse and recycling of water. These practices have the potential to significantly contribute to the sustainable management of natural water resources. Additionally, the reuse of wastewater as an alternative water source and the recovery of resources such as nutrients and energy from wastewater to value-added products are critical actions within the circular economy framework that align with SDG 6 targets (Arana-Landin et al., 2024; Smol et al., 2020; Rebello et al., 2025). The implementation of the circular economy in wastewater/water management has significant environmental and economic benefits, including the reduction of carbon footprints, minimisation of waste generation, enhancement of water system resilience, and contribution to sustainable economic development (Sabău-Popa et al., 2022).
SDG 7	Affordable and clean energy	The efficient use of resources to improve energy efficiency and provision is an important aspect of the circular economy framework, which includes integrating renewable energy sources such as solar, wind, and hydropower. These practices can potentially reduce the dependence on fossil fuels and enhance energy security (Nunes et al., 2023; Kandpal et al., 2024a; Kandpal et al., 2024b; Musa et al., 2024; Olabi, 2019). Waste heat recovery from industrial waste-to-energy networks and industrial processes offers the potential to enhance industrial energy efficiency. Additionally, the use of biogas and anaerobic digestion can significantly increase the share of renewable energy in the global energy mix. This approach is applicable at both the household level and in large- and medium-scale biogas plants for agricultural waste, which has significant, untapped potential (Taskhiri et al., 2015; Schroeder et al., 2019; Cai et al., 2024). As discussed above, circular economy practices promote energy efficiency, integration of renewable energy sources, and support sustainable urban development, which is a viable route to achieving and advancing SDG 7.
SDG 8	Decent work and economic growth	Circular economy practices promote the creation of new job opportunities (also called green jobs) in sectors such as recycling, remanufacturing, and repair. These attributes significantly contribute to decent work and economic growth (Karamustafa et al., 2024; Valverde & Avilés-Palacios, 2021). Furthermore, the circular economy can advance and accelerate innovation and entrepreneurship in sustainable business models, thereby steering economic growth, creating new markets, and generating business opportunities (Voulvoulis, 2022; Gupta et al., 2025).

(Continued)

TABLE 2.1 (*Continued*)
The contribution of the circular economy to achieving United Nations Sustainable Development Goals (SDGs).

SDG	Goal Description	Contribution of the Circular Economy
SDG 9	Industry, innovation and infrastructure	The SDG 9 centres on stimulating inclusive and sustainable industrialisation, advancing innovation, and building resilient infrastructure. Therefore, the circular economy practices emphasise the reuse, sharing, repair, recycling, and refurbishment of existing products and materials; this principle aligns closely with the objectives of SDG 9. Circular economy practices are beneficial to the construction industry through eco-design, industrial synergy, resource optimisation (reduction of resource intensity), and reduced environmental impacts (Khuc et al., 2024; de Oliveira & Oliveira, 2023; Belmonte-Ureña et al., 2021; Puntillo, 2023). Furthermore, circular economy principles, such as waste reduction and sustainable production, are critical drivers of innovation in infrastructure and industrial development. These circular economy approaches must be adopted to promote the development of sustainable infrastructure, advanced manufacturing processes, and innovative technologies (Cervantes Puma et al., 2024).
SDG 11	Sustainable cities and communities	Urban waste reuse, recycling, and reduction, along with improvements in resource efficiency and the promotion of sustainable consumption, address critical city/urban challenges, including housing, infrastructure, and environmental impacts. The integration of circular economy practices into urban planning can enable cities to create inclusive, secure, and resilient environments for their populations. Furthermore, integrating circular economy principles into SDG 11 can heighten the sustainability of cities by encouraging eco-friendly construction materials and energy alternatives. This is crucial in establishing urban sustainability, as it facilitates cities in managing resources efficiently and moderating environmental impacts (Puntillo, 2023; Cervantes Puma et al., 2024; Brandão & Verissimo, 2024; Mpiere & Kamanda, 2025; Mehmood et al., 2025).
SDG 12	Responsible consumption and production	The circular economy framework encourages cities to adopt practices that significantly enhance overall improvement of resource efficiency, reduction of waste generation, and promotion of sustainable practices. Additionally, the circular economy is integral in supporting the emphasis and achievement of responsible consumption and production. They aim to disconnect economic growth from resource constraints. This is achieved by maintaining the highest value and functionality of products, components, and materials (Cervantes Puma et al., 2024; Tsai et al., 2025).

SDG 13	Climate action	The circular economy principles are gradually identified as a critical approach for achieving SDG 13. The relationship between the circular economy and SDG 13 encompasses various strategies and procedures that aim to reduce greenhouse gas emissions and mitigate climate change. Effective waste management and recycling play a vital role in the circular economy by tackling the climate change challenges that cities face. Circular economy practices can lower the carbon footprint by minimising waste generation, optimising resource usage, and promoting the reuse of materials; in doing so, the impact of urbanisation on climate change is reduced (Hajam et al., 2020; Dwivedi et al., 2025; Alnafrah et al., 2025). In conclusion, circular economy methods help reduce greenhouse gas emissions, integrate renewable energy, conserve natural resources, and promote climate-resilient urban development, which significantly contributes to advancing and achieving the broader climate action agenda (Ortiz-de-Montellano et al., 2023; Dwivedi et al., 2025; Hassan et al., 2024; Cervantes Puma et al., 2024).
SDG 14 and 15	Life below water and life on land	Marine pollution reduction is crucial for achieving SDG 14, and this goal can be accomplished by minimising waste and pollution caused by contaminants entering water bodies, which directly benefits marine ecosystems. The circular economy promotes sustainable fishing practices, which in turn help conserve fish stocks and protect marine protected areas (Sampantamit et al., 2020; Valverde & Avilés-Palacios, 2021). Practices such as recycling, reusing, and regenerating natural systems reduce negative environmental impacts on land, including land degradation, and advocate for the conservation of terrestrial ecosystems and sustainable land use. These approaches support SDG 15, which emphasises the protection and restoration of terrestrial ecosystems (Urban & Hametner, 2022; Himshikha et al., 2024; Cervantes Puma et al., 2024). As demonstrated in the literature, the circular economy framework contributes to achieving SDG 14 and SDG 15 by promoting pollution reduction, sustainable resource management, and the conservation of marine and terrestrial biodiversity (Pla-Julián & Guevara, 2019; Delany-Crowe et al., 2019; Valverde & Avilés-Palacios, 2021).

for the circular economy have shown that composting organic waste materials, considering nutrient recovery, reduces environmental impact compared to established waste management methods, such as landfill disposal (Oldfield et al., 2018). Furthermore, waste valorisation promotes responsible consumption and production and lessens pollution and resource depletion (Bhat et al., 2023; Roy et al., 2023). Furthermore, waste valorisation technologies create new markets for waste-derived high-value products, which in turn enhance the economic viability of waste management systems, thereby creating job opportunities and fostering the development of new technologies, ultimately contributing to economic growth (Leder et al., 2020; Nogueira et al., 2023; Halim et al., 2025; Abbate et al., 2025). Several technologies, both mature and emerging, are used in waste valorisation. These include thermochemical conversion, bioconversion, biochemical processes, integrated and hybrid approaches, material recovery, and recycling. Table 2.2 summarises their principles. Notwithstanding the promising benefits of waste valorisation concepts in advancing the circular economy, challenges such as technological readiness, funding availability, and regulatory constraints can potentially limit innovation (Suárez et al., 2023; Wang et al., 2022). Furthermore, integrating waste valorisation into existing waste management systems is complex, posing additional limitations for emerging markets to drive these innovations (Nogueira et al., 2023).

TABLE 2.2
Waste valorisation technological approaches.

Techniques	Pathways and Descriptions
Thermochemical conversion	• **Pyrolysis and Gasification:** In pyrolysis, waste materials are decomposed in the absence of oxygen, while in the gasification process, the waste is thermally decomposed in a controlled oxygen environment (Al-Hammadi & Güngörmüşler, 2025; Ungureanu et al., 2025). Pyrolysis and gasification processes are suitable for the valorisation of biomass, municipal solid waste, and biogenic residues due to their environmental, economic, and energy efficiency performances. They are used to convert waste materials into solid, liquid, and gaseous products, such as syngas, bio-oil, and biochar (Ayub et al., 2023; Al-Hammadi & Güngörmüşler, 2025; Ungureanu et al., 2025).
Biochemical	• **Fermentation** is used to convert organic waste into biofuels (e.g., bioethanol, biogas) and high-value chemicals. Anaerobic digestion is one of the most widely used fermentation processes that occurs in the absence of oxygen during the breakdown of organic matter by bacteria (Sheldon, 2014; Kulkarni & Singh, 2025). • **Bioconversion** utilises microorganisms to transform waste materials, including agricultural residues and food waste, into products such as biofuels, organic acids, and bioplastics, among others (Gueboudji, 2024).

(Continued)

TABLE 2.2 (*Continued*)
Waste valorisation technological approaches.

Techniques	Pathways and Descriptions
Material recovery and recycling	• This process converts various waste materials into valuable products (such as advanced materials, value-added or fine chemicals and energy) to foster a circular economy and ease reliance on raw materials (Arun et al., 2024; Kokkinos et al., 2025). The material recovery and recycling process is crucial for sustainability, as it addresses resource depletion by converting waste into feedstock.
Chemical recycling	• This process involves breaking down waste materials, such as plastics and textiles, into their basic monomers or other valuable chemicals, which are then reused as precursors to produce new materials or chemicals. The chemical processes involved include pyrolysis-based chemical recycling, depolymerisation, hydrothermal process, biocatalytic degradation and solvolysis, among others (Kulkarni & Singh, 2025; Gueboudji, 2024).
Integrated and hybrid approaches	• **Integrating pyrolysis and gasification with anaerobic digestion**: The anaerobic digestion of waste materials raises technical, environmental, and economic concerns. Therefore, to improve its performance and reduce environmental burden, this process is integrated with thermochemical processes such as pyrolysis and gasification. This integrated approach supports the circular economy by converting organic waste into biogas and biochar (Sikarwar et al., 2021; Caiardi et al., 2022). • **Zero-waste valorisation:** This comprehensive approach integrates extraction methodologies, microbial bioconversion, and circular economy principles to convert waste materials into value-added products, generating zero waste residues (Taheri & Hosseini, 2025). The primary challenge of zero-waste valorisation is transitioning from laboratory-scale concepts to industrial-scale applications. This is because the commercialisation of the integrated system is hindered by the complex extraction and bioprocessing phases, which require specialised equipment and high initial capital costs. Furthermore, it has been reported that the economic viability of zero-waste valorisation systems depends on various factors, such as local infrastructure, the availability of consistent waste feedstock, and the market value of the final value-added product (Ribeiro et al., 2020; Lee et al., 2024; Taheri & Hosseini, 2025). • **Biorefining**: This approach incorporates several processes, such as biochemical (enzymatic hydrolysis and fermentation), thermal, and mechanical methods, to maximise the recovery of valuable products from food waste or biomass (Kulkarni & Singh, 2025; Al-Hammadi & Güngörmüşler, 2025) • **Smart bioconversion**: Smart bioconversion is an advanced and intelligent process that integrates nanoengineering, artificial intelligence, and bioconversion approaches (such as microbial, enzymatic, and insect-based systems) to convert organic waste into valuable products. The integration of various innovative approaches enhances effectiveness, sustainability, and economic feasibility (Jayabal & Prabhakar, 2026).

2.5 WASTE VALORISATION AND CIRCULAR ECONOMY PHASES

2.5.1 Formalisation of the Waste Pickers

Informal waste pickers play a crucial role in solid waste management and often operate in parallel with formal waste collection and disposal agents (Moreno-Sánchez & Maldonado, 2006; Masood & Barlow, 2013). Their main tasks include collecting and sorting reusable and recyclable waste materials that can be reincorporated into the economy's production process from streets, dumpsites, or landfills, thus contributing meaningfully to waste reduction and resource recovery (Moreno-Sánchez & Maldonado, 2006; Morais et al., 2022). However, waste pickers frequently encounter sidelining, poor working conditions, and a lack of recognition in formal waste management systems (Morais et al., 2022). Therefore, there is a need to develop or implement strategies to formalise them. Previous studies have reported that improving their working conditions, providing legal recognition, and enhancing their economic stability can be achieved by integrating informal waste pickers into formal systems, such as municipal waste management through contracts, associations, or other formal arrangements (Masood & Barlow, 2013; Kasinja & Tilley, 2018). Additionally, Moreno-Sánchez and Maldonado (2006) stated that applying complementary policies such as subsidies for material recovery and taxes on the use of materials, as well as consumption and disposal taxes, can incentivise the inclusion and integration of waste pickers in formal waste management systems. Formalisation of waste pickers presents both challenges and opportunities. For instance, they often face challenges such as negative public perception, transportation issues, and fluctuating material prices, as well as potential resistance to establishing cooperatives due to concerns about income loss and conflict (Kasinja & Tilley, 2018; Adekunle et al., 2023). Despite these challenges, the formalisation of waste pickers could reduce social inequality, improve income stability, and enhance safety at work, thereby supporting the circular economy by ensuring that valuable materials are recovered, recycled, and reused efficiently (Calderon Marquez et al., 2021; Morais et al., 2022).

2.5.2 Formal Municipal Solid Waste Collection and Pretreatment Before Landfilling

Rapid population growth, booming economy, urbanisation, and high living standards have significantly increased the rate of solid waste generation. Therefore, providing a sustainable and efficient waste management system that considers potential impacts on public health and the environment is critical for most countries (Hemidat et al., 2022; Budihardjo et al., 2023). Therefore, municipal solid waste management is one of the most important solutions to environmental challenges faced by many countries, particularly in developing nations (Hemidat et al., 2022; Derdera & Ogato, 2023). A formal municipal solid waste management system is designed to handle all types of waste materials, encompassing systematic collection and pretreatment processes. This system is devised to minimise environmental impact (reduce landfilling) and promote waste valorisation. These milestones are achieved by applying several key processes, including collection, transfer, resource recovery, recycling, and treatment of solid waste (Hemidat et al., 2022). However, the accumulation of waste in the environment suggests that there are major constraints to solid waste management. Therefore, there is a need to address the technical, operational, and administrative constraints on municipal solid waste management practices in developing countries (Derdera & Ogato, 2023). Incorporating waste management techniques, such as sorting, waste reduction, reuse, recycling, composting, and controlled disposal, is recommended to achieve sustainable integrated municipal solid waste (Abubakar et al., 2022). The collection of waste materials ensures the segregation of recyclables, organics, and hazardous materials. The pretreatment step involves sorting the waste to recover valuable materials and includes biological treatment of organic waste materials using processes such as composting or anaerobic digestion (Xu et al., 2023). Physical and stabilisation processes are used to reduce waste volumes, isolate hazardous waste to prevent contamination, and minimise moisture and odour (Sarker et al., 2025; Sharma et al., 2024). These pretreatment measures are designed to support circular economy principles, decrease landfill burden, extend site lifespan, and reduce greenhouse gas emissions, thus protecting public health, promoting environmental quality, supporting economic productivity, and ensuring sustainability (Goh et al., 2025; Long, 2022).

2.5.3 Stakeholders involved in the CE system

Multiple stakeholders, including policymakers, manufacturers, businesses, government agencies, researchers, and communities, play a central, collaborative role in the circular economy, contributing significantly to promoting sustainability and facilitating the implementation of the circular economic model through their diverse sectoral roles. The involvement of governments and policymakers in establishing regulations, incentives, and standards encourages circular economy practices (Hartley et al., 2023; Provenzano et al., 2024; Munonye, 2025; Mamash et al., 2025). The industry and manufacturers contribute to the redesign and remanufacturing of products to enhance resilience, repairability, and recyclability (Di Stefano et al., 2023; Ishaq et al., 2026). Consumers and communities, on the other hand, play a crucial role in adopting responsible consumption habits and participating in reuse and recycling initiatives, which align well with the principles of the circular economy framework (Tabas et al., 2025; Zhidebekkyzy et al., 2023). The principles of the circular economy are well-supported by academia and research institutions, which continually seek to educate, conduct research, and develop new technologies, providing evidence-based policy recommendations. This is often achieved by embracing waste management and recycling companies that focus on material recovery and creating markets for secondary raw materials (Tabas et al., 2025). Through funding assistance from various financial institutions, circular economy projects are realised (Zhidebekkyzy et al., 2023). Civil society and non-governmental organisations advocate for sustainability, community education, and compliance monitoring, as well as facilitating stakeholder cooperation (Hull et al., 2021). As the world becomes more tech-savvy, the involvement of technology providers in the circular economy enables the incorporation of digital solutions, such as the Internet of Things (IoT), artificial intelligence (AI), and blockchain, to optimise resource flows and enhance transparency across the value chain (Kinnunen et al., 2022; Leder et al., 2020). The roles of each stakeholder in the circular economy are evident, indicating that an interconnected system is necessary to address the challenges faced by waste valorisation advocates, which are crucial steps towards achieving sustainable development through the circular economy (Palafox-Alcantar et al., 2021; Leder et al., 2020).

2.6 CONCLUSION

Waste valorisation offers a pathway to sustainable waste management and resource recovery, which is fundamental to the circular economy model. In addition, the role of waste valorisation in converting waste material into valuable products not only promotes economic development and supports the transition to a sustainable and resource-efficient future as well as efficient waste management practices but also strongly addresses environmental challenges. Various researchers have reported that the potential of waste valorisation in a circular economy can only be realised through collaborative efforts of various stakeholders to address technological and regulatory challenges. The outcomes of such collaboration could transform waste into valuable resources, contributing to both environmental sustainability and economic development. Furthermore, integrating waste valorisation practices with innovative technologies offers a promising approach to achieving a sustainable, circular economy.

REFERENCES

Abbate, S., Centobelli, P., & Di Gregorio, M. (2025). Wine waste valorisation: Crushing the research domain. *Review of Managerial Science*, *19*(3), 963–998.

Aborujilah, A., Wang, H., Niu, Y., Ji, C., & Yang, Y. (2023). The role of circular economy in achieving sustainable development goals (SDGs): An integrative framework. *Estidamaa*, *2023*, 1–8.

Abubakar, I. R., Maniruzzaman, K. M., Dano, U. L., AlShihri, F. S., AlShammari, M. S., Ahmed, S. M. S., Al-Gehlani, W.A.G., & Alrawaf, T. I. (2022). Environmental sustainability impacts of solid waste management practices in the global South. *International Journal of Environmental Research and Public Health*, *19*(19), 12717.

Adekunle, M. M., Sangodoyin, A. Y., & Wahab, B. (2023). Waste picking and the effects of solid waste separation by households in Ibadan metropolis, SW Nigeria. *International Journal of Environmental Studies*, *80*(4), 948–963.

Alaghemandi, M. (2024). Sustainable solutions through innovative plastic waste recycling technologies. *Sustainability*, *16*(23), 10401.

Al-Hammadi, M., & Güngörmüşler, M. (2025). From refuse to resource: Exploring technological and economic dimensions of waste-to-energy. *Biofuels, Bioproducts and Biorefining*, *19*(2), 570–592.

Alnafrah, I., Alqararah, K., & AlJaradin, M. (2025). From linear to circular: Evaluating the effects of circular economy on sustainable development progress in the EU economies. *Journal of Environmental Management*, *382*, 125383.

Aragonés, M. M., & Torralvo, F. A. (2024). How to select the best approach for circular economy assessment? 3D positioning framework, decision support tool and critical analysis for bio-based systems. *Environmental Impact Assessment Review*, *106*, 107493.

Arana-Landin, G., Sigüenza-Tamayo, W., Landeta-Manzano, B., Uriarte-Gallastegi, N., & Laskurain Iturbe, I. (2024, July). Standardizing the management of the circular economy as a mechanism to achieve sustainable development goal 6. In *The International Conference on Industrial Engineering and Industrial Management* (pp. 184–189). Cham: Springer Nature Switzerland.

Arun, M., Barik, D., & Chandran, S. S. (2024). Exploration of material recovery framework from waste A revolutionary move towards clean environment. *Chemical Engineering Journal Advances, 18*, 100589.

Ayub, Y., Zhou, J., Shen, W., & Ren, J. (2023). Innovative valorization of biomass waste through integration of pyrolysis and gasification: Process design, optimization, and multi-scenario sustainability analysis. *Energy, 282*, 128417.

Belmonte-Ureña, L. J., Plaza-Úbeda, J. A., Vazquez-Brust, D., & Yakovleva, N. (2021). Circular economy, degrowth and green growth as pathways for research on sustainable development goals: A global analysis and future agenda. *Ecological Economics, 185*, 107050.

Bhat, R., Sharma, M., Rätsep, R., Malenica, D., & Jõgi, K. (2023). Challenges and prospects of tackling food loss and wastes in the circular economy context. *Sustainable Food Value Chain Development: Perspectives from Developing and Emerging Economies*, 15–36.

Brandão, R., & Verissimo, L. (2024). Circular economy adoption in construction: A pathway to sustainable development and UN SDG 11 achievement. In *An agenda for sustainable development research* (pp. 213–230). Cham: Springer Nature Switzerland.

Budihardjo, M. A., Priyambada, I. B., Chegenizadeh, A., Al Qadar, S., & Puspita, A. S. (2023). Environmental impact technology for life cycle assessment in municipal solid waste management. *Global Journal of Environmental Science & Management (GJESM), 9.*

Cai, Z., Tang, Y., & Lin, J. (2024). Exploring the impact of circular economy practices on ecological footprint, inflation rate, and renewable energy consumption: Evidence from G20 economies. *Environmental Science and Pollution Research, 31*(18), 26536–26554.

Caiardi, F., Belaud, J. P., Vialle, C., Monlau, F., Tayibi, S., Barakat, A., Oukarroum, A., Zeroual, Y., & Sablayrolles, C. (2022). Waste-to-energy innovative system: Assessment of integrating anaerobic digestion and pyrolysis technologies. *Sustainable Production and Consumption, 31*, 657–669.

Calderon Marquez, A. J., Silva de Souza Lima Cano, N., & Rutkowski, E. W. (2021). Inclusion of waste pickers into municipal waste management systems: A comparison between Colombia and Brazil. *The Journal of Environment & Development, 30*(4), 395–425.

Cervantes Puma, G. C., Salles, A., & Bragança, L. (2024). Nexus between urban circular economies and sustainable development goals: A systematic literature review. *Sustainability, 16*(6), 2500.

de Oliveira, C. T., & Oliveira, G. G. A. (2023). What Circular economy indicators really measure? An overview of circular economy principles and sustainable development goals. *Resources, Conservation and Recycling, 190*, 106850.

Delany-Crowe, T., Marinova, D., Fisher, M., McGreevy, M., & Baum, F. (2019). Australian policies on water management and climate change: Are they supporting the sustainable development goals and improved health and well-being? *Globalization and Health, 15*(1), 68.

Derdera, S. E., & Ogato, G. S. (2023). Towards integrated, and sustainable municipal solid waste management system in Shashemane city administration, Ethiopia. *Heliyon, 9*(11).

Di Stefano, C., Elia, S., Garrone, P., & Piscitello, L. (2023). The circular economy as a new production paradigm to enhance resilience of MNEs and the economic system. *AIB Insights, 23*(3), 1–7.

Dwivedi, A., Tripathi, A. R., Kapruwan, V., & Kandpal, V. (2025). Circle of adaption: Circular economy key to adapt to climate change. In *Rethinking resources: Strategies, policies, and partnerships for a circular future* (pp. 255–267). Singapore: Springer Nature Singapore.

Elroi, H., Zbigniew, G., Agnieszka, W. C., & Piotr, S. (2023). Enhancing waste resource efficiency: Circular economy for sustainability and energy conversion. *Frontiers in Environmental Science*, *11*, 1303792.

Fatimah, Y. A., Govindan, K., Murniningsih, R., & Setiawan, A. (2020). Industry 4.0 based sustainable circular economy approach for smart waste management system to achieve sustainable development goals: A case study of Indonesia. *Journal of Cleaner Production*, *269*, 122263.

Ferronato, N., Rada, E. C., Portillo, M. A. G., Cioca, L. I., Ragazzi, M., & Torretta, V. (2019). Introduction of the circular economy within developing regions: A comparative analysis of advantages and opportunities for waste valorization. *Journal of Environmental Management*, *230*, 366–378.

Finamore, M., & Oltean-Dumbrava, C. (2024). Circular economy in construction-findings from a literature review. *Heliyon*, *10*(15).

Ghisellini, P., Cialani, C., & Ulgiati, S. (2016). A review on circular economy: The expected transition to a balanced interplay of environmental and economic systems. *Journal of Cleaner Production*, *114*, 11–32.

Gnekpe, C., & Plantec, Q. (2023). Regulatory push-pull and technological knowledge dynamics of circular economy innovation. *Technological Forecasting and Social Change*, *196*, 122767.

Goh, K. C., Kurniawan, T. A., Othman, M. H. D., Jumaniyozov, K., Aziz, F., Anouzla, A., Ali, I., Khan, M.I., Alsultan, G.A., Zhang, D., & Choo, W. O. (2025). Promoting landfill mining for sustainable resource recovery, circular economy, and climate mitigation: A review. *Frontiers in Environmental Science*, *13*, 1643195.

Gueboudji, Z. (2024). Waste valorization techniques. In *Generation of energy from municipal solid waste: Circular economy and sustainability* (pp. 29–52). Cham: Springer Nature Switzerland.

Guo, J., Ali, S., & Xu, M. (2023). Recycling is not enough to make the world a greener place: Prospects for the circular economy. *Green Carbon*, *1*(2), 150–153.

Gupta, S. K., Sharma, R., Sharma, P., Whig, P., & Hammouch, H. (2025). Circular economy, growth, supply chain, and advantage. *Developments in Environmental Science*, *18*, 3–31.

Hajam, Y. A., Kumar, R., & Kumar, A. (2023). Environmental waste management strategies and vermi transformation for sustainable development. *Environmental Challenges*, *13*, 100747.

Hajam, Y. A., Kumar, R., & Neelam. (2024). Value addition to waste for circular economy and sustainable development. In *Zero waste management technologies* (pp. 137–170). Cham: Springer Nature Switzerland.

Halim, R., Iyer, A., Esteves, S. M., Milliken, S., Kalai, K., Pleissner, D., . . . & Mburu, M. (2025). Is 'waste'an appropriate concept in a sustainable bioeconomy? *Sustainable Chemistry One World*, *5*, 100044.

Hartley, K., Schülzchen, S., Bakker, C. A., & Kirchherr, J. (2023). A policy framework for the circular economy: Lessons from the EU. *Journal of Cleaner Production*, *412*, 137176.

Hassan, S. T., Baloch, M. A., Bui, Q., & Khan, N. H. (2024). The heterogeneous impact of geopolitical risk and environment-related innovations on greenhouse gas emissions: The role of nuclear and renewable energy in the circular economy. *Gondwana Research*, *127*, 144–155.

Hemidat, S., Achouri, O., El Fels, L., Elagroudy, S., Hafidi, M., Chaouki, B., . . . & Guo, J. (2022). Solid waste management in the context of a circular economy in the MENA region. *Sustainability, 14*(1), 480.

Himshikha, Sharma, T., Kaur, T., Singh, A., Mohapatra, A., & Saikia, N. (2024). Regenerative agriculture and sustainable development goals. In *Regenerative agriculture for sustainable food systems* (pp. 135–174). Singapore: Springer Nature Singapore.

Huang, S., Wang, H., Ahmad, W., Ahmad, A., Ivanovich Vatin, N., Mohamed, A. M., . . . & Mehmood, I. (2022). Plastic waste management strategies and their environmental aspects: A scientometric analysis and comprehensive review. *International Journal of Environmental Research and Public Health, 19*(8), 4556.aste landfills in Pakistan. *Renewable and Sustainable Energy Reviews, 44*, 117–131.

Hull, C. E., Millette, S., & Williams, E. (2021). Challenges and opportunities in building circular-economy incubators: Stakeholder perspectives in Trinidad and Tobago. *Journal of Cleaner Production, 296*, 126412.

Hysa, E., Kruja, A., Rehman, N. U., & Laurenti, R. (2020). Circular economy innovation and environmental sustainability impact on economic growth: An integrated model for sustainable development. *Sustainability, 12*(12), 4831.

Ishaq, S., Tanveer, U., Hoang, T. G., & Seuring, S. (2026). Circular design strategies unleashed: Competitiveness and the journey towards circular manufacturing businesses. *Business Strategy and the Environment.* https://doi.org/10.1002/bse.70512

Jansen, B. W., van Stijn, A., Eberhardt, L. C. M., van Bortel, G., & Gruis, V. (2022). The technical or biological loop? Economic and environmental performance of circular building components. *Sustainable Production and Consumption, 34*, 476–489.

Jayabal, R., & Prabhakar, P. (2026). Advanced strategies in food and agricultural waste valorisation: Integrating nanoengineering, artificial intelligence and smart bioconversion for sustainable resource recovery. *Journal of the Science of Food and Agricultur, 106*(8), 4507–4520.

Kandpal, V., Jaswal, A., Santibanez Gonzalez, E. D., & Agarwal, N. (2024a). Policy framework for sustainable energy transition and the circular economy. In *Sustainable energy transition: Circular economy and sustainable financing for environmental, social and governance (ESG) practices* (pp. 217–238). Cham: Springer Nature Switzerland.

Kandpal, V., Jaswal, A., Santibanez Gonzalez, E. D., & Agarwal, N. (2024b). Circular economy principles: Shifting towards sustainable prosperity. In *Sustainable energy transition: Circular economy and sustainable financing for environmental, social and governance (ESG) practices* (pp. 125–165). Cham: Springer Nature Switzerland.

Karamustafa, E. Y., Arsan, B., & Alkan, D. P. (2024). How can the circular economy contribute to the achievement of SDG 8: Decent work and economic growth? In *Transition to the circular economy model: The case of Turkey* (pp. 37–47). Cham: Springer Nature Switzerland.

Karungamye, P. (2024). Energy recovery from solid waste valorisation: Environmental and economic potential for developing countries. *Scientific African, 26*, e02402.

Kasinja, C., & Tilley, E. (2018). Formalization of informal waste pickers' cooperatives in Blantyre, Malawi: A feasibility assessment. *Sustainability, 10*(4), 1149.

Khuc, T. Q., Nguyen, V. T., & Do, S. T. (2024, November). Circular economy: An overview of drivers and benefits in the context of construction industry. In *East Asia-Pacific Conference on Structural Engineering and Construction* (pp. 61–68). Singapore: Springer Nature Singapore.

Kinnunen, P., Karhu, M., Yli-Rantala, E., Kivikytö-Reponen, P., & Mäkinen, J. (2022). A review of circular economy strategies for mine tailings. Cleaner Engineering and Technology, *8*, 100499.

Knäble, D., de Quevedo Puente, E., Pérez-Cornejo, C., & Baumgärtler, T. (2022). The impact of the circular economy on sustainable development: A European panel data approach. *Sustainable Production and Consumption*, *34*, 233–243.

Kokkinos, E., Peleka, E., Tzamos, E., & Zouboulis, A. (2025). Waste valorization technologies in tannery sludge, chromite, and magnesite mining. *Recycling*, *10*(4), 123.

Kulkarni, S., & Singh, S. K. (2025). Sustainable waste valorization of fuels and chemicals into useful resources. *Developments in Environmental Science*, *18*, 67–90.

Kusumo, F., Mahlia, T. M. I., Pradhan, S., Ong, H. C., Silitonga, A. S., Fattah, I. R., . . . & Mofijur, M. (2022). A framework to assess indicators of the circular economy in biological systems. *Environmental Technology & Innovation*, *28*, 102945.

Leder, N., Kumar, M., & Rodrigues, V. S. (2020). Influential factors for value creation within the circular economy: Framework for waste valorisation. *Resources, Conservation and Recycling*, *158*, 104804.

Lee, Z. J., Chung, S. Y., Lee, A. W. L., & Tan, Y. S. (2024). Decision tree approach based on food waste valorisation pathways: A case study on moisture content level of spent coffee grounds in Singapore. *Procedia CIRP*, *122*, 253–258.

Long, H. (2022). Analysis of the key factors of ecological environment protection in the national economic sustainable development goals. *Journal of Environmental and Public Health*, *2022*(1), 3593587.

Mamash, A., Iyiola, K., & Aljuhmani, H. Y. (2025). The role of circular economy entrepreneurship, cleaner production, and green government subsidy for achieving sustainability goals in business performance. *Sustainability*, *17*(9), 3990.

Masood, M., & Barlow, C. Y. (2013). Framework for integration of informal waste management sector with the formal sector in Pakistan. *Waste Management & Research*, *31*(10_suppl), 93–105.

Mehmood, K., Kautish, P., Sithipolvanichgul, J., & Gupta, P. (2025). Circular economy readiness and digital solutions: Redefining urban sustainability through human-centered technological advancements. *Technological Forecasting and Social Change*, *219*, 124242.

Middha, R., Srivastava, N., & Saxena, N. (2024). Chemical principles in waste segregation and recycling. In *Waste management for smart cities* (pp. 1–46). Singapore: Springer Nature Singapore.

Milhem, M., Ateeq, A., Al Astal, A., Alzoraiki, M., Ateeq, R. A., Shafie, D. I., & Santhanamery, T. (2024). Advancing sustainability: Embracing the circular economy for environmental and economic resilience. In *Business sustainability with artificial intelligence (AI): Challenges and opportunities* (Vol. 2, pp. 57–68). Cham: Springer Nature Switzerland.

Mong, G. R., Tan, H., Sheng, D. D. C. V., Kek, H. Y., Nyakuma, B. B., Woon, K. S., . . . & Wong, K. Y. (2024). A review on plastic waste valorisation to advanced materials: Solutions and technologies to curb plastic waste pollution. *Journal of Cleaner Production*, *434*, 140180.

Morais, J., Corder, G., Golev, A., Lawson, L., & Ali, S. (2022). Global review of human waste-picking and its contribution to poverty alleviation and a circular economy. *Environmental Research Letters*, *17*(6), 063002.

Moreno-Sánchez, R. D. P., & Maldonado, J. H. (2006). Surviving from garbage: The role of informal waste pickers in a dynamic model of solid-waste management in developing countries. *Environment and Development Economics*, *11*(3), 371–391.

Morseletto, P. (2020). Targets for a circular economy. *Resources, Conservation and Recycling*, *153*, 104553.

Mpiere, D., & Kamanda, J. K. (2025). Promoting urban sustainability through the local production of ecological paving stones and green coal. *Circular Economy and Sustainability*, 1–14.

Munonye, W. C. (2025). Integrating circular economy principles in business strategies: A policy-driven approach. *Circular Economy and Sustainability*, *5*(3), 1887–1896.

Musa, K., Tufail, S., Erum, N., Said, J., & Mustaffa, A. H. (2024). Impact of the circular economy on ecological footprint: Evidence from Germany. *Environmental Science and Pollution Research*, *31*(44), 56067–56078.

Mustafa, M. F., Liu, Y., Duan, Z., Guo, H., Xu, S., Wang, H., & Lu, W. (2017). Volatile compounds emission and health risk assessment during composting of organic fraction of municipal solid waste. *Journal of Hazardous Materials*, *327*, 35–43.

Nanda, S., & Berruti, F. (2021). Municipal solid waste management and landfilling technologies: A review. *Environmental Chemistry Letters*, *19*(2), 1433–1456.

Navare, K., Muys, B., Vrancken, K. C., & Van Acker, K. (2021). Circular economy monitoring—How to make it apt for biological cycles? *Resources, Conservation and Recycling*, *170*, 105563.

Nogueira, L. A., Lindeløv, B., & Olsen, J. (2023). From waste to market: Exploring markets, institutions, and innovation ecosystems for waste valorization. *Business Strategy and the Environment*, *32*(4), 2261–2274.

Nunes, A. M. M., Coelho Junior, L. M., Abrahão, R., Santos Júnior, E. P., Simioni, F. J., Rotella Junior, P., & Rocha, L. C. S. (2023). Public policies for renewable energy: A review of the perspectives for a circular economy. *Energies*, *16*(1), 485.

O'Connor, K. E. (2021). Microbiology challenges and opportunities in the circular economy. Microbiology, *167*(1), 001026.

Olabi, A. G. (2019). Circular economy and renewable energy. *Energy*, *181*, 450–454.

Oldfield, T. L., White, E., & Holden, N. M. (2018). The implications of stakeholder perspective for LCA of wasted food and green waste. *Journal of Cleaner Production*, *170*, 1554–1564.

Ortiz-de-Montellano, C. G. S., Samani, P., & van der Meer, Y. (2023). How can the circular economy support the advancement of the Sustainable Development Goals (SDGs)? A comprehensive analysis. *Sustainable Production and Consumption*, *40*, 352–362.

Palafox-Alcantar, P. G., Hunt, D. V., & Rogers, C. D. (2021). Current and future professional insights on cooperation towards circular economy adoption. *Sustainability*, *13*(18), 10436.

Panchal, R., Singh, A., & Diwan, H. (2021). Does circular economy performance lead to sustainable development?—A systematic literature review. *Journal of Environmental Management*, *293*, 112811.

Pla-Julián, I., & Guevara, S. (2019). Is circular economy the key to transitioning towards sustainable development? Challenges from the perspective of care ethics. *Futures*, *105*, 67–77.

Priyadarshini, P., & Abhilash, P. C. (2020). Circular economy practices within energy and waste management sectors of India: A meta-analysis. *Bioresource Technology*, *304*, 123018.

Provenzano, M., Pacchera, F., Silvestri, C., & Ruggieri, A. (2024). From vineyard to value: A circular economy approach to viticulture waste. *Resources*, *13*(12), 172.

Provenzano, V., & Seminara, M. R. (2024, May). The nexus between circular economy and sustainable development goals. In *International Symposium: New Metropolitan Perspectives* (pp. 72–82). Cham: Springer Nature Switzerland.

Puntillo, P. (2023). Circular economy business models: Towards achieving sustainable development goals in the waste management sector—Empirical evidence and theoretical implications. *Corporate Social Responsibility and Environmental Management*, *30*(2), 941–954.

Rebello, T. A., Mian, H. R., Chhipi-Shrestha, G., Hewage, K., & Sadiq, R. (2025). Integrating the circular economy concept with one water approach to enhance sustainability in urban water systems: Identification, selection, and ranking of key sustainability indicators. *Clean Technologies and Environmental Policy*, 1–24.

Reddy, V. V., Kumar, P., Rao, A. L. N., Kumar, R., Singh, S., Asha, V., & Kareem, S. H. (2023). Waste to wealth generation: Innovative methodologies in resource utilization and minimization in circular economy. In *E3S Web of Conferences* (Vol. 453, p. 01035). EDP Sciences.

Ribeiro, T. B., Oliveira, A. L., Costa, C., Nunes, J., Vicente, A. A., & Pintado, M. (2020). Total and sustainable valorisation of olive pomace using a fractionation approach. *Applied Sciences*, *10*(19), 6785.

Rocchi, L., Paolotti, L., Cortina, C., Fagioli, F. F., & Boggia, A. (2021). Measuring circularity: An application of modified Material Circularity Indicator to agricultural systems. *Agricultural and Food Economics*, *9*(1), 9.

Rodríguez-Antón, J. M., Rubio-Andrada, L., Celemín-Pedroche, M. S., & Ruíz-Peñalver, S. M. (2022). From the circular economy to the sustainable development goals in the European Union: An empirical comparison. *International Environmental Agreements: Politics, Law and Economics*, *22*(1), 67–95.

Roy, P., Mohanty, A. K., Dick, P., & Misra, M. (2023). A review on the challenges and choices for food waste valorization: Environmental and economic impacts. *ACS Environmental Au*, *3*(2), 58–75.

Sabău-Popa, C. D., Bele, A. M., Dodescu, A. O., & Boloș, M. I. (2022). How does the circular economy applied in the European Union support sustainable economic development? *Sustainability*, *14*(16), 9932.

Saidani, M., Yannou, B., Leroy, Y., Cluzel, F., & Kendall, A. (2019). A taxonomy of circular economy indicators. *Journal of Cleaner Production*, *207*, 542–559.

Sampantamit, T., Ho, L., Van Echelpoel, W., Lachat, C., & Goethals, P. (2020). Links and trade-offs between fisheries and environmental protection in relation to the sustainable development goals in Thailand. *Water*, *12*(2), 399.

Sarker, T. R., Ethen, D. Z., Asha, H. H., Islam, S., & Ali, M. R. (2025). Transformation of municipal solid waste to biofuel and bio-chemicals—a review. *International Journal of Environmental Science and Technology*, *22*(5), 3811–3832.

Schroeder, P., Anggraeni, K., & Weber, U. (2019). The relevance of circular economy practices to the sustainable development goals. *Journal of Industrial Ecology*, *23*(1), 77–95.

Sharma, P., Bano, A., Singh, S. P., Varjani, S., & Tong, Y. W. (2024). Sustainable organic waste management and future directions for environmental protection and techno-economic perspectives. *Current Pollution Reports*, *10*(3), 459–477.

Sheldon, R. A. (2014). Green and sustainable manufacture of chemicals from biomass: State of the art. *Green Chemistry*, *16*(3), 950–963.

Sikarwar, V. S., Pohořelý, M., Meers, E., Skoblia, S., Moško, J., & Jeremiáš, M. (2021). Potential of coupling anaerobic digestion with thermochemical technologies for waste valorization. *Fuel*, *294*, 120533.

Smol, M., Adam, C., & Preisner, M. (2020). Circular economy model framework in the European water and wastewater sector. *Journal of Material Cycles and Waste Management*, *22*(3), 682–697.

Stephan, G. (2022). Circular economy: Illusion or first step towards a sustainable economy: A physico-economic perspective. *Sustainability*, *14*(8), 4778.

Suárez, M. Á., Gambuzzi, E., Disla, J. M. S., Castejón, G., Poggiaroni, G., & Ling, J. (2023). ROOTS-Circular policies for changing the biowaste system. *Open Research Europe*, *3*, 78.

Tabas, A. M., Rehman, M. A., Khitous, F., & Urbinati, A. (2025). Stakeholder and customer engagement in circular economy ecosystems: A systematic literature review and research agenda. *Business Strategy and the Environment*, *34*(1), 402–416.

Taheri, S., & Hosseini, S. S. (2025). Waste not, want not: Comprehensive valorization of fruit and vegetable waste from single-product recovery to zero-waste strategies. *Cleaner Waste Systems*, *11*, 100300.

Taskhiri, M. S., Behera, S. K., Tan, R. R., & Park, H. S. (2015). Fuzzy optimization of a waste-to-energy network system in an eco-industrial park. *Journal of Material Cycles and Waste Management*, *17*(3), 476–489.

Thakker, S., & Narayanamoorthi, R. (2015, March). Smart and wireless waste management. In *2015 International Conference on Innovations in Information, Embedded and Communication Systems (ICIIECS)* (pp. 1–4). IEEE.

Tsai, F. M., Mohamed, A. F., Bui, T. D., Sethanan, K., & Tseng, M. L. (2025). Sustainable development goal 12-responsible consumption and production challenges and opportunities in world regions: A data-driven analysis on circular economy practices. *Journal of Industrial and Production Engineering*, *42*(7), 772–818.

Ungureanu, N., Vlăduţ, N. V., Biriş, S. Ş., Ionescu, M., & Gheorghiţă, N. E. (2025). Municipal solid waste gasification: Technologies, process parameters, and sustainable valorization of by-products in a circular economy. *Sustainability*, *17*(15), 6704.

Urban, P., & Hametner, M. (2022). The economy–environment nexus: Sustainable development goals interlinkages in Austria. *Sustainability*, *14*(19), 12281.

Valverde, J. M., & Avilés-Palacios, C. (2021). Circular economy as a catalyst for progress towards the sustainable development goals: A positive relationship between two self-sufficient variables. *Sustainability*, *13*(22), 12652.

Velenturf, A. P., & Purnell, P. (2021). Principles for a sustainable circular economy. *Sustainable production and consumption*, *27*, 1437–1457.

Voulvoulis, N. (2022). Transitioning to a sustainable circular economy: The transformation required to decouple growth from environmental degradation. *Frontiers in Sustainability*, *3*, 859896.

Wang, X., Li, C., Lam, C. H., Subramanian, K., Qin, Z. H., Mou, J. H., Jin, M., Chopra, S. S., Singh, V., Ok, Y. S., & Yan, J. (2022). Emerging waste valorisation techniques to moderate the hazardous impacts, and their path towards sustainability. *Journal of Hazardous Materials*, *423*, 127023.

Xiong, X., Liu, X., Iris, K. M., Wang, L., Zhou, J., Sun, X., . . . & Tsang, D. C. (2019). Potentially toxic elements in solid waste streams: Fate and management approaches. *Environmental Pollution*, *253*, 680–707.

Xu, P., Shu, L., Li, Y., Zhou, S., Zhang, G., Wu, Y., & Yang, Z. (2023). Pretreatment and composting technology of agricultural organic waste for sustainable agricultural development. *Heliyon*, *9*(5).

Zhidebekkyzy, A., Moldabekova, A., Amangeldiyeva, B., & Šanova, P. (2023). Transition to a circular economy: Exploring stakeholder perspectives in Kazakhstan. *Journal of International Studies*, *16*(3).

Zuberi, M. J. S., & Ali, S. F. (2015). Greenhouse effect reduction by recovering energy from waste landfills in Pakistan. *Renewable and Sustainable Energy Reviews*, *44*, 117–131.

3 Recovery of rare earth elements and platinum group metals

3.1 INTRODUCTION

Rare earth elements (REEs) are divided into two groups based on their ionic radii and atomic masses: light rare earth elements (LREEs) and heavy rare earth elements (HREEs). The LREEs include cerium (Ce), europium (Eu), lanthanum (La), neodymium (Nd), praseodymium (Pr), and promethium (Pm), whereas the HREEs comprise dysprosium (Dy), erbium (Er), gadolinium (Gd), holmium (Ho), lutetium (Lu), scandium (Sc), terbium (Tb), thulium (Tm), ytterbium (Yb), and yttrium (Y). These elements exhibit exceptional properties that enable their extensive use in magnetic, catalytic, electronic, optical, aerospace, and energy storage applications, which explains their widespread incorporation into many consumer products. China is currently the leading producer of REEs, accounting for approximately 70%–90% of global production (Chen et al., 2024). Platinum group metals (PGMs) represent another critical group of metals with extensive applications in catalysis (particularly in the automotive industry), industrial chemical production, petroleum refining, electronics, and medical technologies. The PGMs include iridium (Ir), osmium (Os), palladium (Pd), platinum (Pt), rhodium (Rh), and ruthenium (Ru). South Africa is the dominant producer of PGMs, contributing approximately 85% of global PGM production (Rao & Reddi, 2000).

Owing to their exceptional physicochemical properties and the rapid advancement of technology and renewable energy systems, the demand for REEs and PGMs is expected to increase significantly in the coming decades. For example, global demand for REEs is projected to rise three- to sevenfold by 2040 (O'Connell-Danes et al., 2024). Although REEs are relatively abundant in the Earth's crust, with estimated global reserves of approximately 110 million tonnes according to the United States Geological Survey (USGS), their distribution is highly uneven worldwide (U.S. Geological Survey, 2023). In contrast, PGMs are considerably scarcer, with global mineral reserves estimated at approximately 71 million tonnes as of 2025 (U.S. Geological Survey, 2025). Furthermore, the geographical distribution of PGMs is highly skewed, with nearly 98% of known reserves concentrated in Russia, South Africa, and Zimbabwe (U.S. Geological Survey, 2024). Conversely, the largest consumers of REEs and PGMs are regions such as Europe, North America, and Asia (Zhang et al., 2024). The uneven global distribution of REEs and PGMs has intensified geopolitical competition for control over these strategic resources. Consequently, many countries are exploring alternative materials

DOI: 10.1201/9781003745846-3

and developing technologies to recover REEs and PGMs from spent products and waste streams to mitigate supply risks and balance the growing disparity between demand and supply.

The recovery of REEs and PGMs from end-of-life products offers a viable strategy to alleviate demand–supply pressures while reducing the environmental pollution associated with their primary production. Over the years, these elements have been recovered from a variety of waste streams, including electronic waste (e-waste), acid mine drainage, spent catalysts, wastewater, and coal ash. For example, Xu et al. employed adsorption-based techniques to recover Dy, Eu, and Gd from acid mine drainage (J. Xu et al., 2024). The reported method demonstrated high selectivity and substantial adsorption capacity, both strongly influenced by the adsorbent's physicochemical properties.

The aim of this study was to investigate the sources of REE and PGM contamination in the environment to identify the most suitable sources for their recovery. In addition, the adsorption mechanisms of various adsorbents used for the recovery of REEs and PGMs from water and wastewater were examined to identify functional groups that preferentially bind these metals. This investigation was further supported by an evaluation of recovery efficiencies and adsorption capacities.

3.2 SOURCES OF REES AND PGMS CONTAMINATION

3.2.1 Sources of REEs contamination

Industrial activities, including mining, smelting, and manufacturing, are among the major sources of REE contamination in the environment. Numerous studies have identified mining operations as the most extensively investigated and significant contributors to REE pollution (Godwyn-Paulson et al., 2022; Guo et al., 2019; Tian et al., 2018; Xu et al., 2025; Zhu et al., 2025). For example, Guo et al. reported that mining and smelting activities were the dominant sources of REEs in PM_{10} collected from a mining region in Nandan County, China (Guo et al., 2019). REE concentrations in the sampled areas reached up to 23.54 ng m^{-3}, which were attributed primarily to anthropogenic activities, particularly mining and smelting.

In another study, Zhu et al. assessed REE concentrations in soil, water, and vegetables collected from a mining region in Ganzhou, China (Zhu et al., 2025). Their findings revealed REE concentrations of up to 1915.68 mg kg^{-1} in soil, 197.3 mg kg^{-1} in vegetables, and 4.09 μg L^{-1} in water. Notably, REE concentrations in soil exceeded both provincial and national background levels, supporting the conclusion that mining activities were the primary source of elevated REE contamination in the region.

Wastewater effluents also represent a significant source of REE contamination in the environment (Cao et al., 2025; J. Zhang et al., 2025). For instance, Cao et al. investigated the impact of anthropogenic activities by analysing the influence of wastewater effluents on REE pollution in Jiaozhou Bay, China (Cao et al., 2025).

Their study reported elevated concentrations of REEs in the sampled seawater, which were primarily attributed to wastewater discharges.

Rock weathering is a natural process that can significantly contribute to the contamination of various water bodies with REEs. Liu et al. reported that rock weathering increased REE concentrations in stream water originating from the Jiaping ion-adsorption-type rare earth deposits in Guangxi, China (Liu et al., 2025). Elevated REE levels were observed during the rainy season compared to the dry season, indicating that rainfall enhanced rock weathering and subsequent REE mobilisation.

Similarly, Vesković et al. demonstrated that the weathering of REE-rich rocks contributed to groundwater contamination (Vesković & Onjia, 2025). In an earlier study, Zhaoliang et al. examined 13 weathering profiles of sedimentary rocks, including dolomite, limestone, silicalite, and purple sandstone, to investigate the enrichment and release of REEs during weathering processes (Zhaoliang & Cheng, 2006). Their findings revealed that carbonate and clastic rocks contained high REE concentrations and played a key role in controlling REEs distribution in the Wujiang River.

3.2.2 Sources of PGMs contamination

The automobile industry is considered the primary source of PGM contamination in the environment, as automotive catalytic converters are the largest consumers of PGMs (Shubo et al., 2025; Hughes et al., 2021). Elevated concentrations of PGMs associated with vehicular emissions have been detected in roadside soils (Riga-Karandinos et al., 2006), human hair (Lo Medico et al., 2023), street dust (Ladonin, 2018), water bodies (Bluteau et al., 2025), and ambient air (Sen et al., 2016). Ladonin et al. demonstrated a strong correlation between Rh, Pd, and Pt emissions from motor vehicles and their accumulation in soil and street dust samples collected in Moscow (Ladonin, 2018). Although PGMs are initially deposited in soil and street dust, they can be transported into aquatic systems via surface runoff during rainfall events, where they may pose potential risks to aquatic organisms and human health through the consumption of contaminated water.

Matodzi et al. investigated the occurrence of PGMs in soil, dust, water, and sediment samples collected near a cement manufacturing plant in the Waterberg region of South Africa (Matodzi et al., 2020). The concentrations of PGMs detected in soil and water samples decreased in the order Pt > Ru > Rh > Pd, and the cement plant was identified as the primary source through atmospheric deposition of dust particles. These findings suggest that cement manufacturing facilities may represent significant sources of PGM contamination in the environment, particularly in regions undergoing extensive infrastructural development.

3.3 FUNDAMENTALS AND JUSTIFICATIONS FOR RECOVERING REES AND PGMS

PGMs are extremely rare in the Earth's crust, resulting in relatively limited global reserves estimated at just over 70 million tonnes. The mining and separation of PGMs are also highly expensive and technically challenging. In contrast, REEs, despite their name, are relatively abundant in the Earth's crust, with estimated average concentrations ranging from 130 to 240 μg g^{-1}, which is considerably higher than that of many commonly exploited elements (Balaram, 2019). However, REEs are rarely concentrated in economically exploitable ore deposits due to their wide dispersal and high chemical reactivity, which prevents them from occurring as free metals. This combination of dispersal and chemical reactivity makes the extraction and separation of REEs difficult and costly. The scarcity of PGMs, the complex extraction of REEs, and the increasing global demand for these metals have prompted a surge of research focused on developing techniques to recover PGMs and REEs from end-of-life products and mining waste.

In addition to their scarcity, PGMs and REEs are unevenly distributed globally. As previously discussed, South Africa, Russia, and Zimbabwe hold the largest PGM reserves, whereas China dominates REE production. Leveraging its technological capabilities, China has effectively established a near-monopoly in the separation and purification of REEs, a process that is technically challenging due to the chemical properties and dispersed nature of these elements. Any disruption in supply from these limited regions, whether due to geopolitical tensions or conflicts, could have significant global repercussions. The impact is expected to be particularly severe given the global shift towards electric vehicles, fuel cells, wind turbines, and automotive catalytic converters, all of which rely heavily on PGMs and REEs. Recent events, such as China's reduction of REE exports to conserve its reserves, illustrate this risk. In response to the uneven distribution and growing demand, many countries and industries have intensified efforts to recover these metals from end-of-life products, including electronics and automotive catalytic converters. Additionally, researchers are increasingly exploring the recovery of REEs and PGMs from mining wastewater to mitigate supply–demand imbalances and reduce environmental contamination.

Environmental pollution with PGMs and REEs is a growing concern due to their increasing demand and widespread use. End-of-life products containing these metals often enter the environment, where they accumulate in agricultural soil and water systems, either directly or through runoff. Additionally, some industries and mining operations release untreated or insufficiently treated wastewater, further contaminating aquatic ecosystems. Recovering REEs and PGMs from wastewater is therefore essential to mitigate their potential toxic effects on aquatic organisms and humans who rely on contaminated water sources.

3.4 ADSORPTIVE RECOVERY OF REES AND PGMS FROM WATER/WASTEWATER

3.4.1 Adsorptive recovery of REEs

The recovery of REEs and PGMs from water and wastewater is essential to address supply–demand imbalances, the uneven global distribution of reserves and production, and environmental contamination with its associated risks. Among various recovery techniques, adsorption is one of the most widely employed due to its high efficiency, simplicity, cost-effectiveness, low waste generation, and potential for adsorbent reuse. A key advantage of adsorption is the ability to explore diverse adsorbent materials and combinations to optimise recovery efficiency for REEs and PGMs.

For example, Xu et al. fabricated β-cyclodextrin-based nanosponges (β-CD/PVA-SA NSs) for the adsorptive recovery of Dy, Eu, and Gd from acid mine drainage (J. Xu et al., 2024). The adsorbent exhibited high adsorption capacities and removal efficiencies in simulated wastewater, as summarised in Table 3.1. When applied to real acid mine drainage, β-CD/PVA-SA NSs achieved approximately 99% removal efficiency for 14 REEs present, while exhibiting minimal interaction with competing ions such as Li, Mg, and Ca. X-ray photoelectron spectroscopy (XPS) analysis indicated that REEs coordinated with oxygen-containing functional groups (C=O, O=C–C, and O–H/C–O–C) on the adsorbent, elucidating the adsorption mechanism. The exceptional performance of β-CD/PVA-SA NSs can therefore be attributed to the synergistic combination of β-cyclodextrin, polyvinyl alcohol, and sodium alginate, which are rich in oxygen moieties. This finding also highlights the strong affinity of REEs for oxygen-containing functional groups.

In another study, Xu et al. employed ZIF-8, a metal-organic framework (MOF), as an adsorbent for the selective recovery of Dy, Gd, and Y from mining wastewater (Wu et al., 2024). ZIF-8 exhibited high adsorption capacities for these target REEs in simulated wastewater, as summarised in Table 3.1. When applied to real mining wastewater, the adsorbent achieved removal efficiencies exceeding 92% while selectively targeting REEs (Wu et al., 2024). The adsorption mechanism involved coordination of metal ions to nitrogen groups and Zr–O–H functional groups of the ZIF-8 framework, along with electrostatic interactions that facilitated binding to the material's surface. Approximately 40% of the adsorbed REEs were recovered, with the relatively low recovery attributed to strong metal-sorption site binding. The recovered REEs can be reused in various applications, reducing environmental waste and promoting the principles of circular economy and green chemistry.

In a notable study, Zhang et al. achieved the selective recovery of La and Ce from mining wastewater using magnetic zeolite synthesised from rare earth tailings (J. Zhang et al., 2025). Rare earth tailings are a by-product generated during REE processing. As shown in Table 3.1, the adsorbent demonstrated high removal efficiencies and substantial adsorption capacities. When applied to real acid-mine wastewater, recoveries of Ce and La exceeded 90%, which is particularly impressive given the matrix complexity and competing ions. Utilising waste rare earth tailings as a raw material for adsorbent synthesis makes this process both cost-effective and environmentally sustainable, as it generates a value-added product from waste while preventing the disposal of potentially hazardous materials into the environment.

3.4.2 Adsorptive recovery of PGMs

PGMs have also attracted significant attention for adsorptive recovery from water and wastewater. Guo et al. employed a green and efficient biosorption approach to recover

TABLE 3.1
Adsorptive recovery of REEs and PGMs from water/wastewater using various adsorbents.

Adsorbent	Elements	Sample	Recovery Performance (%)	Adsorption Capacity (mg g^{-1})	Reference
		Rare earth elements			
β-CD/PVA-SA NSs	Eu, Dy, and Gd	Acidic mine drainage	>95	222, 217, and 204	J. Xu et al. (2024)
Poly(ortho amino-phenol) resin	Nd and Er	Wastewater	63 & 73	38.0 and 47.7	Algarni et al. (2025)
DGCN-K	Nd, Dy, and Lu	Water	-	146, 181, and 204	Cai et al. (2025)
Magnetic zeolite	Ce and La	Mining wastewater	80 & 70	99.9 and 123	J. Zhang et al. (2025)
PCDP-M-SHM	Gd, Eu, Nd, and Y	Wastewater	>96	40.9–64.3	Nkinahamira et al. (2021)
ZIF-8	Dy, Gd, and Y	Mining wastewater	-	723, 630, and 610	Wu et al. (2024)
Ec-FeNPs	Ce, Er, Dy, Gd, Ho, La, Lu, Nd, Pr, Sm, Tb, Tm, and Y	Mining wastewater	>90	-	X. Xu et al. (2024)
SA/PVA	Ce, Er, Dy, Gd, La, Nd, Pr, Sm, Tb, and Y	Mining wastewater	27.6–52.6	-	Y. Zhang et al. (2025)
		Platinum group metals			
Zr-MSA@HLS	Pd	Metallurgical wastewater	84.3	217	Li et al. (2025)
EPS	Pt and Pd	Water	-	1402 and 1992	Guo et al. (2026)
BTMSPA-SG	Pt and Pd	Water	>98	48.5 and 29.7	Mphahlele et al. (2024)
DTMSP-BT-SG	Pt and Pd	Water		471 and 382	Mphahlele et al. (2025)
AT-C	Pd	Industrial wastewater	96.6	179	Zhang et al. (2023)
Yeast-functionalised bentonite	Pt	Water	98.5	0.255	Mosai et al. (2020)

Pt and Pd from aqueous solutions (Guo et al., 2026). In their study, the rhizobacterium *Pseudomonas putida* was induced by Pt and Pd to secrete extracellular polymeric substances (EPS), which were subsequently extracted and used as biosorbents. EPS is rich in active functional groups, including amino, carboxyl, and phosphate moieties, which serve as sorption sites for Pt and Pd. The EPS biosorbent exhibited exceptionally high maximum adsorption capacities for Pt and Pd in simulated wastewater, exceeding 1000 mg g^{-1} (Table 3.1). XPS analysis revealed that EPS coordinated with Pt and Pd through various functional groups of amino acids, including C=O/O–C–O, C–O/C–N, $-NH_3^+$, –NH/–NH_2, and C=O/P=O. EPS is abundant, easily produced in high yields, and cost-effective, making it a promising alternative for the efficient recovery of PGMs for reuse in applications such as automotive catalysis. Given its environmental and economic advantages, greater attention should be directed towards biosorption as a green, cost-effective, efficient, and waste-free method for recovering PGMs and other precious metals.

Activated carbon has long played a key role in adsorbing various metal ions due to its abundant active sorption sites, which depend on the carbon source. Zhang et al. functionalised activated carbon with 2-aminothiazole to recover Pd from industrial wastewater (Zhang et al., 2023). The functionalisation introduced sulfur atoms to promote selective adsorption of Pd, which exhibits a high affinity for sulfur-containing functional groups. The resulting adsorbent demonstrated high recovery efficiency and adsorption capacity, as shown in Table 3.1. When applied to real industrial wastewater, Pd recovery reached 96.6%, confirming the suitability of this adsorbent for mitigating Pd pollution and recovering this valuable metal for reuse. XPS analysis revealed that the successful adsorption of Pd was primarily due to the synergistic coordination of C–S and C=O functional groups with the metal, whereas C=N and –NH groups played no significant role, likely due to their low affinity for Pd. This observation highlights the critical importance of selecting an appropriate adsorbent and appropriately functionalising it to achieve selective recovery of PGMs and REEs from wastewater, particularly in complex matrices containing multiple competing ions.

Mosai et al. recovered Pt from water using yeast-functionalised bentonite, prepared from spent yeast from brewing and widely available bentonite (Mosai et al., 2020). The waste-derived adsorbent achieved high recovery efficiencies, as summarised in Table 3.1. Importantly, the study included a comprehensive cost analysis covering adsorbent preparation and the recovery of Pt and other PGMs, including Ir, Os, Pd, Rh, and Ru. The analysis revealed that up to USD 57 000 could be generated as profit from the recovered PGMs (Mosai et al., 2020). This finding is significant, as it demonstrates the practical economic potential of circular economy strategies, particularly for the recovery of REEs and PGMs from waste streams.

3.5 CHALLENGES AND PROSPECTS OF RECOVERY TECHNIQUES

Notably, there are relatively few studies demonstrating the complete recovery of PGMs and REEs from water and wastewater, followed by their desorption from the adsorbent into individual metal ions for reuse in various applications.

Most research has focused on the removal of PGMs and REEs from waste, an important step for mitigating environmental contamination and reducing exposure of living organisms to toxic metals. However, the goal is not only to remove these metals but also to recover and separate them, thereby supporting circular economy principles and addressing the imbalance between high demand and limited supply.

Several studies report high removal efficiencies but poor recovery performance from the adsorbents. For example, Wu et al. recovered only approximately 40% of the adsorbed REEs from ZIF-8 post-adsorption (Wu et al., 2024). This limitation underscores the need for further investigation into cost-effective and environmentally friendly solvent systems for metal desorption. The challenge is often attributed to the strong coordination bonds between the metal ions and functional groups of the adsorbent, which can prevent efficient elution, particularly when adsorbed metal ions form complexes with ligands or substitute existing metal centres in the material. Consequently, careful selection of both the adsorbent and the eluting solvent is crucial to achieve efficient and selective recovery of REEs and PGMs.

Waste-derived nanomaterials are expected to play a pivotal role in enabling greener and more cost-effective recovery of REEs and PGMs from wastewater. This approach offers dual environmental benefits: it prevents waste from entering the environment and enables value-added nanomaterials to recover REEs and PGMs from wastewater before they can contaminate surface waters. Additionally, because the production of these nanomaterials is cost-effective, their use can increase the economic value of recovered precious metals.

In parallel, the use of microorganisms and their secreted products in biosorption represents a promising, low-cost, and environmentally friendly alternative for REE and PGM recovery. The potential of biosorption was demonstrated by the effective recovery of Pt and Pd using EPS secreted by *Pseudomonas putida* (Guo et al., 2026). EPS-based biosorption outperformed several conventional adsorbents for Pt and Pd recovery, highlighting its promising prospects for the sustainable recovery of REEs and PGMs from wastewater.

Future studies should incorporate comprehensive economic assessments of REE and PGM recovery techniques, including estimates of the profits from recovered metals. Such analyses will help industries select cost-effective, efficient recovery methods while promoting sustainable practices. For example, future iterations of Table 3.1 could include a column detailing the costs associated with each adsorbent and recovery method, enabling practical and direct comparisons. Considering costs is essential not only for industrial decision-making but also for informing policy development and promoting large-scale adoption of sustainable recovery technologies worldwide.

3.6 CONCLUSION

This study demonstrated that environmental pollution by REEs and PGMs originates from multiple sources, including mining and processing activities, manufacturing plants (e.g., phosphate fertiliser and cement production), and wastewater discharges. It was further shown that REEs and PGMs can be effectively recovered from wastewater, which is critical for mitigating environmental contamination, addressing supply–demand imbalances, and alleviating the effects of uneven global distribution of these metals. Adsorptive recovery using various materials exhibited high performance, with removal efficiencies exceeding 98% in some cases. Notably, one scenario indicated that up to USD 57 000 could be generated in profit from recovered PGMs using just 1.0 g of adsorbent. These findings highlight that the recovery of REEs and PGMs not only protects the environment but also offers significant economic benefits.

REFERENCES

Algarni, F., Elshehy, E. A., Kashif, M., & El-Said, W. A. (2025). Synthesis of a high surface area porous poly(ortho amino-phenol) resin for effective adsorption of rare earth elements Nd(III) and Er (III) ions. *Journal of the Taiwan Institute of Chemical Engineers*, *173*.

Balaram, V. (2019). Rare earth elements: A review of applications, occurrence, exploration, analysis, recycling, and environmental impact. *Geoscience Frontiers*, *10*(4), 1285–1303.

Bluteau, G., Ponton, D. E., Rosabal, M., & Amyot, M. (2025). Biodynamics and environmental concentrations of the platinum group elements in freshwater systems. *Environmental Science and Technology*, *59*(12), 6203–6213.

Cai, H., Rong, M., Liang, S., Wang, Q., Meng, Q., Ni, S., Liu, H., & Yang, L. (2025). Microenvironment modulation of two-dimensional defective carbon nitride with extended nanoporous furnished oxygen-enriched sites for efficient recovery of rare earth elements. *Journal of Hazardous Materials*, *489*.

Cao, A., Fu, Y., Wang, J., Xue, L., He, W., & Liu, Q. (2025). Impact of human activities on rare earth elements in coastal regions: Anthropogenic gadolinium sources, budget, and transport. *Water Research*, *287*.

Chen, P., Ilton, E. S., Wang, Z., Rosso, K. M., & Zhang, X. (2024). Global rare earth element resources: A concise review. *Applied Geochemistry*, 175.

Godwyn-Paulson, P., Jonathan, M. P., Rodríguez-Espinosa, P. F., & Rodríguez-Figueroa, G. M. (2022). Rare earth element enrichments in beach sediments from Santa Rosalia mining region, Mexico: An index-based environmental approach. *Marine Pollution Bulletin*, *174*.

Guo, G., Song, B., Lei, M., & Wang, Y. (2019). Rare earth elements (REEs) in PM_{10} and associated health risk from the polymetallic mining region of Nandan County, China. *Human and Ecological Risk Assessment*, *25*(3), 672–687.

Guo, Q., Li, S., Yin, B., Zhang, X., Leng, Z., Huang, Y., Lin, W., Yan, P., Dai, W., & Sun, S. (2026). Target-metal stress-induced synthesis of extracellular polymeric substances by Pseudomonas putida enhances the adsorption of trace platinum and palladium from wastewater. *Separation and Purification Technology*, *387*.

Hughes, A. E., Haque, N., Northey, S. A., & Giddey, S. (2021). Platinum group metals: A review of resources, production and usage with a focus on catalysts. *Resources*, 10.

Ladonin, D. V. (2018). Platinum-group elements in soils and street dust of the southeastern administrative district of Moscow. *Eurasian Soil Science*, *51*, 268–276.

Li, F., Zhang, Y., Zheng, F., Dai, M., Zhou, Z., Yang, F., Bai, H., Wang, S., & Yang, X. (2025). Surface-functionalized loofah sponge biocomposite: An eco-friendly and cost-effective sorbent for palladium recovery from metallurgical wastewater. *Separation and Purification Technology*, *377*.

Liu, J., Han, Z., Dong, C., Wei, X., & Chen, Y. (2025). Geochemical survey of stream sediments and stream water for ion-adsorption type rare earth deposits (IAREDs): A pilot study in Jiaping IARED, Guangxi, South China. *Minerals, 15.*

Lo Medico, F., Varrica, D., & Alaimo, M. G. (2023). Occurrence of palladium and platinum in human scalp hair of adolescents living in urban and industrial sites. *Science of the Total Environment, 892.*

Matodzi, V., Legodi, M. A., & Tavengwa, N. T. (2020). Determination of platinum group metals in dust, water, soil and sediments in the vicinity of a cement manufacturing plant. *SN Applied Sciences, 2.*

Mosai, A. K., Chimuka, L., Cukrowska, E. M., Kotzé, I. A., & Tutu, H. (2020). Removal of platinum (IV) from aqueous solutions with yeast-functionalised bentonite. *Chemosphere, 239.*

Mphahlele, M. R. R., Mosai, A. K., Tutu, H., & Kotzé, I. A. (2024). Advancements in sustainable platinum and palladium recovery: Unveiling superior adsorption efficiency and selectivity with a novel silica-anchored acylthiourea adsorbent. *RSC Advances, 14,* 7507–7516.

Mphahlele, M. R. R., Mosai, A. K., Tutu, H., & Kotzé, I. A. (2025). Efficient recovery of platinum and palladium by fixed-bed column adsorption using acylthiourea- and amine-modified silica. *Journal of Sustainable Metallurgy, 11,* 4334–4346.

Nkinahamira, F., Alsbaiee, A., Wang, Y., Yang, X., Chen, T. Y., Cao, M., Feng, M., Sun, Q., & Yu, C. P. (2021). Recovery and purification of rare earth elements from wastewater and sludge using a porous magnetic composite of β-cyclodextrin and silica doped with PC88A. *Separation and Purification Technology, 266.*

O'Connell-Danes, J. G., Ozen Ilik, B., Hull, E. E., Ngwenya, B. T., Morrison, C. A., & Love, J. B. (2024). A simple supramolecular approach to recycling rare earth elements. *ACS Sustainable Chemistry and Engineering, 12,* 9301–9305.

Rao, C. R. M., & Reddi, G. S. (2000). Platinum group metals (PGM): Occurrence, use and recent trends in their determination. *TrAC Trends in Analytical Chemistry, 19,* 565–586.

Riga-Karandinos, A. N., Saitanis, C. J., & Arapis, G. (2006). First study of anthropogenic platinum group elements in roadside top-soils in Athens, Greece. *Water, Air, and Soil Pollution, 172,* 3–20.

Sen, I. S., Mitra, A., Peucker-Ehrenbrink, B., Rothenberg, S. E., Tripathi, S. N., & Bizimis, M. (2016). Emerging airborne contaminants in India: Platinum group elements from catalytic converters in motor vehicles. *Applied Geochemistry, 75,* 100–106.

Shubo, A., Sun, S., Tu, G., Liu, R., Xiao, F., Shi, R., Sui, C., & Yu, K. (2025). Iron capture mechanism for harmless recovering platinum group metals from spent automobile catalyst. *Environmental Technology (United Kingdom), 46*(10), 1666–1678.

Tian, S., Liang, T., Li, K., & Wang, L. (2018). Source and path identification of metals pollution in a mining area by PMF and rare earth element patterns in road dust. *Science of the Total Environment, 633,* 958–966.

U.S. Geological Survey. (2023). Mineral commodity summaries 2023. *U.S. Geological Survey,* 1–210.

U.S. Geological Survey. (2024). Mineral commodity summaries 2024. *U.S. Geological Survey,* 1–210.

U.S. Geological Survey. (2025). Mineral commodity summaries 2025 (ver. 1.2, March 2025). *U.S. Geological Survey,* 1–212.

Vesković, J., & Onjia, A. (2025). Influencing factors of groundwater 238U, 232Th, 40K, and rare earth element contamination: Insights from the two-dimensional Monte Carlo simulation of radiological risks. *Marine Pollution Bulletin, 213.*

Wu, R., Hong, B., Xue, C., Chen, Zhibiao, & Chen, Zuliang. (2024). ZIF-8 Used for the selective recovery of heavy rare earth elements from mining wastewater. *Environmental Science and Technology, 58*(22), 9612–9623.

Xu, D. X., Cheng, X. P., Huang, C. L., Liu, X. R., Xie, C. D., Qiu, R. L., Liu, C., Guo, M. N., Morel, J. L., Liu, W. S., & Tang, Y. T. (2025). Recovery of rare earth elements from sediments affected by mining activities. *Journal of Hazardous Materials, 496.*

Xu, J., Wang, D., Yang, J., He, J., Yang, T., Zhao, X., Liu, X., Zhang, J., Zhang, L., & Zhao, Y. (2024). Efficient recovery of rare earth elements from acidic wastewater by a green β-CDs-functionalized nanosponge. *Small, 20.*

Xu, X., Weng, X., Owens, G., & Chen, Z. (2024). Iron nanoparticles synthesized using Euphorbia cochinchinensis leaf extracts exhibited highly selective recovery of rare earth elements from mining wastewater: Exploring the origin of high selectivity. *Journal of Hazardous Materials, 480.*

Zhang, J., Cheng, J., Xu, G., Hou, X., Yu, Z., Liu, W., Zhang, Y., Chen, J., Yang, J., Wu, M., Yang, W., Li, S., Zhang, M., Chen, L., Li, Z., Zhang, Jiaxia, Wu, X., Wang, X., Xiao, J., & Xie, X. (2025). Selective recovery of rare earth elements from mine wastewater using the magnetic zeolite prepared by rare earth tailings. *Chemical Engineering Journal, 511.*

Zhang, L., Li, B., Shao, P., Zhou, X., Li, D., Hu, Z., Dong, H., Yang, L., Shi, H., & Luo, X. (2023). Selective capture of palladium from acid wastewater by thiazole-modified activated carbon: Performance and mechanism. *Environmental Research, 238.*

Zhang, S., He, X., Ding, Y., Shi, Z., & Wu, B. (2024). Supply and demand of platinum group metals and strategies for sustainable management. *Renewable and Sustainable Energy Reviews*, 204.

Zhang, Y., Cai, W., & Chen, Z. (2025). Sodium alginate/polyvinyl alcohol for the recovery of rare earth elements from mining wastewater: Mechanism and properties. *Separation and Purification Technology, 354.*

Zhaoliang, S., & Cheng, Y. (2006). Enrichment and release of rare earth elements during weathering of sedimentary rocks in Wujiang catchments Southwest China. *Journal of Rare Earths, 24*, 491–496.

Zhu, L., Huang, J., Hu, G. H., Wang, Q., Huang, H., Wang, S., Wu, C., Sun, Z., Fang, Y., Hao, M., & Xiong, L. (2025). Pollution and health risk assessment of rare earth elements in soil, water and vegetables from an ion-adsorption rare earth mining area in Ganzhou, China. *Environmental Geochemistry and Health, 47.*

4 Classification of nanomaterials derived from waste

4.1 INTRODUCTION

4.1.1 Background and importance of nanomaterials from waste

Nanomaterials are defined as materials with at least one dimension in the nanometre range (1–100 nm) (Baig et al., 2021). At this scale, matter exhibits unique and often enhanced physical, chemical, and biological properties due to the dominance of surface atoms and the influence of quantum effects. Their nanoscale dimensions also allow for diverse morphologies, including spherical, tubular, hollow, spiral, and irregular structures (Ealia & Saravanakumar, 2017; Joudeh & Linke, 2022; Mekuye & Abera, 2023). These structural and size-dependent features give rise to remarkable functionalities, for instance, the tensile strength of carbon nanotubes, the electrical conductivity of graphene, and the semiconducting and catalytic activity of metal oxide nanoparticles (Fatima et al., 2020; Iordache et al., 2023; Wang et al., 2023). Such properties have enabled nanomaterials to transform sectors ranging from renewable energy and electronics to medicine, construction, and environmental remediation, as shown in Figure 4.1.

Although promising, traditional techniques for synthesising nanomaterials often rely on high-quality chemicals, advanced reagents, and energy-intensive methods. While they are effective, they are costly, environmentally unsustainable, and produce secondary pollutants, as well as depleting non-renewable resources (Omran & Baek, 2022). In response to these challenges, researchers are increasingly utilising waste as a green feedstock for synthesising nanomaterials. Agricultural wastes, plastic wastes, industrial wastes, and fossil-fuel-based materials are readily available and cost-effective and often contain valuable building blocks such as carbon, silica-based materials, and metals needed for producing nanostructures (Nath et al., 2023; Deivayanai et al., 2024; Flores-Contreras et al., 2024; Li et al., 2025).

The production of high-value nanomaterials from waste materials not only eliminates the problem of waste disposal but also reduces dependence on virgin materials, lowers production costs, and increases resource efficiency (Omran & Baek, 2022; Deivayanai et al., 2024). The valorisation of waste as nanomaterials largely aligns with the principles of green chemistry and sustainable development, as well as those of a circular economy (Omran & Baek, 2022; Flores-Contreras et al., 2024). Rather than dumping materials once used as waste, this approach highlights the importance of recycling, recovery, and reuse, thereby converting waste streams into technologically valuable products (Lizundia et al., 2022; Deivayanai et al., 2024).

DOI: 10.1201/9781003745846-4

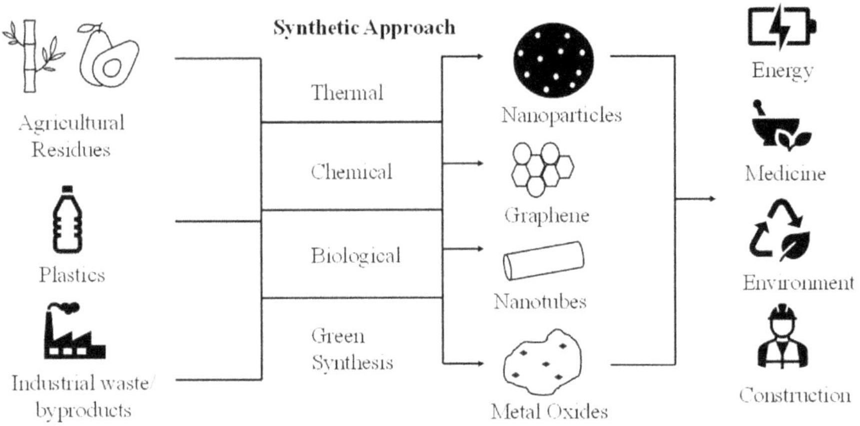

FIGURE 4.1 Illustrating the transformation of various waste sources into nanomaterials via different synthetic methods (thermal, chemical, biological, and green synthesis) and their applications in energy, medicine, environment, and construction.

Notably, this dual focus on waste minimisation and advanced material production ensures that both environmental preservation and technical innovation are pursued simultaneously.

4.2 OVERVIEW OF NANOMATERIALS AND WASTE UTILISATION

4.2.1 Definition and Unique Properties of Nanomaterials

Nanomaterials exhibit unique structural features that cause their behaviour to differ significantly from bulk materials, mainly due to their high surface-to-volume ratio and the influence of quantum confinement (Khan et al., 2019; Joudeh & Linke, 2022). These nanoscale effects impart unusual optical, electronic, and catalytic properties. For example, gold nanoparticles display vibrant red or purple colours (markedly different from those of bulk gold), while titanium dioxide (TiO_2) nanoparticles exhibit superior photocatalytic efficiency, enabling the breakdown of organic pollutants under sunlight (Khan et al., 2019; Tee et al., 2021; Chauke et al., 2025).

Due to these properties, nanomaterials have become crucial in a wide range of applications. In the energy sector, they are used to improve the performance of lithium-ion batteries, fuel cells and hydrogen production catalysts (Xu et al., 2020; Lu et al., 2020). In medicine and biotechnology, they serve as carriers for targeted drug delivery, diagnostic imaging agents, and antimicrobial coatings (Jana & Seth, 2020; Toro-Córdova et al., 2020; Woźniak et al., 2022). Environmental applications include the purification of water, the removal of air pollutants, and the remediation of contaminated soils (Cao et al., 2021; Mukhopadhyay et al., 2022; Elhenawy et al., 2024). In construction, nanomaterials enhance the strength, durability, and functionality of composites. Their usefulness across such diverse fields underscores the growing global demand for nanomaterials and the need for sustainable production pathways.

4.2.2 Waste as a resource used to produce nanomaterials

Traditional methods of nanomaterial production typically rely on high-purity chemical precursors and energy-intensive processes, which increase costs and environmental impacts. In recent years, however, waste streams have emerged as abundant, low-cost, and underutilised alternatives for nanomaterial synthesis. This pattern shift is directly linked to the global sustainability agenda, which emphasises circular economy approaches in which waste is re-envisioned as a valuable resource rather than a disposal problem.

Agricultural residues are among the most promising precursors. Biomass such as rice husks, coconut shells, corn cobs, wheat waste, grass, and sugarcane bagasse are rich in lignocellulose, carbon, and silica (Akhayere et al., 2019; Yaqueen et al., 2025). When subjected to controlled thermal or chemical treatments, these residues can yield nanostructures such as carbon nanotubes, graphene, and silica nanoparticles. For instance, rice husk ash, which contains more than 90% amorphous silica, is an inexpensive raw material for producing nanosilica (Costa & Paranhos, 2018; Askaruly et al., 2020), while coconut shells can be carbonised and activated to generate carbon nanomaterials with high surface area (Tsai & Jiang, 2018; Sujiono et al., 2022). This not only diverts massive amounts of agricultural by-products from open burning or landfills but also creates nanomaterials for applications ranging from energy storage to catalysis.

Additionally, synthetic waste streams also hold significant potential. Plastics, one of the most persistent pollutants in terrestrial and marine ecosystems, can be thermally decomposed via pyrolysis or catalytic cracking to produce carbonaceous nanostructures such as carbon dots, graphene-like nanosheets, and porous carbons (Choi et al., 2022; Abdelfatah et al., 2024; Mohamed et al., 2025). Such upcycling of plastics offers a technological pathway to reduce plastic pollution while simultaneously generating high-value nanomaterials for sensors, batteries, and biomedical applications.

Industrial by-products are another valuable source. Fly ash, generated in millions of tonnes annually from coal-fired power plants, is composed largely of silica, alumina, and trace metal oxides (Yadav & Fulekar, 2020; Alterary & Marei, 2021). Through processes such as alkali activation or hydrothermal treatment, fly ash can be converted into zeolites, nanosilica, and other functional nanostructures with applications in adsorption, catalysis, energy storage, and construction (Boycheva et al., 2020; Li et al., 2022; Yadav et al., 2023). Similarly, petroleum coke (a carbon-rich by-product of oil refining) can serve as a precursor to graphene, carbon nanofibres, and other advanced carbon nanomaterials when processed at high temperatures or through chemical activation (Saha et al., 2021; Cabral et al., 2024).

This approach is not only resource-efficient but also timely, given the rising volumes of global waste. According to the World Bank, annual municipal solid waste generation is estimated to exceed 2.01 billion tonnes worldwide and is projected to grow substantially with ongoing population growth, urbanisation, and industrialisation (Yakah et al., 2023). Ineffective waste management contributes to environmental degradation, greenhouse gas emissions, and health risks (Ferronato & Torretta, 2019; Kristanto & Koven, 2019). Thus, converting these waste streams into nanomaterials then represents a dual benefit, as it addresses mounting environmental challenges while creating high-value products that can feed into diverse industries.

Treating waste as a raw material for nanotechnology, civilisations can simultaneously move towards cleaner production systems, reduce dependency on virgin resources, and advance the principles of a circular economy. This strategy not only aligns with sustainability goals but also offers economic opportunities through the value-added use of materials that are otherwise considered liabilities.

4.2.3 Advantages of waste-derived nanomaterials

The synthesis of nanomaterials from waste offers multiple advantages. Economically, it reduces the cost of nanomaterial production as raw materials are often inexpensive or sometimes even carry a negative value, such as when industries must pay for waste disposal. Environmentally, it reduces landfill burden and lowers greenhouse gas emissions associated with the burning process. It also promotes resource efficiency by reducing demand for virgin materials such as graphite or silica.

Additionally, many waste-derived synthesis processes are aligned with green-chemistry principles, avoiding hazardous reagents and operating under relatively mild conditions. For example, hydrothermal treatment of biomass can yield carbon dots without the need for toxic solvents, while alkaline extraction of silica from rice husks is far less energy-intensive than producing silica from mined quartz. Compared with conventional methods, waste-derived synthesis routes are often simpler, more sustainable, and scalable, positioning them as promising alternatives for the future of nanomaterial production.

4.3 RECENT PROGRESS AND MILESTONE ACHIEVEMENTS ON THE SYNTHESIS OF NANOMATERIALS

4.3.1 Recent advances

In the last decade, advances in synthesis methods have greatly accelerated progress in this field. Researchers have increasingly adopted green synthesis approaches that minimise the use of hazardous chemicals, reduce energy consumption, and utilise renewable or benign solvents. For example, hydrothermal and solvothermal methods using water or ethanol as solvents have become popular for producing carbon dots and metal oxide nanoparticles from biomass and food waste (Monday et al., 2021; Ndlwana et al., 2021; Bilge & Sınağ, 2022). Similarly, pyrolysis and gasification of plastics have been optimised to yield carbon nanomaterials with controlled morphologies and properties (Ren et al., 2022).

The investigation of waste streams has diversified in recent years. Beyond traditional sources like biomass and fly ash, attention has recently shifted to electronic waste, which contains precious metals such as platinum, gold, silver, and copper (Li et al., 2019). Sophisticated extraction and reduction techniques (pyrometallurgical and hydrometallurgical) have enabled the recovery of these metals at the nanoscale, as shown in Figure 4.2, enabling applications in electronics, sensors, and catalysis (Li et al., 2019; Tunali et al., 2021). Likewise, industrial sludge and textile waste

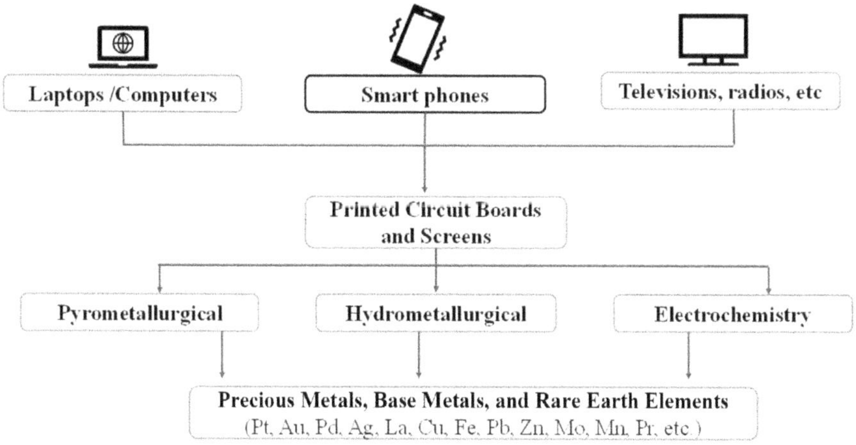

FIGURE 4.2 Recovery of metals from electronic waste via chemical processing routes. Printed circuit boards and screens are treated using pyrometallurgical, hydrometallurgical, or electrochemical methods to extract precious metals (Pt, Au, Pd, Ag), base metals (Cu, Fe, Pb, Zn, Mo, Mn), and rare earth elements (La, Pr, etc.).

(particularly cotton and cotton blends) have been investigated as potential sources for producing polymer-based nanomaterials such as cellulose nanocrystals (CNCs), as well as hybrid nanomaterials (Teixeira et al., 2025). These expansions highlight the adaptability of waste-to-nanomaterial strategies to a wide range of feedstocks.

Breakthrough studies have demonstrated the high performance of waste-derived nanomaterials in real-world applications. For instance, carbon dots synthesised from polyethene terephthalate (PET) waste have been employed as fluorescent probes for bio-imaging (Wu et al., 2022; Arpita et al., 2023). Zeolites derived from fly ash have shown strong adsorption capacity for heavy metals and organic pollutants in wastewater treatment (Buzukashvili et al., 2024; Kumari et al., 2024). High-impact publications and patents in recent years reflect growing industrial interest, particularly in applications related to environmental remediation and energy storage (Motalib Hossain et al., 2024).

4.3.2 Emerging trends

One of the most promising trends in the field is the integration of artificial intelligence (AI) and machine learning (ML) into the synthesis of nanomaterials. These tools are being applied to optimise synthesis conditions, predict material properties and design waste-to-material conversion processes with greater efficiency (Iravani et al., 2024; Tao et al., 2021; Tripathy et al., 2024). Such approaches reduce the need for trial-and-error experimentation, thereby accelerating the development of high-performance nanomaterials.

Another emerging trend is the scaling up of production. While laboratory-scale demonstrations have been successful for many years, recent work has shifted towards pilot-scale and industrial-scale synthesis. For example, pilot-scale production of zeolitic materials from coal fly ash has already been demonstrated for nutrient adsorption and fertiliser use (e.g., zeolite type X synthesis with strong adsorption performance) (Bonetti et al., 2021). Companies have also begun exploring commercial production of graphene from petroleum coke, validated by recent work showcasing scalable, cost-effective graphene manufacture for perovskite solar cell electrode (Alkhatani et al., 2025) and the recent strategic partnership King Abdulaziz City for Science and Technology (KACST), the Oil Sustainability Program (OSP), and the Center of Excellence for Energy Research (CEER) aimed at industrialising this route (KACST, 2025). Similarly, activated carbon is produced at an industrial scale from biomass. For example, Rotocarb in South Africa manufactures activated carbon from macadamia nut shells. While in India, the Council of Scientific and Industrial Research (CSIR-CMERI) recently developed a pilot plant for biomass-derived activated carbon for wastewater treatment (Rotocarb, 2016; CSIR-CMERI, 2024).

There is also increasing emphasis on multifunctionality and hybridisation. Researchers are now combining waste-derived nanomaterials with other functional materials to create composites that offer synergistic properties. For instance, carbon nanofibres derived from biomass can be integrated with metal oxides to improve supercapacitor performance (Ruiz-Montoya et al., 2021; Gao et al., 2024), while hybrid composites from mixed waste streams can serve as advanced photocatalysts for environmental remediation (Boruah et al., 2020).

Altogether, these advances or emerging trends indicate that waste-derived nanomaterials are moving from the realm of academic curiosity towards mainstream industrial adoption. They are no longer seen merely as a niche approach to waste management, but rather as a cornerstone of sustainable nanotechnology development.

4.4 CLASSIFICATION OF NANOMATERIALS DERIVED FROM WASTE AND SYNTHETIC METHODS

The classification of nanomaterials synthesised from waste is crucial for understanding their origins, chemical nature, and potential applications. Waste-derived nanomaterials can be broadly grouped by the type of waste source used for synthesis or by the type of nanomaterial obtained. Each approach offers distinct insights: while classification by waste source emphasises feedstock availability and waste valorisation strategies, classification by nanomaterial type highlights the physicochemical properties and technological potential of the resulting materials.

In practice, these two perspectives are closely linked. The chemical composition of the waste strongly influences the class of nanomaterials obtained, while the synthesis route determines their quality, morphology, and functional performance. For example, carbon-rich feedstocks such as biomass or plastics favour the production of carbon-based nanostructures, whereas inorganic-rich residues, such as fly ash, yield silica or metal oxide nanomaterials. Table 4.1 summarises the major waste sources, the corresponding nanomaterials, synthesis routes, and applications compiled from recent studies.

Several of these pathways are already being explored industrially, as discussed in Section 4.3, where examples such as Rotocarb, CSIR-CMERI, and KACST initiatives highlight the potential for scaling waste-derived nanomaterials

The transformation of waste into nanomaterials relies on diverse synthesis techniques, each with unique strengths and limitations. The choice of method depends

TABLE 4.1
Classification of waste-derived nanomaterials.

Waste Source	Nanomaterial Type	Synthesis Methods	Applications	Reference
Biomass	Carbon-based nanomaterials (carbon nanotubes, graphene), metal oxides	Carbonisation, hydrothermal treatment	Energy storage, catalysis and biomedical uses	Yan et al. (2025); Wang et al. (2022)
Plastic waste	Carbon dots, graphene-like materials	Pyrolysis, chemical vapour deposition (CVD)	Sensors, energy storage, environmental remediation	Ma et al. (2024); Arpita et al. (2023)
Fly ash	Silica nanoparticles, zeolites, metal oxides	Chemical extraction, thermal processing	Construction materials, adsorbents, and catalysis	Lin et al. (2016); Yadav et al. (2023(b))
Petroleum coke	Carbon-based nanomaterials (graphene, carbon nanofibres)	High-temperature processing, electrochemical methods	Energy storage, catalysis	Alkhatani et al. (2025); Cabral et al. (2024)
Other industrial waste (e.g., industrial textile waste; e-waste)	Polymer-based nanomaterials, metal nanoparticles	Chemical extraction, hydrothermal, pyrometallurgical	Electronics, environmental remediation, advanced composites	

on the waste composition, the desired nanomaterial properties, and the process scalability. For example, pyrolysis is highly effective for converting plastic waste into carbon nanomaterials, while hydrothermal treatment offers precise control over the morphology of biomass-derived products. Table 4.2 provides a comparative overview of these synthesis methods, including their advantages, limitations, and scalability.

4.5 APPLICATIONS OF WASTE-DERIVED NANOMATERIALS

Waste-derived nanomaterials exhibit unique physicochemical properties, such as large surface area, tunable porosity, and diverse surface functionalities that enable their use across a broad spectrum of fields. Their sustainable origin adds value by reducing raw material demand and closing resource loops. In the following, the major application domains are described in detail.

4.5.1 Environmental applications

One of the most prominent uses of waste-derived nanomaterials is in water purification and pollution remediation. Carbon-based nanostructures (graphene, carbon nanotubes, activated carbons) synthesised from agricultural residues or plastic wastes provide abundant adsorption sites and functional groups that efficiently capture emerging contaminants and heavy metals such as lead, arsenic, and cadmium (Mubarik et al., 2021; Kumar, 2023). Similarly, metal oxide nanoparticles from fly ash, like titanium dioxide or zinc oxide, exhibit high photocatalytic activity, allowing the degradation of dyes, pharmaceuticals and other persistent organic pollutants under visible or UV light (Gadore & Ahmaruzzaman, 2021; Gadore et al., 2024). These characteristics make them excellent candidates for incorporation into filtration membranes, packed-bed reactors, and photocatalytic treatment units for industrial wastewater.

Air-quality improvement is another critical area. Porous carbons and doped metal oxides can be embedded in air-filtration membranes or coatings to trap volatile organic compounds (VOCs) and fine particulate matter ($PM_{2.5}$) ($PM_{2.5}$ are particles ≤ 2.5 μm that can penetrate deep into the lungs and contribute to heart and lung diseases) (Henning et al., 2021; Zehra et al., 2024; Agrawal & Nawaz, 2025). Their high specific surface area and surface chemistry enable rapid adsorption of gaseous pollutants, such as formaldehyde and benzene. Recent research also indicates the potential of nanostructured catalysts from waste to convert harmful nitrogen oxides (NOx) into benign nitrogen gas (Chen et al., 2023; Graul et al., 2024; Shi et al., 2024). This can help in reducing smog formation in urban settings.

Beyond remediation, these nanomaterials can serve as real-time environmental sensors to detect pollutants. Carbon dots and doped graphene derived from plastic waste can be engineered to possess fluorescent or electrochemical sensing capabilities, enabling the monitoring of water and air contaminants at trace levels.

TABLE 4.2
Comparison of synthesis methods for waste-derived nanomaterials.

Method	Advantages	Limitations	Scalability	Reference
Carbonisation	Simple process; suitable for biomass and carbon-rich wastes	Requires high temperatures, possible emissions	Moderate	Khandaker et al. (2025); Bai et al. (2024); Oh et al. (2025)
Hydrothermal/ Solvothermal	Green, controlled morphology, versatile	Time-intensive, high-pressure systems are needed	Moderate to high	Mehmandoust et al. (2022)
Pyrolysis	Effective for plastics; yields carbon nanomaterials	Energy-intensive; secondary gas emissions	High	Belbessai et al. (2022); Dai et al. (2023)
Chemical vapour deposition (CVD)	Produces high-quality nanomaterials	Expensive; complex operation	Moderate	Tripathi et al. (2017); Hatta et al. (2017)

Note: Scalability is the practicality of moving from a small experiment to large-scale industrial production while maintaining performance and economic viability.

4.5.2 Electrochemical sensors for environmental monitoring

Electrochemical sensing is a rapidly advancing field in which waste-derived nanomaterials serve as electrode modifiers to improve sensitivity, selectivity, and stability in the detection of toxic pollutants. The high surface area, tunable porosity, and rich surface functionalities of these nanostructures (such as oxygenated groups on activated carbon or heteroatom-doped graphene) facilitate efficient electron transfer and provide abundant active sites for analyte adsorption (Onfray & Thiam, 2023). As a result, electrodes modified with these materials exhibit low detection limits and rapid response times (Onfray & Thiam, 2023; Bressi et al., 2024), making them ideal for monitoring trace concentrations of heavy metals and emerging organic contaminants in water, soil, and industrial effluents.

For heavy-metal detection, carbon-based nanomaterials derived from agricultural or plastic waste have been widely employed to sense lead (Pb^{2+}), cadmium (Cd^{2+}), mercury (Hg^{2+}), and arsenic (As^{3+}) (Bressi et al., 2024; Pabel et al., 2024; Sairaman & Young-Joon, 2025). When incorporated onto glassy carbon or screen-printed electrodes, biomass-derived graphene, carbon nanodots, or carbon nanotubes enhance electron transport and provide functional groups that bind strongly with metal ions (Bressi et al., 2024; Pabel et al., 2024; Sairaman & Young-Joon, 2025). Coupling these nanomaterials with bimetallic nanoparticles (e.g., Au, Pt, Cu, Ag) or conducting polymers further improves catalytic activity and anti-fouling properties, enabling real-time monitoring of drinking water and industrial discharge (Campuzano et al., 2019; Meng et al., 2024).

Electrochemical sensors based on waste-derived nanomaterials are also proving highly effective for emerging organic contaminants, such as pharmaceuticals, pesticides, endocrine disruptors, and phenolic compounds. Nitrogen-doped graphene or carbon dots produced from food or plastic waste offer tailored surface chemistry that promotes π–π interactions and hydrogen bonding with organic molecules (Kuang et al., 2023; Sooryakanth et al., 2025). These interactions increase sensitivity in differential pulse voltammetry or square wave voltammetry, allowing the detection of compounds such as bisphenol A, diclofenac, and triclosan at parts-per-billion levels. In addition, combining carbon materials with metal oxides, such as ZnO, TiO_2, or Fe_3O_4, to form hybrid nanocomposites exhibits synergistic electrocatalytic effects (Zhang et al., 2024).

The scalability and low cost of producing these nanomaterials from waste give electrochemical sensors a distinct advantage for deployment in resource-limited or decentralised monitoring systems. Portable, disposable, or even wearable devices built with waste-derived nanostructured electrodes can provide continuous, on-site analysis without the need for sophisticated laboratory infrastructure. By integrating these sensors with wireless or IoT platforms, real-time data on heavy metal and organic pollutant levels can support early warning systems, regulatory compliance, and community-driven environmental protection (Yuan et al., 2018; Nagabooshanam et al., 2019; Ajakwe et al., 2025). This emerging application highlights not only the technological potential of waste-to-nanomaterial strategies but also their critical role in safeguarding water and ecosystem health.

4.5.3 Energy applications

The energy sector benefits significantly from the electrochemical, catalytic, and conductive properties of waste-derived nanomaterials. Their hierarchical porosity and high conductivity support the development of next-generation energy storage technologies. Biomass- or petroleum-coke-derived graphene and carbon nanofibres function as high-performance electrodes for lithium-ion, sodium-ion, and potassium-ion batteries, as well as for supercapacitors (Ferdous et al., 2024; Benoy et al., 2025). The interconnected pore networks enhance ion transport, while graphitic domains enable rapid charge–discharge cycles and long-term stability (Lei et al., 2024; Benoy et al., 2025). Such characteristics reduce dependence on mined graphite, lowering both cost and environmental impact. Waste-derived nanomaterials also excel as catalysts for renewable energy conversion. Transition-metal oxides obtained from fly ash or industrial sludge, such as nickel, cobalt, or iron oxides, exhibit strong activity in the hydrogen evolution reaction (HER) (Altalhi et al., 2022), oxygen evolution reaction (OER) (Lu et al., 2023), and oxygen reduction reaction (ORR) (Yuan & Dai, 2016). These processes underpin water electrolysis, fuel cells, and metal–air batteries. Doping carbon materials with heteroatoms, such as nitrogen or sulfur, derived from agricultural residues or plastics, further enhances their catalytic performance and durability, providing a sustainable alternative to expensive PGMs (Zago et al., 2024; Fatima et al., 2025).

4.5.4 Biomedical applications

Biomedical technologies have embraced waste-derived nanomaterials for their biocompatibility and functional versatility. Carbon dots synthesised from food waste or natural fibres exhibit bright, tunable fluorescence, and low toxicity, making them ideal for bioimaging and biosensing applications (Choi et al., 2020; Divyabharathi et al., 2025). These nanomaterials enable non-invasive imaging of tissues and real-time monitoring of biological processes, providing cost-effective alternatives to conventional fluorescent probes (Etefa & Dejene, 2025).

In drug-delivery systems, waste-derived nanostructures such as mesoporous silica or functionalised graphene can be loaded with therapeutic agents and engineered for targeted, controlled release. Their large surface area, adjustable pore size, and ease of surface modification enable efficient encapsulation and delivery of chemotherapeutics, antibiotics, or genetic material (Zhao & Liu, 2014; Porrang et al., 2021; Chen et al., 2024). This precision reduces side effects and enhances treatment efficacy. Moreover, certain waste-derived metal oxide nanoparticles, including silver and zinc oxide, display strong antimicrobial activity against bacteria and fungi. When incorporated into wound dressings, medical device coatings, or dental materials, these nanomaterials help prevent infections and biofilm formation (Das et al., 2019; Díaz-Muñoz et al., 2025; Thakur et al., 2025). The use of readily available waste streams to produce such biomedical-grade materials offers a sustainable route to affordable healthcare solutions, particularly for resource-limited settings.

4.5.5 Industrial and construction applications

In the industrial sector, waste-derived nanomaterials enhance the strength, durability, and functionality of composites. Fly ash-based silica nanoparticles and nano-zeolites can be incorporated into cement and concrete to improve compressive strength, reduce permeability, and increase resistance to chemical attack (Ali et al., 2021; Peng et al., 2023). This not only upgrades the performance of construction materials but also reduces the carbon footprint associated with traditional cement production. Polymer-based nanomaterials derived from textile or plastic waste serve as lightweight, high-strength reinforcement agents in polymers, paints, and coatings. Their nanoscale dispersion improves mechanical properties, abrasion resistance, and thermal stability of industrial coatings and packaging materials. For example, graphene derived from petroleum coke can be incorporated into anticorrosion coatings (Kulyk et al., 2022; Verma et al., 2023), significantly extending the lifetime of metal infrastructure exposed to harsh environments. By converting difficult-to-manage waste streams into high-value functional materials, industries can lower production costs, reduce environmental impact, and move towards circular economy models.

4.6 CHALLENGES AND LIMITATIONS ASSOCIATED WITH SYNTHETIC APPROACHES AND APPLICATIONS OF NANOMATERIALS

Despite the promise of waste-derived nanomaterials, their widespread adoption faces several technical, environmental, and economic challenges. Understanding these challenges is essential for guiding future research, ensuring safety and fostering industrial-scale implementation.

4.6.1 Technical challenges

A major technical challenge is the variability in waste composition, which can directly impact the reproducibility and quality of the resulting nanomaterials. Agricultural residues, plastic waste, and industrial by-products can differ significantly in moisture content, elemental composition, and impurity levels depending on geographic origin, seasonal factors, or upstream processing. Such inconsistencies can lead to unpredictable particle sizes, morphologies, or functional group distributions in the final nanomaterials, making it difficult to ensure uniform performance across batches. Another key issue is the scalability and reproducibility of synthesis methods. While laboratory-scale processes such as hydrothermal treatment or pyrolysis have demonstrated success, scaling them up to continuous industrial operations introduces new variables, including heat and mass transfer limitations, energy demands, and reactor design complexities (Hossain et al., 2025; Oluremi, 2025). Maintaining consistent temperature and pressure profiles in large-scale systems is challenging and can result in lower yields or heterogeneous product quality (Hossain et al., 2025). Furthermore, many green synthesis routes that excel in small batches may require modification or hybridisation with conventional techniques to achieve a commercially viable amount without compromising the environmental benefits.

4.6.2 Environmental and safety concerns

Although these nanomaterials originate from recycled waste, their potential toxicity cannot be overlooked. Nanoparticles possess high reactivity and can interact with biological systems in unpredictable ways, including generating reactive oxygen species, crossing cell membranes, and bioaccumulating in tissues (Kumah et al., 2023). Metal-based nanomaterials, such as those containing nickel or cobalt, may pose risks to aquatic life and human health if not properly contained or disposed of after use (Kanwal et al., 2019; Sengul & Asmatulu, 2020). Comprehensive toxicological studies and standardised safety assessments remain limited, leaving uncertainties around long-term exposure (Kumah et al., 2023). A related challenge involves the life cycle assessment (LCA) of waste-derived nanomaterials. While converting waste into high-value products aligns with circular economy principles, the full environmental footprint (covering energy consumption, chemical inputs, emissions, and end-of-life disposal) must be carefully evaluated (Abdelbasir et al., 2020; Noman et al., 2022). For instance, high-temperature pyrolysis or chemical vapour deposition may offset some of the sustainability benefits if powered by fossil fuels or generate hazardous by-products. Rigorous LCAs are needed to ensure that these materials truly reduce net environmental impacts compared to conventional nanomaterials (Abdelbasir et al., 2020).

4.6.3 Economic and regulatory barriers

Economic feasibility is another significant limitation. The cost of processing waste into nanomaterials can remain high, particularly when feedstock pretreatment, purification, or sophisticated equipment is required, especially from underdeveloped regions (Hossain et al., 2025; Oluremi, 2025). Competing with established, large-scale suppliers of conventional nanomaterials from well-developed regions often demands substantial capital investment and optimised supply chains (Hossain et al., 2025). Without clear market incentives or government subsidies, start-ups and small enterprises may struggle to achieve profitability (Oluremi, 2025).

Regulatory frameworks governing the production and use of nanomaterials are still evolving in many regions (Chávez-Hernández et al., 2024). Clear guidelines on classification, labelling, and safe handling are often lacking or inconsistent across jurisdictions (Chávez-Hernández et al., 2024; Ghosh & Kumar, 2024). This regulatory uncertainty can delay commercialisation, complicate international trade, and discourage investors. Harmonising global standards and establishing risk-based, science-driven policies will be critical to fostering confidence among manufacturers, end users, and the public (Ghosh & Kumar, 2024).

4.7 FUTURE PERSPECTIVES

A major frontier lies in developing greener, more efficient synthesis methods. Emerging approaches such as microwave-assisted hydrothermal processing, low-temperature plasma treatments, and enzyme-mediated conversions can significantly reduce energy input and chemical usage. Combining these techniques with renewable energy sources, such as solar-powered furnaces or electrochemical reactors driven by green electricity, could lower the carbon footprint of large-scale production while improving product consistency. Future research should focus on exploring new waste sources and types of nanomaterials. Underutilised feedstocks, such as marine biomass, textile residues, and construction debris, offer untapped potential for the production of unique nanostructures with tailored properties. Investigating multi-component wastes, where mixed plastics or composite sludges serve as precursors, may yield hybrid nanomaterials with enhanced functionality. Another critical direction is enhancing the multifunctional applications of nanomaterials. For example, designing a single material capable of simultaneous pollutant adsorption, photocatalytic degradation, and antimicrobial activity could reduce the need for multiple treatment steps in environmental remediation. Advanced computational modelling and ML tools can help predict structure–property relationships, enabling the rational design of such versatile materials.

4.8 CONCLUSION

The investigation of waste-derived nanomaterials has advanced remarkably, demonstrating how discarded resources can be transformed into high-value products with applications across environmental, energy, biomedical, and industrial sectors. This chapter outlined classification frameworks based on both waste source and nanomaterial type, surveyed a range of synthesis methods, and highlighted a wide spectrum of applications, from pollution remediation and energy storage to electrochemical sensing and structural composites. At the same time, it emphasised the technical, environmental, economic, and regulatory challenges that must be addressed to realise full-scale deployment.

Waste-derived nanomaterials demonstrate the principles of sustainable development and the circular economy. By converting agricultural residues, plastics, fly ash, petroleum coke, and other industrial by-products into functional nanostructures, these technologies reduce landfill burdens, lower greenhouse gas emissions, and create new economic opportunities. Their unique physicochemical properties (large surface area, tunable porosity, and customisable surface chemistry) make them powerful tools for tackling pressing global issues such as water scarcity, clean energy generation, and advanced manufacturing. Researchers must continue to develop greener, scalable synthesis methods and explore underutilised waste streams. Industry and government should collaborate to establish standards, provide incentives, and integrate these materials into commercial supply chains. By aligning scientific innovation with supportive policies and market adoption, waste-derived nanomaterials can move from promising laboratory discoveries to essential building blocks of a more sustainable and resilient future.

REFERENCES

Abdelbasir, S. M., McCourt, K. M., Lee, C. M., & Vanegas, D. C. (2020). Waste-derived nanoparticles: Synthesis approaches, environmental applications and sustainability considerations. *Frontiers in Chemistry*, *8*, 782.

Abdelfatah, A. M., Hosny, M., Elbay, A. S., El-Maghrabi, N., & Fawzy, M. (2024). From waste to worth: Upcycling plastic into high-value carbon-based nanomaterials. *Polymers, 17*(1), 63.

Agrawal, S., & Nawaz, T. (2025). A closed-loop sustainable method for the fabrication of electrospun nanofiber using food waste for filtration of particulate matters (PMs) and volatile organic matter (VOCs) from air. *Separation and Purification Technology, 358*, 130416.

Ajakwe, I. U., Ajakwe, S. O., Lee, J. M., & Kim, D. S. (2025). Internet-of-Things-blockchain integration in environmental pollution monitoring data management: Trends and techniques. *International Journal of Environmental Science and Technology*, 1–20.

Akhayere, E., Essien, E. A., & Kavaz, D. (2019). Effective and reusable nano-silica synthesized from barley and wheat grass for the removal of nickel from agricultural wastewater. *Environmental Science and Pollution Research*, 26(25), 25802–25813.

Ali, D., Sharma, U., Singh, R., & Singh, L. P. (2021). Impact of silica nanoparticles on the durability of fly ash concrete. *Frontiers in Built Environment, 7*, 665549.

Alkhatani, M., Almutairi, M. M., Alalwan, M. H., Algharash, H. K., Aldossary, R. A., Alzahrani, Y. A., . . . & Almalki, M. (2025). Petroleum coke-derived graphene as a low-cost electrode material for efficient perovskite solar cells. *Nanoscale Advances, 7*(15), 4710-4720.

Altalhi, T., Mezni, A., Ibrahim, M. M., Refat, M. S., Gobouri, A. A., Safklou, A. M., Mousli, A. M., Attia, M. S., Boruah, P. K., Das, M. R., & Ryl, J. (2022). Cathodic activation of titania-fly ash cenospheres for efficient electrochemical hydrogen production: A proposed solution to treat fly ash waste. *Catalysts, 12*(5), 466.

Alterary, S. S., & Marei, N. H. (2021). Fly ash properties, characterization and applications: A review. *Journal of King Saud University-Science, 33*(6), 101536.

Arpita, Kumar, P., Kataria, N., Narwal, N., Kumar, S., Kumar, R., Khoo, K. S., & Show, P. L. (2023). Plastic waste-derived carbon dots: Insights of recycling valuable materials towards environmental sustainability. *Current Pollution Reports, 9*(3), 433–453.

Askaruly, K., Azat, S., Sartova, Z., Yeleuov, M., Kerimkulova, A., & Bekseitova, K. (2020). Obtaining and characterization of amorphous silicafrom rice husk. *Journal of Chemical Technology and Metallurgy, 55*(1), 88–97.

Bai, Y. L., Zhang, C. C., Rong, F., Guo, Z. X., & Wang, K. X. (2024). Biomass-derived carbon materials for electrochemical energy storage. *Chemistry—A European Journal, 30*(23), p.e202304157.

Baig, N., Kammakakam, I., & Falath, W. (2021). Nanomaterials: A review of synthesis methods, properties, recent progress and challenges. *Materials Advances, 2*(6), 1821–1871.

Belbessai, S., Benyoussef, E. H., Gitzhofer, F., & Abatzoglou, N. (2022). Carbon nanomaterial production using waste plastic pyrolysis over a new catalyst made from mining residues: Effect of plastic type. *Chemical Engineering Journal Advances, 12*, 100424.

Benoy, S. M., Das, A. K., Sarmah, D., Pawar, M., Shelke, M. V., & Saikia, B. K. (2025). In situ solid-state synthesis of nitrogen-enriched porous carbon nanosheets from petroleum coke for lithium-ion hybrid capacitors. *Energy & Fuels, 39*(21), 10053–10069.

Bilge, S., & Sınağ, A. (2022). TiO2/core-shell structured carbon support materials derived from hydrothermal carbonization of waste masks biomass: A green photocatalyst. *Inorganic Chemistry Communications, 144*, 109911.

Bonetti, B., Waldow, E. C., Trapp, G., Hammercshmitt, M. E., Ferrarini, S. F., Pires, M. J., Estevam, S. T., & Aquino, T. F. (2021). Production of zeolitic materials in pilot scale based on coal ash for phosphate and potassium adsorption in order to obtain fertilizer. *Environmental Science and Pollution Research, 28*(3), 2638–2654.

Boruah, P. K., Yadav, A., & Das, M. R. (2020). Magnetic mixed metal oxide nanomaterials derived from industrial waste and its photocatalytic applications in environmental remediation. *Journal of Environmental Chemical Engineering*, *8*(5), 104297.

Boycheva, S., Marinov, I., Miteva, S., & Zgureva, D. (2020). Conversion of coal fly ash into nanozeolite Na-X by applying ultrasound assisted hydrothermal and fusion-hydrothermal alkaline activation. *Sustainable Chemistry and Pharmacy*, *15*, 100217.

Bressi, V., Celesti, C., Ferlazzo, A., Len, T., Moulaee, K., Neri, G., Luque, R., & Espro, C. (2024). Waste-derived carbon nanodots for fluorimetric and simultaneous electrochemical detection of heavy metals in water. *Environmental Science: Nano*, *11*(3), 1245–1258.

Buzukashvili, S., Sommerville, R., Rowson, N. A., & Waters, K. E. (2024). An overview of zeolites synthesised from coal fly ash and their potential for extracting heavy metals from industrial wastewater. *Canadian Metallurgical Quarterly*, *63*(1), 130–152.

Cabral, R. L. B., Galvão, E. R. V. P., Fechine, P. B. A., Galvão, F. M. F., & do Nascimento, J. H. O. (2024). A minireview on the utilization of petroleum coke as a precursor for carbon-based nanomaterials (CNMs): Perspectives and potential applications. *RSC Advances*, *14*(28), 19953–19968.

Campuzano, S., Pedrero, M., Yáñez-Sedeño, P., & Pingarrón, J. M. (2019). Antifouling (bio) materials for electrochemical (bio) sensing. *International Journal of Molecular Sciences*, *20*(2), 423.

Cao, J. J., Huang, Y., & Zhang, Q. (2021). Ambient air purification by nanotechnologies: From theory to application. *Catalysts*, *11*(11), 1276.

Chauke, N. M., Ngqalakwezi, A., & Raphulu, M. (2025). Transformative advancements in visible-light-activated titanium dioxide for industrial wastewater remediation. *International Journal of Environmental Science and Technology*, 1–32.

Chávez-Hernández, J. A., Velarde-Salcedo, A. J., Navarro-Tovar, G., & Gonzalez, C. (2024). Safe nanomaterials: From their use, application and disposal to regulations. *Nanoscale Advances*, *6*(6), 1583–1610.

Chen, S. Y., Jian, J. Y., & Lin, H. M. (2024). Functionalization of rice husk-derived mesoporous silica nanoparticles for targeted and imaging in cancer drug delivery. *Journal of the Science of Food and Agriculture*, *104*(4), 2120–2129.

Chen, T., Ren, S., Chen, L., Chen, Z., Li, X., Wang, M., & Yang, J. (2023). Catalysts prepared from solid wastes for efficient removal of NOx in NH3-SCR process: A review. *Catalysis Today*, *420*, 114175.

Choi, J., Yang, I., Kim, S. S., Cho, S. Y., & Lee, S. (2022). Upcycling plastic waste into high value-added carbonaceous materials. *Macromolecular Rapid Communications*, *43*(1), 2100467.

Choi, Y., Zheng, X. T., & Tan, Y. N. (2020). Bioinspired carbon dots (biodots): Emerging fluorophores with tailored multiple functionalities for biomedical, agricultural and environmental applications. *Molecular Systems Design & Engineering*, *5*(1), 67–90.

Costa, J. A. S., & Paranhos, C. M. (2018). Systematic evaluation of amorphous silica production from rice husk ashes. *Journal of Cleaner Production*, *192*, 688–697.

Council of Scientific and Industrial Research-Central Mechanical Engineering Research Institute (CSIR-CMERI). (2024). *Batch-scale production of high-quality activated carbon from biomass wastes for wastewater treatment.* India Science, Technology & Innovation. Retrieved September 10, 2025, from https://www.indiascienceandtechnology.gov.in/research/batch-scale-production-high-quality-activated-carbon-biomass-wastes-waste-water-treatment-initiative

Dai, L., Karakas, O., Cheng, Y., Cobb, K., Chen, P., & Ruan, R. (2023). A review on carbon materials production from plastic wastes. *Chemical Engineering Journal*, *453*, 139725.

Das, G., Patra, J. K., Debnath, T., Ansari, A., & Shin, H. S. (2019). Investigation of antioxidant, antibacterial, antidiabetic and cytotoxicity potential of silver nanoparticles synthesized using the outer peel extract of Ananas comosus (L.). *PloS One*, *14*(8), p.e0220950.

Deivayanai, V. C., Karishma, S., Thamarai, P., Kamalesh, R., Saravanan, A., Yaashikaa, P. R., & Vickram, A. S. (2024). Innovations in plastic remediation: Catalytic degradation and machine learning for sustainable solutions. *Journal of Contaminant Hydrology*, *267*, 104449.

Díaz-Muñoz, L. L., Castañeda-Aude, J. E., Landín-Sandoval, V. J., Lozano-Álvarez, J. A., & Escárcega-González, C. E. (2025). Valorization and harnessing citrus waste: Focus on green synthesis of metal nanoparticles with antimicrobial applications and other potential uses. *Food Reviews International*, 1–46.

Divyabharathi, R., Kalidasan, B., Sakthi Suriya Raj, J. S., Chinnasamy, S., Pandey, A. K., Giri, J., & Fatehmulla, A. (2025). Harnessing agro-wastes for sustainable nanomaterials: The promise of biofluorescent carbon dots. *Environmental Progress & Sustainable Energy*, *44*(1), p.e14525.

Ealia, S. A. M., & Saravanakumar, M. P. (2017, November). A review on the classification, characterisation, synthesis of nanoparticles and their application. In *IOP conference series: Materials science and engineering* (Vol. 263, No. 3, p. 032019). IOP Publishing.

Elhenawy, S., Khraisheh, M., AlMomani, F., Al-Ghouti, M., Selvaraj, R., & Al-Muhtaseb, A. A. (2024). Emerging nanomaterials for drinking water purification: A new era of water treatment technology. *Nanomaterials*, *14*(21), 1707.

Etefa, H. F., & Dejene, F. B. (2025). Applications of green carbon dots in personalized diagnostics for precision medicine. *International Journal of Molecular Sciences*, *26*(7), 2846.

Fatima, M., Ahmad, F., Noor, S., Shafique, S., Ibrahim, A. B., Habib, M. A., Khan, M. I., & Shanableh, A. (2025). Application of carbon-based electrocatalysts derived from waste materials in fuel cells and electrolyzers. *Sustainable Energy & Fuels*, *9*(13), 3523–3549.

Fatima, R., Warsi, M. F., Zulfiqar, S., Ragab, S. A., Shakir, I., & Sarwar, M. I. (2020). Nanocrystalline transition metal oxides and their composites with reduced graphene oxide and carbon nanotubes for photocatalytic applications. *Ceramics International*, *46*(10), 16480–16492.

Ferdous, A. R., Shah, S. S., Shah, S. N. A., Johan, B. A., Al Bari, M. A., & Aziz, M. A. (2024). Transforming waste into wealth: Advanced carbon-based electrodes derived from refinery and coal by-products for next-generation energy storage. *Molecules*, *29*(9), 2081.

Ferronato, N., & Torretta, V. (2019). Waste mismanagement in developing countries: A review of global issues. *International Journal of Environmental Research and Public Health*, *16*(6), 1060.

Flores-Contreras, E. A., González-González, R. B., Pablo Pizaña-Aranda, J. J., Parra-Arroyo, L., Rodríguez-Aguayo, A. A., Iñiguez-Moreno, M., González-Meza, G. M., Araújo, R. G., Ramírez-Gamboa, D., Parra-Saldívar, R., & Melchor-Martínez, E. M. (2024). Agricultural waste as a sustainable source for nanoparticle synthesis and their antimicrobial properties for food preservation. *Frontiers in Nanotechnology*, *6*, 1346069.

Gadore, V., & Ahmaruzzaman, M. (2021). Fly ash-based nanocomposites: A potential material for effective photocatalytic degradation/elimination of emerging organic pollutants from aqueous stream. *Environmental Science and Pollution Research*, *28*(34), 46910–46933.

Gadore, V., Mishra, S. R., Sillanpää, M., & Ahmaruzzaman, M. (2024). Emerging applications of waste fly ash for remediation of environmental contaminants: A high-value and sustainable approach towards utilization of waste materials. *Nanotechnology for Environmental Engineering*, *9*(2), 207–237.

Gao, Y., Huang, Y., Xin, X., Jiang, T., Zong, M., & Xu, J. (2024). Biomass modified carbon nanofibers use for flexible supercapacitor electrodes with thermal conductive performances. *Journal of Energy Storage*, *99*, 113382.

Ghosh, M., & Kumar, R. (2024). Regulatory issues in nanotechnology. In *Nanotechnology theranostics in livestock diseases and management* (pp. 765–788). Singapore: Springer Nature Singapore.

Graul, T., Martínez, M. G., Qiu, Y., Fourgeaud, F., Zeng, K., & Nzihou, A. (2024). NOx decomposition using Ni-and Fe-loaded biocarbon catalysts. *Applied Catalysis O: Open*, *194*, 206982.

Hatta, M. N. M., Hashim, M. S., Hussin, R., Aida, S., Kamdi, Z., Ainuddin, A. R., & Yunos, M. Z. (2017, October). Synthesis of carbon nanostructures from high density polyethylene (HDPE) and polyethylene terephthalate (PET) waste by chemical vapour deposition. In *Journal of Physics: Conference Series* (Vol. 914, No. 1, p. 012029). IOP Publishing.

Henning, L. M., Abdullayev, A., Vakifahmetoglu, C., Simon, U., Bensalah, H., Gurlo, A., & Bekheet, M. F. (2021). Review on polymeric, inorganic, and composite materials for air filters: From processing to properties. *Advanced Energy and Sustainability Research*, *2*(5), 2100005.

Hossain, F., Hoque, M. E., Khan, A. A., & Arifuzzaman, M. (2025). Lignocellulosic waste-derived nanomaterials: Types and applications in wastewater pollutant removal. *Water*, *17*(16), 2426.

Iordache, M., Oubraham, A., Sorlei, I. S., Lungu, F. A., Capris, C., Popescu, T., & Marinoiu, A. (2023). Noble metals functionalized on graphene oxide obtained by different methods—new catalytic materials. *Nanomaterials*, *13*(4), 783.

Iravani, S., Khosravi, A., Zare, E. N., Varma, R. S., Zarrabi, A., & Makvandi, P. (2024). MXenes and artificial intelligence: Fostering advancements in synthesis techniques and breakthroughs in applications. *RSC Advances*, *14*(49), 36835–36851.

Jana, S., & Seth, M. (2020). Nanomaterials based superhydrophobic and antimicrobial coatings. *NanoWorld Journal*, *6*(2), 26–28.

Joudeh, N., & Linke, D. (2022). Nanoparticle classification, physicochemical properties, characterization and applications: A comprehensive review for biologists. *Journal of Nanobiotechnology*, *20*(1), 262.

Kanwal, Z., Raza, M. A., Manzoor, F., Riaz, S., Jabeen, G., Fatima, S., & Naseem, S. (2019). A comparative assessment of nanotoxicity induced by metal (silver, nickel) and metal oxide (cobalt, chromium) nanoparticles in Labeo rohita. *Nanomaterials*, *9*(2), 309.

Khan, I., Saeed, K., & Khan, I. (2019). Nanoparticles: Properties, applications and toxicities. *Arabian Journal of Chemistry*, *12*(7), 908–931.

Khandaker, T., Islam, T., Nandi, A., Anik, M. A. A. M., Hossain, M. S., Hasan, M. K., & Hossain, M. S. (2025). Biomass-derived carbon materials for sustainable energy applications: A comprehensive review. *Sustainable Energy & Fuels*, *9*(3), 693–723.

King Abdullah City for Science and Technology (KACST), Oil Sustainability Program (OSP), & Center of Excellence for Energy Research (CEER). (2025, July 07). *KACST, OSP and CEER pioneer cost-effective graphene production from petroleum coke*. USD Analytics. Retrived September 10, 2025, from https://www.usdanalytics.com/blog/kacst-osp-and-ceer-pioneer-cost-effective-graphene-production-from-petroleum-coke

Kristanto, G. A., & Koven, W. (2019). Estimating greenhouse gas emissions from municipal solid waste management in Depok, Indonesia. *City and Environment Interactions*, *4*, 100027.

Kuang, T., Jin, M., Lu, X., Liu, T., Vahabi, H., Gu, Z., & Gong, X. (2023). Functional carbon dots derived from biomass and plastic wastes. *Green Chemistry*, *25*(17), 6581–6602.

Kulyk, B., Freitas, M. A., Santos, N. F., Mohseni, F., Carvalho, A. F., Yasakau, K., Fernandes, A. J., Bernardes, A., Figueiredo, B., Silva, R., & Tedim, J. (2022). A critical review on the production and application of graphene and graphene-based materials in anti-corrosion coatings. *Critical Reviews in Solid State and Materials Sciences*, *47*(3), 309–355.

Kumah, E. A., Fopa, R. D., Harati, S., Boadu, P., Zohoori, F. V., & Pak, T. (2023). Human and environmental impacts of nanoparticles: A scoping review of the current literature. *BMC Public Health*, *23*(1), 1059.

Kumar, S. (2023). Carbon based nanomaterial for removal of heavy metals from wastewater: A review. *International Journal of Environmental Analytical Chemistry*, *103*(19), 7961–7978.

Kumari, S., Chowdhry, J., Kumar, M., & Garg, M. C. (2024). Zeolites in wastewater treatment: A comprehensive review on scientometric analysis, adsorption mechanisms and future prospects. *Environmental Research*, *260*, 119782.

Lei, Y. J., Zhao, L., Lai, W. H., Huang, Z., Sun, B., Jaumaux, P., Sun, K., Wang, Y. X., & Wang, G. (2024). Electrochemical coupling in subnanometer pores/channels for rechargeable batteries. *Chemical Society Reviews*, *53*(8), 3829–3895.

Li, G., Li, M., Zhang, X., Cao, P., Jiang, H., Luo, J., & Jiang, T. (2022). Hydrothermal synthesis of zeolites-calcium silicate hydrate composite from coal fly ash with co-activation of Ca (OH) 2-NaOH for aqueous heavy metals removal. *International Journal of Mining Science and Technology*, *32*(3), 563–573.

Li, N., Li, Z., Cui, Y., Liang, L., Peng, W., Cheng, Z., Yan, B., & Chen, G. (2025). Chemical upcycling of biodegradable plastic waste: A critical review. *Green Chemistry*, *27*(30), 9000–9015.

Li, Z., Diaz, L. A., Yang, Z., Jin, H., Lister, T. E., Vahidi, E., & Zhao, F. (2019). Comparative life cycle analysis for value recovery of precious metals and rare earth elements from electronic waste. *Resources, Conservation and Recycling*, *149*, 20–30.

Lin, L., Lin, Y., Li, C., Wu, D., & Kong, H. (2016). Synthesis of zeolite/hydrous metal oxide composites from coal fly ash as efficient adsorbents for removal of methylene blue from water. *International Journal of Mineral Processing*, *148*, 32–40.

Lizundia, E., Luzi, F., & Puglia, D. (2022). Organic waste valorisation towards circular and sustainable biocomposites. *Green Chemistry*, *24*(14), 5429–5459.

Lu, J., Guo, J., Song, S., Yu, G., Liu, H., Yang, X., & Lu, Z. (2020). Preparation of Ag nanoparticles by spark ablation in gas as catalysts for electrocatalytic hydrogen production. *RSC Advances*, *10*(63), 38583–38587.

Lu, Q., Han, Q., Wang, X., Wei, C., Guan, X., Qu, C., & Li, J. (2023). High-value utilization of Cr-containing sludge: Eco-friendly and ultra-low-cost electrocatalyst for efficient OER in alkaline media from Cr-containing sludge. *Journal of Environmental Management*, *346*, 119020.

Ma, Z., Wang, L., Wang, Z., Cui, H., Tang, E., Hou, H., Xu, G., Song, T., & Wang, Q. (2024). Recent advances of plastic waste-derived carbon materials for energy storage, environmental remediation and organic synthesis applications. *ChemCatChem*, *16*(23), e202401072.

Mehmandoust, M., Li, G., & Erk, N. (2022). Biomass-derived carbon materials as an emerging platform for advanced electrochemical sensors: Recent advances and future perspectives. *Industrial & Engineering Chemistry Research*, *62*(11), 4628–4635.

Mekuye, B., & Abera, B. (2023). Nanomaterials: An overview of synthesis, classification, characterization and applications. *Nano Select*, *4*(8), 486–501.

Meng, W., Han, X., Han, R., Zhang, X., Zeng, X., Duan, J., & Luo, X. (2024). A highly stable electrochemical sensor with antifouling and antibacterial capabilities for mercury ion detection in seawater. *Analytica Chimica Acta*, *1309*, 342685.

Mohamed, M. G., Mansour, N. A., Mousa, A. A., & Mazrouaa, A. M. (2025). Applications of solid waste-derived carbon nanomaterials in packaging material. In *Waste derived carbon nanomaterials* (Vol. 2, pp. 1–16). American Chemical Society.

Motalib Hossain, M. A., Ker, P. J., Tiong, S. K., Indra Mahlia, T. M., & Hannan, M. A. (2024). Recent progress in nanomaterials of battery energy storage: A patent landscape analysis, technology updates and future prospects. *Nanotechnology Reviews*, *13*(1), 20230215.

Mubarik, S., Qureshi, N., Sattar, Z., Shaheen, A., Kalsoom, A., Imran, M., & Hanif, F. (2021). Synthetic approach to rice waste-derived carbon-based nanomaterials and their applications. *Nanomanufacturing*, *1*(3), 109–159.

Mukhopadhyay, R., Sarkar, B., Khan, E., Alessi, D. S., Biswas, J. K., Manjaiah, K. M., Eguchi, M., Wu, K. C., Yamauchi, Y., & Ok, Y. S. (2022). Nanomaterials for sustainable remediation of chemical contaminants in water and soil. *Critical Reviews in Environmental Science and Technology*, *52*(15), 2611–2660.

Nagabooshanam, S., Roy, S., Mathur, A., Mukherjee, I., Krishnamurthy, S., & Bharadwaj, L. M. (2019). Electrochemical micro analytical device interfaced with portable potentiostat for rapid detection of chlorpyrifos using acetylcholinesterase conjugated metal organic framework using Internet of things. *Scientific Reports*, *9*(1), 19862.

Nath, P. C., Ojha, A., Debnath, S., Sharma, M., Sridhar, K., Nayak, P. K., & Inbaraj, B. S. (2023). Biogeneration of valuable nanomaterials from agro-wastes: A comprehensive review. *Agronomy*, *13*(2), 561.

Ndlwana, L., Raleie, N., Dimpe, K. M., Ogutu, H. F., Oseghe, E. O., Motsa, M. M., Msagati, T. A., & Mamba, B. B. (2021). Sustainable hydrothermal and solvothermal synthesis of advanced carbon materials in multidimensional applications: A review. *Materials*, *14*(17), 5094.

Newman Monday, Y., Abdullah, J., Yusof, N. A., Abdul Rashid, S., & Shueb, R. H. (2021). Facile hydrothermal and solvothermal synthesis and characterization of nitrogen-doped carbon dots from palm kernel shell precursor. *Applied Sciences*, *11*(4), 1630.

Noman, E. A., Al-Gheethi, A., Al-Sahari, M., Mohamed, R. M. S. R., Crane, R., Ab Aziz, N. A., & Govarthanan, M. (2022). Challenges and opportunities in the application of bioinspired engineered nanomaterials for the recovery of metal ions from mining industry wastewater. *Chemosphere*, *308*, 136165.

Oh, Y. S., Seo, S. W., Yang, J. J., Jeong, M., & Ahn, S. (2025). Eco-friendly, biomass-derived materials for electrochemical energy storage devices. *Coatings*, *15*(8), 915.

Oluremi, D. (2025). *Industrial waste as a source for graphene nanosheets*. Federal University of Technology, Ondo, Nigeria.

Omran, B. A., & Baek, K. H. (2022). Valorization of agro-industrial biowaste to green nanomaterials for wastewater treatment: Approaching green chemistry and circular economy principles. *Journal of Environmental Management*, *311*, 114806.

Onfray, C., & Thiam, A. (2023). Biomass-derived carbon-based electrodes for electrochemical sensing: A review. *Micromachines*, *14*(9), 1688.

Pabel, M. Y., Yasmin, S., Shaikh, M. A. A., & Kabir, M. H. (2024). Electronic waste derived reduced graphene oxide supported silver nanoparticles for the electrochemical sensing of trace level arsenite in aqueous medium. *Sensors and Actuators A: Physical*, *366*, 115028.

Peng, X., Wang, Q., & Wu, J. (2023). Effect of nanosilica on the strength and durability of cold-bonded fly ash aggregate concrete. *Sustainability*, *15*(21), 15413.

Porrang, S., Rahemi, N., Davaran, S., Mahdavi, M., & Hassanzadeh, B. (2021). Preparation and in-vitro evaluation of mesoporous biogenic silica nanoparticles obtained from rice and wheat husk as a biocompatible carrier for anti-cancer drug delivery. *European Journal of Pharmaceutical Sciences*, *163*, 105866.

Ren, S., Xu, X., Hu, K., Tian, W., Duan, X., Yi, J., & Wang, S. (2022). Structure-oriented conversions of plastics to carbon nanomaterials. *Carbon Research*, *1*(1), 15.

Rotocarb. (2016). *About us*. Rotocarb Activated Carbons. Retrieved September 10, 2025, from https://rotocarb.co.za/about-us/

Ruiz-Montoya, J. G., Quispe-Garrido, L. V., Gómez, J. C., Baena-Moncada, A. M., & Gonçalves, J. M. (2021). Recent progress in and prospects for supercapacitor materials based on metal oxide or hydroxide/biomass-derived carbon composites. *Sustainable Energy & Fuels*, *5*(21), 5332–5365.

Saha, S., Lakhe, P., Mason, M. J., Coleman, B. J., Arole, K., Zhao, X., Yakovlev, S., Uppili, S., Green, M. J., & Hule, R. A. (2021). Sustainable production of graphene from petroleum coke using electrochemical exfoliation. *NPJ 2D Materials and Applications*, *5*(1), 75.

Sairaman, S., & Young-Joon, K. (2025). Biochar-derived electrochemical sensors: A green route for trace heavy metal detection. *Chemosensors*, *13*(8), 278.

Sengul, A. B., & Asmatulu, E. (2020). Toxicity of metal and metal oxide nanoparticles: A review. *Environmental Chemistry Letters*, *18*(5), 1659–1683.

Shi, Y., Zhao, T., Du, J., Feng, B., Tang, Q., Shan, S., Zhang, M., Wang, Q., & Zhao, J. (2024). Low temperature NH3 selective catalytic reduction of NOx over CeOx-biochar catalyst prepared by air oxidation. *Journal of the Energy Institute*, *112*, 101484.

Sooryakanth, B., Rajan, M. R., Sudhabose, S., Arulalan, S., & Venkatesan, S. (2025). Sustainable conversion of chitosan waste into nitrogen-doped graphene quantum dots: Green synthesis pathways and biomedical potential. *Exon*, *2*(3), 173–199.

Sujiono, E. H., Zabrian, D., Zharvan, V., & Humairah, N. A. (2022). Fabrication and characterization of coconut shell activated carbon using variation chemical activation for wastewater treatment application. *Results in Chemistry*, *4*, 100291.

Tao, H., Wu, T., Aldeghi, M., Wu, T. C., Aspuru-Guzik, A., & Kumacheva, E. (2021). Nanoparticle synthesis assisted by machine learning. *Nature Reviews Materials*, *6*(8), 701–716.

Tee, S. Y., Ye, E., Teng, C. P., Tanaka, Y., Tang, K. Y., Win, K. Y., & Han, M. Y. (2021). Advances in photothermal nanomaterials for biomedical, environmental and energy applications. *Nanoscale*, *13*(34), 14268–14286.

Teixeira, M. A., Barreiros, M. P., Felgueiras, H. P., Fangueiro, R., & Ferreira, D. P. (2025). Unlocking the potential of nanocellulose from textile waste: A pathway to nanocomposite applications. *Cellulose*, *32*(1), 29–57.

Thakur, P., Omar, R. A., Talreja, N., Chauhan, D., & Ashfaq, M. (2025). Waste-derived Ca and Zn-based bimetallic (Ca/Zn) nanorods encapsulated chitosan-based haemostatic dressing bandage: A step towards waste to bandages. *Journal of Industrial and Engineering Chemistry*, *143*, 327–338.

Toro-Córdova, A., Sanz, B., & Goya, G. F. (2020). A concise review of nanomaterials for drug delivery and release. *Current Nanoscience*, *16*(3), 399–412.

Tripathi, P. K., Durbach, S., & Coville, N. J. (2017). Synthesis of multi-walled carbon nanotubes from plastic waste using a stainless-steel CVD reactor as catalyst. *Nanomaterials*, *7*(10), 284.

Tripathy, A., Patne, A. Y., Mohapatra, S., & Mohapatra, S. S. (2024). Convergence of nanotechnology and machine learning: The state of the art, challenges and perspectives. *International Journal of Molecular Sciences*, *25*(22), 12368.

Tsai, W. T., & Jiang, T. J. (2018). Mesoporous activated carbon produced from coconut shell using a single-step physical activation process. *Biomass Conversion and Biorefinery*, *8*(3), 711–718.

Tunali, M., Tunali, M. M., & Yenigun, O. (2021). Characterization of different types of electronic waste: Heavy metal, precious metal and rare earth element content by comparing different digestion methods. *Journal of Material Cycles and Waste Management*, *23*(1), 149–157.

Verma, A., Singh, H., Mahton, Y., Saha, P., Tiwary, C. S., & Bhattacharya, J. (2023). Use of waste to wealth process derived sustainable silica-rich graphene analogues to provide enhanced corrosion resistance properties for coatings on carbon steel, exposed to marine environments. *Surface and Coatings Technology*, *464*, 129420.

Wang, J., Khan, M., Hussain, A., Khan, I., Nawaz, A., Ragab, A. H., Sayqal, A., Lei, T., & Zada, A. (2023). Carbon foam composites containing carbon nanotubes and graphene oxide as additives for enhanced mechanical, thermal, electrical and catalytic properties. *Journal of Materials Research and Technology*, *24*, 608–622.

Wang, Y., Liu, Y., Chen, Z., Zhang, M., Liu, B., Xu, Z., & Yan, K. (2022). In situ growth of hydrophilic nickel-cobalt layered double hydroxides nanosheets on biomass waste-derived porous carbon for high-performance hybrid supercapacitors. *Green Chemical Engineering*, *3*(1), 55–63.

Woźniak, M., Płoska, A., Siekierzycka, A., Dobrucki, L. W., Kalinowski, L., & Dobrucki, I. T. (2022). Molecular imaging and nanotechnology—emerging tools in diagnostics and therapy. *International Journal of Molecular Sciences*, *23*(5), 2658.

Wu, Y., Ma, G., Zhang, A., Gu, W., Wei, J., & Wang, R. (2022). Preparation of carbon dots with ultrahigh fluorescence quantum yield based on PET waste. *ACS Omega*, *7*(42), 38037–38044.

Xu, X., Xiong, F., Meng, J., Wang, X., Niu, C., An, Q., & Mai, L. (2020). Vanadium-based nanomaterials: A promising family for emerging metal-ion batteries. *Advanced Functional Materials*, *30*(10), 1904398.

Yadav, V. K., Amari, A., Gacem, A., Elboughdiri, N., Eltayeb, L. B., & Fulekar, M. H. (2023(b)). Treatment of fly-ash-contaminated wastewater loaded with heavy metals by using fly-ash-synthesized iron oxide nanoparticles. *Water*, *15*(5), 908.

Yadav, V. K., Amari, A., Wanale, S. G., Osman, H., & Fulekar, M. H. (2023). Synthesis of floral-shaped nanosilica from coal fly ash and its application for the remediation of heavy metals from fly ash aqueous solutions. *Sustainability*, *15*(3), 2612.

Yadav, V. K., & Fulekar, M. H. (2020). Advances in methods for recovery of ferrous, alumina and silica nanoparticles from fly ash waste. *Ceramics*, *3*(3), 384–420.

Yakah, N., Samavati, M., Akuoko Kwarteng, A., Martin, A., & Simons, A. (2023). Prospects of waste incineration for improved municipal solid waste (MSW) management in Ghana—a review. *Clean Technologies*, *5*(3), 997–1011.

Yan, Y., Sun, W., Wei, Y., Liu, K., Ma, J., & Hu, G. (2025). Review of biomass-derived carbon nanomaterials—From 0D to 3D—For supercapacitor applications. *Nanomaterials*, *15*(4), 315.

Yaqueen, M. N., Kuity, A., & Ahmed, M. A. (2025). Sustainable nano silica production from agricultural residues: A review of green synthesis techniques and performance in asphalt applications. *International Journal of Pavement Research and Technology*, 1–18.

Yuan, S. J., & Dai, X. H. (2016). An efficient sewage sludge-derived bi-functional electrocatalyst for oxygen reduction and evolution reaction. *Green Chemistry*, *18*(14), 4004–4011.

Yuan, Y., Feng, S., Alahi, M. E. E., Nag, A., Afsarimanesh, N., Zhang, H., & He, S. (2018). Development of an internet of things based electrochemical microfluidic system for free calcium detection. *Applied Sciences*, *8*(8), 1357.

Zago, S., Scarpetta-Pizo, L. C., Zagal, J. H., & Specchia, S. (2024). PGM-free biomass-derived electrocatalysts for oxygen reduction in energy conversion devices: Promising materials. *Electrochemical Energy Reviews*, *7*(1), 1.

Zehra, F., Taushiba, A., Dwivedi, S., Rajanikanth, B. S., Nagar, P. K., & Lawrence, A. (2024). A comprehensive review on removal of VOCs from air using fly ash as an adsorbent. *Journal of Air Pollution and Health*, *9*(1), 97–122.

Zhang, H., Sun, H., Huang, S., Lan, J., Li, H., & Yue, H. (2024). Biomass-derived carbon materials for electrochemical sensing: Recent advances and future perspectives. *Critical Reviews in Analytical Chemistry*, 1–26.

Zhao, X., & Liu, P. (2014). Biocompatible graphene oxide as a folate receptor-targeting drug delivery system for the controlled release of anti-cancer drugs. *RSC Advances*, *4*(46), 24232–24239.

5 Synthesis of nanomaterials

5.1 INTRODUCTION

Solid waste management in developing countries is a real environmental and social concern, as the most commonly used treatment option is final disposal in open dump sites or unsuitable sanitary landfills (Navarrete et al., 2011; Tejaswini et al., 2022). The main concern is what will happen to the waste in the landfills over time. Valorising waste offers a sustainable approach to waste management while simultaneously generating value-added products for diverse applications (Tuck et al., 2012). This involves the process of converting waste materials, by-products, or unwanted outputs into valuable products or energy sources, effectively turning waste into a resource (Capanoglu et al., 2022; Ferronato et al., 2019). If this is successful, there will be many benefits, like preserving the environment, advancing the circular economy, reducing the amount of waste sent to landfills, and creating a more sustainable future by recovering valuable materials and energy from what was previously considered worthless (Gueboudji, 2024; Tawo & Mbamalu, 2025).

Recycling waste materials into high-value products, such as steel, fuels, and ceramics, is a markedly feasible process, which has inspired researchers to synthesise various nanomaterials from waste resources for numerous potential applications (Munoz-Sandoval, 2014). Nanomaterials, such as graphene, nanoparticles, and carbon nanotubes (CNTs), among others, have been priority materials for research in nanotechnology (Etxebarria et al., 2012). Graphene and CNTs, with their robust mechanical, electronic, physical, and chemical properties, are playing increasingly significant roles in supercapacitors, transistors, transparent photo-electronic devices, and sensors (Deng et al., 2016; Mathura et al., n.d.). A variety of synthesis methods have been used to produce nanomaterials from waste. This includes pyrolysis, ball milling, sol-gel, hydrothermal, calcination, and co-precipitation methods (Chen et al., 2014; Tursunov, 2014). Although nanomaterials of high quality can be produced through these conventional methods, a crucial challenge remains to achieve environmentally friendly production on a large scale at a low cost (Devi et al., 2021; Zahid et al., 2018). The synthesis of nanomaterials from waste is likely to reduce solid waste and lower fabrication costs simultaneously. More importantly, transforming waste into valuable products is a significant step forward in recycling.

5.2 NANOMATERIALS DERIVED FROM FOOD AND AGRICULTURAL WASTE

5.2.1 Sugarcane

Sugarcane-based products are inherently rich in elements such as silicon, carbon, and nitrogen (Anitha et al., 2024; Iwuozor et al., 2024; Mod et al., 2024). As such,

DOI: 10.1201/9781003745846-5

these are ideal precursors for a wide range of applications. One appealing approach is to transform them into high-interest nanomaterials that can be employed in several prominent applications (Mod et al., 2024). Among nanomaterials, sugarcane products based on silica nanoparticles (SNPs), carbon dots (CDs), metal/metal oxide-based NPs, nanocellulose, cellulose nanofibres (CNFs), and nano-biochar are increasingly reported. By manipulating experimental conditions and selecting suitable starting precursors and elements, it is possible to design these nanomaterials with highly desirable properties suited to specific applications. Sugarcane products have been utilised in numerous studies, either alone or in combination with other materials, to produce various NPs. Most studies reported using bagasse as a precursor, while sugarcane juice is used as a chemical catalyst or a controlling agent. The use of sugarcane leaves for the synthesis of nanomaterials has also been successfully demonstrated. A variety of nanomaterials, including silica, carbon, and metal species, can be synthesised from sugarcane products, as discussed in detail in the forthcoming sections.

5.2.1.1 Silica nanomaterials from sugarcane

Sugarcane has microscopic phytolith structures, essentially silica nanoparticles, that provide structural strength to tissues (Negrão & Driemeier, 2022; Roselló et al., 2015). These phytoliths can be a rich source of silica for extraction from sugarcane plants. Silica nanoparticles exhibit exceptional physical, chemical, electronic, and optical properties, characterised by a high surface area and small particle size (Liu et al., 2018). Numerous studies have used sugarcane bagasse for the synthesis of silica nanomaterials. Two primary synthesis methods have been reported in the literature. The first method is direct calcination, in which bagasse is calcined to form silica nanoparticles (Falk et al., 2019; September et al., 2023). The second method is the sol-gel method. In this method, a strong base, such as sodium silicate, is used to extract silica, and silicon is removed as silicates, forming a gel under controlled pH conditions (Ni'mah et al., 2023). The major difference between these reported synthesis methods is that the calcination method produces biocompatible, biogenic silica nanomaterials (Seroka et al., 2022). The silica nanoparticles synthesised by the sol-gel method lose their biogenic properties and therefore have limited use in life science fields (Ni'mah et al., 2023). Figure 5.1 illustrates the calcination method and the sol-gel synthesis process of silica nanomaterials derived from sugarcane. In the synthesis of silica nanomaterials, the initial step, common across synthesis methods, is washing and grinding bagasse. These steps remove basic physical impurities and create a more homogenous texture for the subsequent reactions. Acid leaching is another purification step in which the bagasse/bagasse ash powder is leached with a strong acid, such as hydrochloric or sulphuric acid (Seroka et al., 2022). Pressurised thermal conditions (using an autoclave) have also been used to increase the efficacy of this reaction. The purpose of this step is to remove all metal ion impurities (Na, Ca, Mg, Fe, K, Mn, and Al) and also to initiate hydrolysis of the structural organic components, including lignin, cellulose, and hemi-cellulose (September et al., 2023). Following this, the material is thoroughly cleansed with distilled water and dried at 80°C. The step is also important in determining the crystallinity and surface area of the synthesised SNPs.

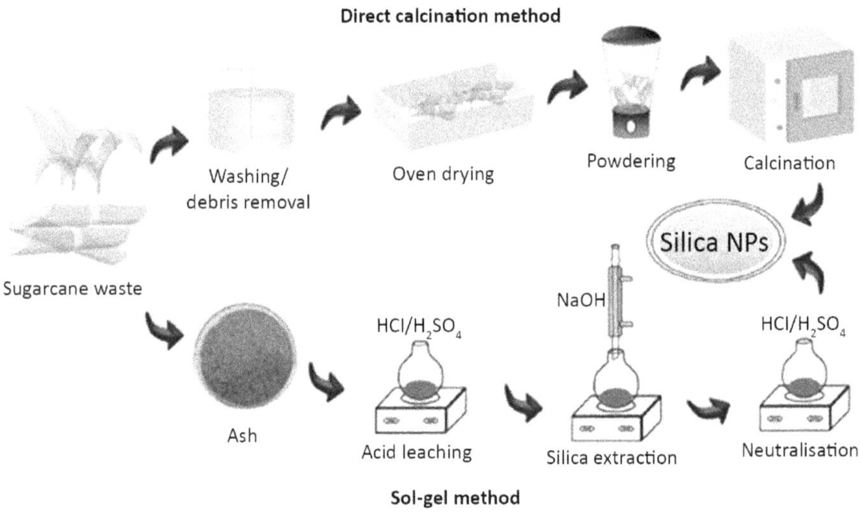

FIGURE 5.1 Calcination and sol-gel procedure

In a study conducted by Falk, a comparative analysis was performed among three different calcination-based routes for synthesising silica nanomaterials using both sugarcane bagasse and sugarcane leaves as silica precursors (Falk et al., 2019). Following the initial washing, the raw material in the first route was further pressed into a finer powder before undergoing calcination. In the second route, after milling, the powder underwent acid leaching before calcination. Lastly, in the third route, as a slight variation from the second, the acid-leaching step was placed after the calcination step. The study revealed that the leaching step in either succession improves the silica content in both bagasse (95%) and leaf (88%) ashes, while also enhancing the total surface area of silica nanoparticles compared to direct calcination. Falk et al. also conducted a similar study comparing two methods for synthesising silica nanoparticles. The comparison was between the sol-gel method and direct calcination of bagasse ash followed by leaching. In the sol-gel method, the bagasse ash was first leached with HCl, and then added to a NaOH solution to form sodium silicate. The formed solution was then subjected to a "gelling stage" by titration with HCl. The obtained exogels were then filtered, dried, and characterised. The silica nanoparticles obtained via the sol-gel process exhibited a higher purity, uniformity (size averaging around 10 nm), and amorphous character in comparison to the conventional calcination method, while the latter was more useful in preserving the "biogenic" nature of the synthesised silica nanomaterials with less uniformity in size, shape, and distribution. The sol-gel method has also been used in several publications owing to its ease of production and the homogeneous nature of the synthesised silica nanomaterials.

Vaibhav and colleagues made a comparative study to determine the potential of various agricultural wastes in producing SNPs. Extraction of silica was performed from four sources: rice husk, bamboo leaves, sugarcane bagasse, and groundnut shell. After removing the moisture, each sample was incinerated at 900°C, followed by dissolution of the silica via alkali leaching with 1 M NaOH. Then, the solution was filtered, treated with H_2SO_4, and dried. The resultant silica was highly pure in all samples. Sugarcane bagasse was found to be one of the best agricultural sources to produce high-grade silica nanomaterials (Vaibhav et al., 2015).

Hamad et al. also reported the synthesis of silica nanomaterials from sugarcane bagasse for use as an additive in industrial drilling fluids, enhancing their properties. This was achieved by thermal treatment in a muffle furnace, followed by sol-gel synthesis of nanosilica. They also confirmed the biodegradable nature of the synthesised particles. The properties of the obtained silica nanomaterials were similar to most silica nanomaterials reported in the literature (Hamad et al., 2021).

5.2.1.2 Carbon-based nanomaterials from sugarcane

Agro-industrial residues and biomass waste, including sugarcane bagasse, are ideally suited for the fabrication of carbon-based nanomaterials, such as CDs, carbon nanomaterials, and graphene oxide. CDs are one of the most versatile nanomaterials, comprising nanocrystalline regions of sp^2-hybridised graphitic carbon (Malpani et al., 2022; Padhan et al., 2025). Their small size and large surface area, coupled with improved photostability, high quantum yield, biocompatibility, and biodegradability, make them well-suited for various applications (R. Kumar et al., 2025). There are two main approaches to synthesising CDs from sugarcane. These approaches are bottom-up and top-down. The bottom-up approach uses a passivation agent to agglomerate smaller molecules into larger CDs (Gaur et al., 2024). The top-down approach, on the other hand, can use degradable precursors to synthesise low-cost, biocompatible CDs. In line with these two approaches, various methods have been used to synthesise CDs, including wet-chemical, thermal, electrochemical, and microwave-mediated methods. The following methods are discussed below.

Thambiraj and co-workers employed the chemical oxidation method to synthesise carbon quantum dots (CQDs) from sugarcane bagasse pulp. The bagasse pulp was air-dried and burned at 60°C, followed by the extraction of CQDs using toluene as the solvent. These CQDs also contained carbonyl and hydroxyl functional groups on their surfaces, aiding their application in sensors, adsorption, and drug-delivery devices (Thambiraj & Shankaran, 2016). In another study, Hashmi and colleagues modified the top-down approach to synthesise carbon dots. Using sugarcane bagasse as a precursor, the authors dialysed the supernatant against ultra-pure water to obtain the CD residues, which were then freeze-dried. The carbon dots obtained were monodispersed and spherical. Regarding the mechanism, the authors theorise that the bagasse particles undergo an initial hydrothermal breakage (under high-temperature, high-pressure conditions), where oxidation of water leads to their breakdown into smaller products and the loss of other impurities. These then follow a "bottom-up" (again, under hydrothermal conditions) approach to re-polymerise into CDs (Hashmi et al., 2020). Other researchers have reported the preparation of carbon dots from sugarcane products. Sim and co-workers used sugarcane juice as a precursor via the hydrothermal route to produce carbon dots. In the first step, the mixture of sugarcane juice and ethanol was autoclaved, followed by multiple washes (using dichloromethane and acetone) and centrifugation steps to produce CDs. These were then conjugated with graphitic carbon nitride ($g\text{-}C_3N_4$) to form a novel nanocomposite for the degradation of bisphenol A (BPA) under natural sunlight. The performance of the nanocomposites was 3.5-fold higher than that of pristine g-C_3N_4, confirming the beneficial role of CDs in enhancing BPA degradation (Sim et al., 2018).

Graphene oxide (GO) could also be derived from sugarcane products. Hashmi and co-workers devised a simple, single-step synthesis for GO from three agro-waste products, including sugarcane bagasse, orange peels, and rice bran, and subsequently compared the physicochemical properties of the resulting GO materials. The carbonaceous precursors were mixed with ferrocene and heated in a muffle furnace at 300°C to obtain GO NPs from the respective precursors. The authors reported an average particle size of 4 nm for bagasse and approximately 2.5 nm when used in a composite with two other agro residues, making it a cost-effective and environmentally benign alternative for GO nanomaterial synthesis (Hashmi et al., 2020). A similar synthesis of GO from bagasse was also demonstrated by Poonsawat, who used simple oxidation in a muffle furnace with ferrocene catalyst (Poonsawat et al., 2019). Table 5.1 summarises other reports for the synthesis of nanomaterials from sugarcane. The table indicates that bagasse and juice are commonly used parts of sugarcane for the synthesis of nanomaterials.

5.2.2 Nanomaterials Derived from Tea Waste

Tea is the most widely consumed beverage worldwide and has numerous health benefits. Although there are many different types of tea, black tea and green tea account for approximately 98% of the world's total tea production. Tea waste production involves withering, crushing, fermentation, drying, and finally packaging. All the waste generated during this production line is referred to as tea waste. Tea production generates substantial waste that cannot be effectively utilised for value creation. This waste contains many different components, including protein, fibre, caffeine, and polyphenols (V. Kumar et al., 2025). Due to its rich composition, it can be revalorised for different purposes, such as the synthesis of nanomaterials. Pruned tea leaves and tea residues are rich in lignin and holocellulose, which are rich in carboxyl, hydroxyl, phenolic hydroxyl, oxy groups, and oxygen-/heteroatoms-containing groups. It has been reported that thermochemical technologies, such as pyrolysis, can transform waste tea leaves into carbonaceous nanomaterials (Mahal & Al-Mutlaq, 2023). The formed carbonaceous nanomaterials usually exhibit relatively large surface areas, porous structures, and abundant functional groups (Nille et al., 2021). Pruned tea leaves have a high content of nitrogen functionalities (amino acids and caffeine), which can be introduced into biochar, making it beneficial as a nitrogen-rich biomass precursor for preparing contaminant adsorbents (Nille et al., 2021).

Several synthesis methods have been explored for converting tea waste into activated carbon, a common nanomaterial derived from waste. These methods include pyrolysis, hydrothermal carbonisation (HTC), and gasification. Pyrolysis is a biomass thermal conversion process performed with a complete absence of oxygen at temperatures above 400°C (Seth et al., 2025). In recent years, hydrothermal carbonisation has been particularly attractive for biochar production. HTC is a moderate thermal conversion reaction with a relatively low heating temperature (180°C–350°C) performed in subcritical water with autogenous pressures (Periakaruppan et al., 2021).

TABLE 5.1
Sugarcane product-derived nanomaterials, their properties, and application perspectives.

Sugarcane Material	Type of Nanomaterial	Synthesis Method	Description of Nanomaterials	Reference
Bagasse	Silica	Thermal method	Biocompatible mesoporous and core-shell	Falk et al. (2019)
Bagasse	Silica	Sol-gel	Amorphous, micro-structured	Shange et al. (2025)
Bagasse	Graphene oxide	Calcination	3D porous structure	Thangaraj et al. (2023)
Sugarcane juice	Carbon dot	Capping agent	Synthesis of graphite carbon nitride	Parveen et al. (2024)
Sugarcane juice	Silver nanoparticles	Capping agent	Synthesis of bio-composite between spherical silver nanoparticles and chitosan	Yong et al. (2020)
Sugarcane juice	Silver-doped silver chloride	Capping agent, halide-supplying agent	Synthesis of hybrid Ag@AgCl plasmonic nanoparticles (NPs)	Kulkarni and Bhanage (2014)

It was reported from literature that the specific surface area of pyrolytic carbons is generally larger than that of hydrothermal carbons (Arundhathi et al., 2024). These examples illustrate why pyrolysis is widely used to synthesise nanomaterials from tea waste. Furthermore, the surface area of pyrolysed products is generally larger than that obtained via the HTC process, consistent with other reports (Debnath et al., 2021). Chemical agents have been widely used to modify nanomaterials, aiming to achieve enhanced surface properties, introduce functional groups, and create novel structures (Çakmak et al., 2024). The physicochemical properties of nanomaterials are influenced by the source of tea waste, the activation agents, the thermal conversion approaches, and the preparation conditions, including temperature and time (Nasar, 2023). Table 5.2 summarises several synthesis methods and conditions for tea waste-derived activated nanocarbon reported in the literature. According to Table 5.2, pyrolysis is the most utilised method for converting tea into nanomaterials. Fewer studies utilised hydrothermal carbonisation.

Ehab and Shayma reported the fabrication of carbon nanoflakes from domestic waste tea leaves. The morphology of the products showed the observation of carbon-flake nanomaterials having thin graphitic outer mono shells with a diameter of about 35 nm and different monolith sizes of about 15–180 nm (Mahal & Al-Mutlaq, 2023). In another study reported by Darshana and colleagues, they conducted a study where silver nanoparticles were derived from waste tea (Rajput et al., 2020).

Viera and colleagues reported a study in which zinc oxide (ZnO) nanomaterial was synthesised from green tea (*Camellia sinensis*) aqueous extract, providing an eco-friendly alternative route. To obtain the ZnO nanomaterial, a zinc(II) nitrate precursor solution was prepared in distilled water at room temperature. In an open glass flask, an aqueous green tea extract was slowly added to the zinc(II) precursor solution. The reaction mixture was then stirred continuously. Finally, ZnO was obtained after the material was completely dried at 60°C and calcined at 500°C for 4 h. The results obtained suggested that the chelation and stabilisation properties of the polyphenols present in the green tea extract play a role in forming the ZnO nanomaterial (Vieira et al., 2024).

5.2.3 Nanomaterials derived from potato peelings

Potatoes are a widely cultivated crop and a vital source of nutrition for humans. Potato peelings are a waste product that is completely discarded, leading to rotting, microbial growth, and unpleasant odours (Vescovo et al., 2025). However, other researchers have demonstrated that potato peelings are a valuable material with potential economic and chemical interest due to their antioxidant, antibacterial, apoptotic, chemopreventive, and anti-inflammatory properties (Vescovo et al., 2025). Potato plants are rich in carbon-based compounds, including protein, lipids, and starch. Thus, turning leftover potato peels into biochar and nanomaterials is an excellent approach to utilise a waste product. It has been discovered that nanomaterials, such as activated nanocarbon derived from potato peelings, are effective at removing contaminants from water, particularly chromium ions from tannery effluents (Jimenez-Champi et al., 2023). Its surface chemistry and morphology, including functional groups such as hydroxyl, carbonyl, amine, and aromatic groups, as well

TABLE 5.2
Summary of several synthesis methods and conditions for tea waste-derived activated nanocarbon reported in previous studies.

Type of tea waste	Synthesis method	Synthesis conditions	Product	Reference
Tea industry waste	Pyrolysis	Tea waste: $ZnCl_2$ = 1:1; 700°C, 4 h	Activated nanocarbon	Wu et al. (2024)
Green tea processing waste	Pyrolysis	Dried at 105°C, 20 h; Activated; Carbonised at 700°C	Activated nanocarbon	C. Zhou et al. (2024)
Tea waste from the factory	Pyrolysis	Tea waste: H_2SO_4 = 1:1.5; 200°C, 24 h; Soaked in 1% $NaHCO_3$	Activated nanocarbon	Setianingsih (2025)
Waste tea leaves after infusion	Pyrolysis	Pyrolysed at 300°C, 2 h	Activated nanocarbon	Wong et al. (2018)
Factory-rejected tea	Hydrothermal carbonisation	200°C, 5 h	Activated nanocarbon	Zheng et al. (2024)
Pruned tea branches	Hydrothermal carbonisation	Tea waste: $ZnCl_2$ = 3:1; 5% Urea, HTC for 2 h	Activated carbon	Li et al. (2021)
Tea factory waste	Chemical energy activation, microwave energy	Tea waste: H_3PO_4 = 1:3; 900 W Microwave heated for 30 s; Carbonised for 1 h	Activated nanocarbon	Yagmur et al. (2008)

as its porosity and high surface area, are responsible for its high adsorption capacity (Tejaswini et al., 2022). Following is a brief discussion of various nanomaterials derived from potato peeling waste.

In a study conducted by Raigond, nanocrystals were derived from potato peelings and used for water treatment. This was achieved through two methods: direct hydrolysis and enzyme pretreatment. These methods broke down the starch granules into nanocrystals. However, the enzyme pretreatment process was shorter than that of acid hydrolysis. Nanocrystals of optimal size were obtained with both methods, with sizes ranging from 10 to 50 nm. The derived nanocrystals were disk-like platelets. Cellulose nanocrystals were also derived from cellulosic material in the potato peel. Cellulose was isolated from peel waste with alkali treatment. Furthermore, cellulose nanocrystals were prepared from potato peel and cellulose microcrystalline using acid hydrolysis. Microscopic images revealed that the aqueous suspension of cellulose nanocrystals derived from potato peel consisted of single, rod-shaped particles, whereas those derived from cellulose microcrystalline were rod-like nanoparticles, agglomerated into bundles, including some rods in single units that were well separated (Raigond et al., 2018),.

Deepa reported a study in which silver nanoparticles generated from potato peelings were applied for wastewater treatment. Their research established an ecologically acceptable method for generating AgNPs from potato peelings, a type of leftover kitchen waste considered zero-value material. Potato peels were used to generate AgNPs with sizes of 20–60 nm. The biosynthesised nanoparticles were found to be spherical, as indicated by FESEM image analysis. It was concluded that the synthesis of AgNPs from potato waste will support food waste management, which in turn will help resolve environmental issues and minimise waste, thereby boosting the economy in developing countries and opening an alternative path to sustainable economic growth. This work enabled an environmentally friendly, cost-effective route to material production (Deepa et al., 2023)

Afolabi reported a study in which carbon nanotubes were synthesised from potato peels via controlled pyrolysis and evaluated for their potential as eco-friendly adsorbents for crude oil remediation. Potato peels were carbonised at 300°C, and the results revealed that the CNTs exhibited a maximum adsorption capacity of 9538.17 mg g^{-1} (Afolabi et al., 2025).

Parkhe reported a study in which potato peels were used to synthesise poly-L-lysine (PLL)-functionalised gold nanoparticles (PLL@AuNP). The method offered an eco-friendly, Green-Chemistry approach. PLL, a water-soluble cationic polymer, was used to functionalise gold nanoparticles, making them suitable for antibacterial, antioxidant, and photocatalytic applications. To synthesise PLL@AuNPs from potato peelings, the prepared extract was added to an aqueous $HAuCl_4 \cdot 3H_2O$ precursor solution at room temperature with constant stirring. The reduction of Au^{3+} to Au^0 was confirmed by a colour change from light yellow to cherry red. Additionally, 0.1% PLL was added to a 10 ml AuNPs solution and shaken vigorously to obtain surface-functionalised AuNPs, i.e., PLL@AuNPs. The results revealed that PLL@AuNPs particles were nearly spherical, with a particle size of approximately 20 nm. This confirmed that the gold nanomaterial was successfully formed (Parkhe et al., 2023)

Osman explored the use of activated carbon (AC) and CNTs from potato peel waste. Different ACs were synthesised via two activation steps: first, with phosphoric acid (designated PP) and then with potassium hydroxide (designated PK). The AC produced after the two activation steps exhibited a good surface area of 833 $m^2 g^{-1}$ with a pore volume of 0.44 $cm^3 g^{-1}$. This study helped advance the circular economy by effectively upcycling and valorising lignocellulosic biomass waste, such as potato peel waste, into high surface area AC and, subsequently, multi-walled carbon nanotubes (MWCNTs). Consequently, MWCNTs were prepared from the produced AC by mixing it with a nitrogen-based material, melamine, and an iron precursor, iron(III) oxalate hexahydrate. This study concluded that such an approach will make the synthesis of value-added materials more sustainable, more eco-friendly, and less expensive (Osman Ahmed et al., 2019).

In another study conducted by Meraj and workers, CQDs were derived from potato peelings. The synthesis was carried out via a highly efficient one-step hydrothermal method. The hydrothermal process promotes carbonisation and the formation of CQDs from the organic compounds in the potato peel extract. This was conducted in an autoclave sealed in an oven, heated to 180°C, and maintained at this temperature continuously for 18 h. The BET surface area of the CQDs was measured to be 5.878 m^2/g, with a pore size of 6.52 nm, confirming that the desired nanomaterial had been successfully formed. The synthesised CQDs were used to remove the MB dye from wastewater (Meraj et al., 2024).

5.2.4 Banana peelings

Banana peels are a rich source of polymers like lignin, hemi-cellulose, and pectin, and they also contain polyphenolic compounds (quercetin, catechin, epicatechin, gallic acid, ferulic acid, gallocatechin, and chrysin) and flavonoids, which act as a ligand and form a coordinate bond with the metal ion (Orsuwan et al., 2016). This is indicated in Figure 5.2. The extract of banana peel is used as both a reducing and a capping agent due to the presence of secondary metabolites, such as flavonoids, polyphenols, and alkaloids (Issara et al., 2024). The various functional groups present in these metabolites could be used for the synthesis of nanoparticles (Pereira et al., 2021). Banana extracts are primarily used to extract nanoparticles. Motivated by the unique characteristics of the banana peel, various kinds of nanoparticles, such as metal and metal oxide nanoparticles, including gold, silver, zinc, and titanium dioxide, were reported.

Traditionally, these nanoparticles were synthesised by using toxic chemicals and capping agents like polyvinyl chloride, anionic surfactants, which are hazardous, expensive, and require high-temperature conditions (Ibrahim, 2015; Issara et al., 2024). Thus, the novel and facile approach of synthesising NPs using banana peel extract is safer, cheaper, environmentally friendly, and does not require high-temperature conditions. The available literature indicates that a natural green method for nanoparticle formulation from waste banana peels for the synthesis of various metal nanoparticles, such as silver, gold, and titanium dioxide, has been reported. The synthesis methods are discussed below.

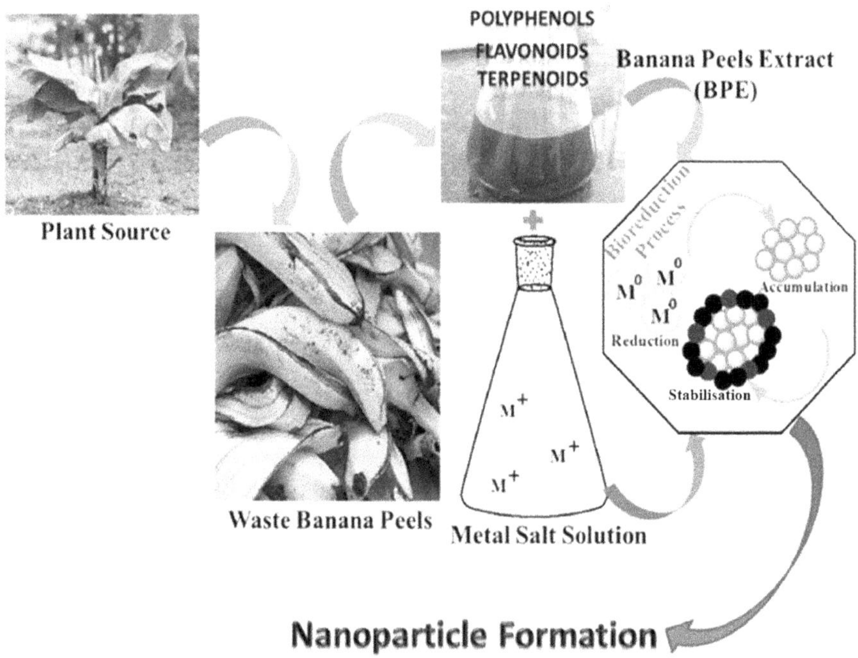

FIGURE 5.2 Synthesis pro metal ions from banana peelings.

Source: Chugh & Kaur (2022)

Ibrahim reported a method for synthesising silver nanoparticles using banana peelings. In the procedure, 6.8 mg of dry ash extract of banana peelings was dissolved in an $AgNO_3$ solution. The mixture was then incubated at 30°C in the dark to avoid photocatalysis of the $AgNO_3$ under the fixed reaction conditions. The obtained AgNPs were used in a waste-treatment method to reduce microbial load (Ibrahim, 2015).

Orsuwan and workers prepared composite films of AgNPs nanomaterials from banana using the casting method. Banana powder (4 g) was then added to distilled water, which was heated to 90°C with stirring until completely dissolved, a process that took approximately 20 min. Then, glycerol was added as a plasticiser to achieve final silver concentrations of 0.5, 1.0, and 2.0 mM. An aqueous solution of $AgNO_3$ was added dropwise to the prepared solution, and the mixture was stirred continuously for approximately 4 h at 90°C. The silver ions were reduced to metallic AgNPs using banana powder, which contains amino acids from proteins, terpenoids, and phenolic hydroxyl groups from tannins. These banana/AgNP films can be used as edible coatings or in food packaging to enhance the shelf life of packaged foods (Orsuwan et al., 2016).

Ondiek developed a method for synthesising gold nanoparticles, in which chloroauric acid ($HAuCl_4$) was dissolved in distilled water using banana peelings. This mixture contained approximately 10 mg of banana peel ash in approximately 2 mL of the chloroauric acid solution (1 mM), which was further incubated at 80°C for 3 min in a water bath. The various changes in the reaction conditions enabled the synthesis of a range of NPs with bright colours, exhibiting typical UV-visible spectra (Ondiek, 2016).

Hammed and colleagues reported the preparation of TiO_2 NPs from banana peel extract. TiO_2 has a molar mass of 79.87 g/mol and a density of 4.2 g cm^{-3}. The bulk TiO_2 particles had an average size of 550 nm. 5 mL of deionised distilled water was used as the solvent to prepare a 100 mg mL^{-1} solution. Then, only 5 ml of this was added dropwise to 50 ml of banana peel extract with continuous stirring at 300 rpm for about an hour on a hot plate using a magnetic stirrer at room temperature (Hameed et al., 2019).

5.2.5 Corn cob

Corn cob is a waste material obtained from the incineration of corn cobs with high silica (SiO_2) content (Sulaiman et al., 2022). The applications of silica extracted from corn cob ash include serving as a supplementary cementitious material in concrete, as a source of high-purity silica, and as a precursor for producing silica nanoparticles with potential industrial uses (Owoeye et al., 2015; Sulaiman et al., 2022). Amorphous silica extracted from corn cob ash can be utilised in various water purification processes due to its capabilities in adsorption, filtration, ion exchange, and disinfection (Sulaiman et al., 2022).

The high surface area of silica synthesised from corn cob waste improves its capacity to bind and extract contaminants from water. Corn cobs are a plentiful agricultural waste that contains a significant amount of silica (Pieła et al., 2020). Due to their unique properties, studies on the use of silica from maize cobs have demonstrated its effectiveness in removing contaminants from water. Water pollutants and impurities can be effectively removed by exploiting silica particles' ability to form hydrogen bonds with water molecules. Salami et al., (2022) have established that magnetic silica derived from corn cobs is effective at extracting contaminants from water, particularly rhodamine B. As a result, converting corn cob waste into low-cost magnetic silica is another effective way to modify it. Water treatment with silica derived from maize cobs is a viable solution.

5.2.6 Egg shells

One of the agricultural wastes that has garnered attention recently is eggshells. Every day, many eggshells are generated worldwide as biowaste. Eggshell waste is mostly generated from households, restaurants, and bakeries. Pure calcium carbonate, with low porosity, is the major component of eggshells and has the potential to be turned into useful products (Weerakoon et al., 2024). Nano-calcium oxide can be synthesised from a waste egg shell by the sol-gel method (Habte et al., 2019). Due to its hierarchical and porous structure, the eggshell membrane is used to synthesise magnetic $CuFe_2O_4$ nanomaterials with multifunctional properties, including catalytic and antibacterial activities, for industrial water treatment (Ismael et al., 2024). Other synthesis reports on nanomaterials from eggshells are discussed below.

In a study reported by Zhang and co-workers, eggshell membrane (ESM), a type of agricultural waste with a unique hierarchical and porous structure, was identified as a promising bio-template for the successful fabrication of ESM-derived magnetic

$CuFe_2O_4$ nanocomposites (abbreviated as ESM/$CuFe_2O_4$). The nanomaterial exhibited unique morphological features, with 3D coral-like, layered, multi-channel architectures that were distinct from those of pure $CuFe_2O_4$. The nanomaterial was synthesised using a citrate combustion method, in which $Cu(NO_3)_2{\cdot}3H_2O$ and $Fe(NO_3)_3{\cdot}9H_2O$ were dissolved in deionised water and mixed thoroughly. Then, the ESM powder was dispersed into the above mixture and stirred for 1 h, after which a specific amount of citric acid was added. Copper nitrate, ferric nitrate, and citric acid were dosed in a molar ratio of 1:2:3.2. The resulting homogeneous suspension was stirred at 90°C for approximately 4 h to form a sticky gel, then transferred to the oven and dried at 100°C for 12 h. After that, the resulting foam sample was ground into uniform powder and collected in a crucible, followed by calcination at 500°C for 2 h at a rate of 5°C min^{-1} The obtained product was defined as magnetic ESM/$CuFe_2O_4$ (Zhang et al., 2021).

In another study by Adaikalam et al. (2024), CaO nanomaterial derived from eggshells was synthesised and evaluated for its effectiveness in photocatalytic pollution remediation, electrocatalytic activity, optical sensing, and antibacterial activity. Unlike other techniques, this calcination process does not require any chemical reagents because the $CaCO_3$ in eggshells is highly pure. Calcium oxide nanoparticles were prepared by subjecting waste eggshells (ES) to high-temperature calcination, and the synthesised CaO nanoparticles were characterised for their structural, morphological, chemical, optical, and other properties. The CaO was synthesised as follows: The washed eggshells were dried at room temperature for 2 days and then ground into a fine powder using a mixer grinder. The obtained eggshell powder was calcined at 900°C for 1 h. Subsequently, it was cooled to around 100°C, and the resulting CaO powder was crushed in a mortar and pestle to a fine nanoscale powder. The resulting CaO NPs were collected and preserved for further characterisation and use. The synthesis procedure reported by Zia and Adaikalam was similar, and it is illustrated in Figure 5.3.

5.3 NANOMATERIALS OBTAINED FROM INDUSTRIAL WASTE

5.3.1 Tyre waste

Worldwide, roughly 1 billion scrap tyres are generated every year (Gomez et al., 2025). A significant portion is landfilled or openly burned, contributing to air pollution and environmental degradation (Magagula et al., 2023; Tsang, 2013). For that reason, the disposal of scrap rubber in landfills is banned in most developed countries (Turgut & Yesilata, 2008). Thus, there is interest in finding sustainable solutions for managing rubber tyre waste. Generally, tyres are composed of 45%–47% natural rubber, 21%–22% carbon black, 12%–25% metals, 5%–10% textile, 6%–7% additives, and 1%–2% zinc, ~1% sulfur (Abdelbasir et al., 2020). NPs were also obtained from recycled tyres via ball milling for 5 consecutive hours. The method yielded rubber ash particles smaller than 500 nm; when aided with silicon waste, the particle size was further reduced to <50 nm. As waste tyres are composed principally of carbon (approximately 81.2 wt%), they are considered a promising source of carbon for various potential applications.

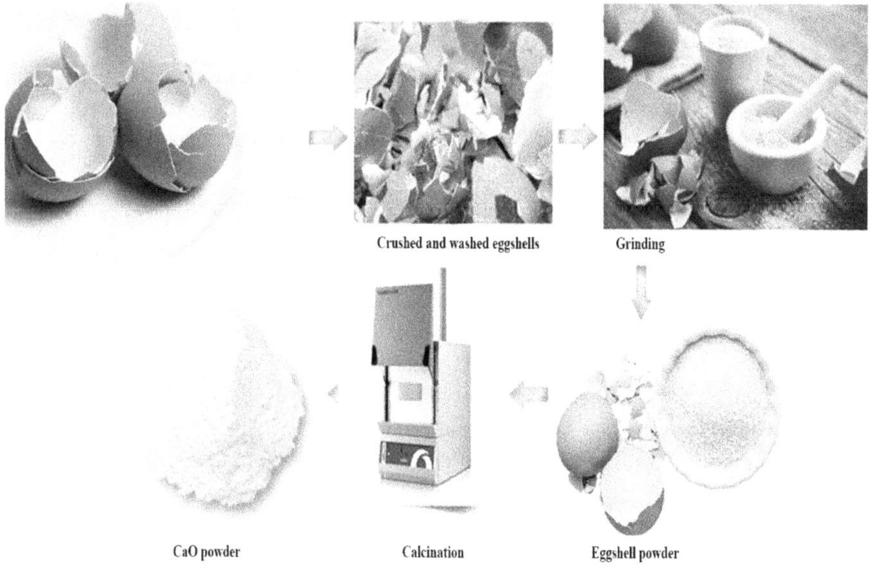

FIGURE 5.3 Synthesis procedure of CaO nanomaterial (Adaikalam et al., 2024).

Maroufi and co-workers synthesised high-value carbon nanoparticles (CNPs) from waste tyre rubber (WTR), using a high-temperature approach. The transformation of WTRs was carried out at 1550°C over various reaction times (5 s to 20 min). The total energy utilisation for the process was acceptable as the decomposition of WTR and the formation of CNPs occur quickly and completely (Maroufi et al., 2017). Another facile and low-cost method for preparing carbon black nanoparticles from waste tyres was reported (Thathola et al., 2024). This method involved thermal transformation under high temperatures (1000°C) and self-pressurised conditions, producing carbon nanomaterials (~22 nm) with a high yield (approximately 81%) under optimised conditions. The chain-like agglomerated nanoparticles exhibited good thermal stability and conductivity, and chemical analysis indicated that they were partially oxidised (C, 84.9%; O, 4.9%; S, 10.2%). Many studies have reported the use of unconventional hydrocarbon waste as low-cost starting materials, such as tyres and plastics, for the synthesis of materials useful for water treatment.

In a study conducted by Mwafy, multi-walled carbon nanotubes (MWCNTs) were synthesised from tyres by the chemical vapour deposition technique. XRD, FESEM, EDX, Raman spectroscopy, and TGA were used to evaluate the synthesised MWCNTs. XRD analysis revealed a diffraction pattern consistent with a graphite structure, indicating crystalline materials with a well-ordered atomic arrangement. Meanwhile, SEM microscopy revealed a tube diameter of approximately 40 nm. Raman spectroscopy showed that the IG/ID ratio equals 1.02, which means that the presence of defects in the skeleton of CNTs due to the distortion process happened on the graphene layers, whereas TGA showed that the effect of thermal combustion of purified CNT started at 500°C with a mass loss of about 95%Mwafy, 2020).

Miz-Fernandez also presented the results of the synthesis and morphological characterisation of carbon nanostructures obtained from the decomposition of residual solid from waste tyres (RSWT) in quartz tubes under reduced pressure (1.33 Pa) at 900°C for 15 min. The synthesis yielded two phases: a fragmented, black, solid powder composed of MWCNTs, onion-type fullerenes, spheres, and a very bright, dark metallic film. Analysis by microscopy (SEM and TEM) showed that the MWCNTs had an average diameter of approximately 25 nm and a length greater than 100 nm, while the diameter of the onion-type fullerenes was found to be 8 nm. The nanospheres ranged in diameter from 500 nm to 1.5 μm, and some had a metallic core surrounded by carbon layers (Mis-Fernández et al., 2012).

Gómez et al. reported the development of a simple and inexpensive method for the preparation of carbon black nanomaterial from waste tyres. Under optimal conditions, the yield of recovered carbon black nanomaterials (~22 nm) was approximately 81%. The nanomaterials presented good thermal stability and conductivity, and formed chain-like agglomerates; chemical composition analysis and solubility tests indicated that they are partly oxidised (Gómez-Hernández et al., 2019).

Sharaf Addin also investigated the use of waste tire-derived char for the synthesis of (CNMs) through microwave-assisted pyrolysis. Waste tyre rubber crumbs were pyrolysed at 600°C for 20 min to produce char, which was subsequently purified and utilised as a precursor for CNMs synthesis. CNMs were synthesised using tire-derived char and ferrocene as a catalyst at varying microwave power levels (600–1000 W), reaction times (10–40 min), and catalyst-to-precursor ratios (25–100 wt%). Comprehensive characterisation techniques, including FTIR, XRD, Raman spectroscopy, TEM, and FESEM, revealed that carbon nanofibres (CNFs) synthesised at 1000 W under optimal conditions exhibited bamboo-shaped, tubular structures with diameters ranging from 425 to 881 nm. XRD confirmed the graphitic nature of the CNFs, while Raman spectroscopy indicated a moderate degree of graphitisation (I_D/I_G = 1.61). TEM analysis confirmed a tip-growth mechanism, in which residual iron-based catalyst nanoparticles were embedded at the tips of the CNFs, promoting their elongation. FESEM analysis revealed residual iron-based catalyst particles embedded within the CNFs, underscoring the need to further refine synthesis conditions to enhance purity. These CNFs, with their porous structure and high surface area, hold potential for adsorption-based applications such as wastewater treatment and pollutant removal (Sharaf Addin et al., 2025)

From the discussions above, it is evident that various nanomaterials can be synthesised from waste tyres. This includes carbon nanomaterials, carbon nanofibres, carbon black, and multi-walled carbon nanotubes, among others. Researchers have employed various synthesis methods to produce nanomaterials from waste tyres, including pyrolysis. Therefore, recycling waste tyres is a viable alternative to reduce the large amount of waste tyres and decrease their negative environmental impact, producing high-quality nanomaterials for use in batteries, sensors, electronic devices, catalysis, pigments, concrete, and plastics, among many other applications.

5.3.2 Battery waste

Over 50 million units of waste lead batteries are produced annually in countries such as China and Japan (Hou et al., 2024). Given the toxicological effects of lead exposure and its bioaccumulation potential, recycling these batteries is a responsible approach that helps avoid ecological and public health problems (Dente et al., 2025). Lead pastes are made of metal oxide, dioxide, and sulfate forms that are difficult to recover and reuse because of their insolubility (Van der Kuijp et al., 2013). However, pyrometallurgical processes can be employed to retrieve lead from spent lead-acid battery paste. For example, Xiaowu and co-workers developed a synthetic method to produce Pb nanopowder from simulated spent lead paste via pyrolysis. The authors produced submicron particles of lead oxide by a thermal process coadjuvanted with poly(N-vinyl-2-pyrrolidone), which helped decrease particle size (Jie et al., 2022). While the technique overcomes difficulties associated with the insolubility of lead paste in aqueous media, pyrometallurgical processes may have unfavourable consequences due to their high-energy requirements and the undesirable emissions of air pollutants such as sulfur dioxide and lead dust (Yang et al., 2020).

Similarly, a large percentage of transportable batteries in industrialised countries are made of zinc–manganese (Zn–Mn) nanomaterials. China alone produces more than 15 billion units every year. Exposure to toxic concentrations of Zn can impair calcium uptake and cause metabolic deficiencies in humans and other organisms (Matović et al., 2011). Thus, recovering Zn from recycled Zn–Mn batteries is a convenient approach for enabling the reuse of the metal in other applications while minimising its uncontrolled release into the environment. Synthesis of Zn nanomaterials, such as NPs, nanofibres, and flaky NPs, using recycled Zn–Mn batteries as a starting material has been actively investigated in recent years. Lv and colleagues demonstrated a facile and highly efficient method for separating zinc from spent batteries. Samples of recycled zinc-cathodes were put in a corundum crucible and subjected to a vacuum pressure of 1 Pa, heating temperature of 800°C, and condensing temperature 200°C. The simple technique can attain zinc separation efficiency up to 99.68%. By following the separation process with nitrogen gas flow into the heating chamber, structured nanoparticles with different morphologies, including hexagonal prisms, fibres, and sheets, can be obtained (Lv et al., 2024).

5.3.3 Mine tailings (acid mine drainage)

Mining waste can generate environmental pollution, including acid mine drainage (AMD). AMD is hazardous to its surroundings and can pollute both surface and groundwater. Nanomaterials are advanced materials used in all fields of human activity and development. However, their production is still expensive and may pollute the environment due to the chemicals used and/or the energy required for their synthesis. A smart solution could be to use mining waste and AMD to produce nanomaterials with properties similar to those of nanomaterials derived from clean chemicals. Thus, both wastes will be valorised, and useful materials will be produced and applied to extract metals. Over the past 10–15 years, with the emergence of nanoscience and nanotechnologies, the number of studies on the production of NPs from mining waste has also increased. Mainly, iron (Fe) mining waste has attracted the scientists' attention probably due to two reasons: (i) diverse, unique properties of iron-based nanoparticles and especially of the magnetite nanoparticles (MNPs), and (ii) due to the mineral beneficiation technology, usually this waste is rich in Fe that could be used as a cheap precursor for preparing MNPs.

5.3.3.1 Magnetite nanoparticles from AMD

The synthesis of MNPs from AMD started 15–17 years ago. It was discovered that ferric iron from AMD pumped out from abandoned underground coal mines. Hydrogen peroxide was added to the raw AMD to oxidise the eventually available Fe^{2+} to Fe^{3+}. Then, the pH of the AMD was raised from its natural level of 2.6 to 3.5–4.0 by adding 4 M NaOH. Iron was precipitated as ferric hydroxide/oxyhydroxide. At pH 3.5, other metals remained in the solution. The precipitate was separated from AMD by centrifugation. The ferric precipitate was resolubilised and used as a source of ferric iron. Reagent-grade $FeSO_4{\cdot}7H_2O$ was the ferrous iron source. A solution containing Fe^{3+} and Fe^{2+} was prepared under an $N_2(g)$ atmosphere to prevent oxidation, then mixed for a specified time to remove dissolved oxygen. Furthermore, the pH was raised to 9.5 by adding a 6.4 M NH_4OH solution. NPs were allowed to grow for 30 min under an N_2 atmosphere. The black precipitate (magnetite NPs with a size of 10–15 nm) was separated from the solution with the aid of an external magnetic field and washed. The prepared MNPs were of the same quality as those prepared with chemical reagents (pure iron salts) under similar conditions.

5.4 POLYSTYRENE WASTE AND POLYETHYLENE TEREPHTHALATE WASTE

Nowadays, our planet suffers from the accumulation of plastic products, which can cause significant environmental harm by polluting the air, water, and land. Plastic is one of the most significant environmental hazards. Plastic is a non-biodegradable material, and several toxic chemicals leach from it, seeping into soil, water, plants, and animals. This leads to cancer and other serious ailments, aquatic life endangerment, and environmental pollution. The use of waste plastic for the synthesis of various nanomaterials can not only protect humans and the environment from the dangers of plastic but also yield beneficial substances for other purposes. Different nanomaterials can be synthesised from waste plastics, such as polyvinyl chloride, which serves as the carbon source for the fabrication of MoC_2 nanoparticles. These nanoparticles are used in various fields, including water purification, gas storage, and electrocatalysis for the generation of sustainable hydrogen. The following section discusses some of the reported nanomaterials synthesised from plastic waste.

In a study reported by Zhou and co-workers, fibrous carbon nanomaterials were prepared from waste PET bottles. This process was highly efficient, facile, and catalyst-free in preparing fibrous carbon nanomaterial with promising hydrophobic and electrical properties. The fibrous carbon nanomaterial prepared can be used in both sheet and powder forms, providing a versatile surface for preparing novel carbon-based composites with significant optical properties and conductivity. The prepared carbon nanomaterial derived from waste PET has also been utilised in fabricating a strain sensor with good durability for detecting pollutants in water bodies (X. Zhou et al., 2020).

Essawy and colleagues described an economical, solvent-free process that converts polyethylene terephthalate (PET) bottle waste into carbon nanostructure materials via thermal dissociation in a closed system under autogenic pressure, along with additives and/or catalysts, which can act as cluster nuclei for carbon nanostructure materials such as fullerenes and carbon nanotubes. This study successfully produced and controlled the microstructure of various carbon nanoparticles from PET waste by optimising preparation parameters, including time, additives, and catalyst amounts. The main novelty of their work was the acquisition of high-value carbon-based nanomaterials from PET waste using a simple, solvent-free chemical technique. Thus, the prepared materials are promising, cost-effective, and eco-friendly and may be used in various applications, including water purification (El Essawy, Konsowa, et al., 2017).

In another study by El Essawy and co-workers, they synthesised graphene from PET bottle waste. The synthesised graphene was characterised using SEM, TEM, BET, and FTIR. The graphene was then used for removing dyes from water. The adsorption of methylene blue (MB) and acid blue 25 (AB25) onto graphene was studied, and the effects of contact time and adsorbent dosage were optimised. The Response Surface Methodology (RSM) was also applied to investigate the effect of three variables (dye concentration, time, and temperature) and their interaction on the removal efficiency. Adsorption kinetics is favoured a pseudo-second-order model, and isotherms are favoured by the Langmuir and the Freundlich isotherm models, respectively. Thermodynamic parameters indicate that the dye adsorption is spontaneous and endothermic. Plastic waste can be converted into valuable carbon-based nanomaterials for adsorbing organic contaminants from aqueous solutions (El Essawy, Ali, et al., 2017).

Chingakham Chinglenthoiba and co-workers reported the synthesis and characterisation of a known metal-organic framework (MOF) using terephthalic acid (TPA) recovered from the hydrolysis of PET bottle waste. This approach adds value to the existing large amounts of bottle waste in the environment. Fully characterised zinc-TPA MOF (MOF-5) was used for the extraction and removal of engineered polyvinyl chloride (PVC) and polymethylmethacrylate (PMMA) nanoparticles from water with a high efficiency of 97% and 95%, respectively. Kinetic and isotherm models for the adsorption of polymer nanoparticles (PNPs) on the MOF surface were investigated to understand the mechanism. The Q_{max} values for PVC and PMMA NPs were 56.65 mg g^{-1} and 33.32 mg g^{-1}, respectively. MOF-5 was characterised before and after adsorption of PNPs on the surface of MOF-5 using a range of techniques. The MOF-5 with PNPs on its surface was successfully regenerated and reused for the adsorption and removal of polymer nanoparticles, yielding consistent results across five extraction cycles. As a proof-of-concept, MOF-5 was also used to remove plastic particles from commercially available body scrub gel solutions. Such methods and materials are necessary to mitigate the health hazards associated with emerging micro- and nanoplastic pollutants in the environment (Chinglenthoiba et al., 2024).

While still discussing MOFs, another research was conducted where a novel green approach, the waste PET bottle material has effectively been used as the starting precursor instead of terephthalic acid for the synthesis of five terephthalate-based nanoporous trivalentMOFs namely MIL-47, MIL-53(Cr), MIL-53(Al), MIL-53(Ga), and MIL-101(Cr). The optimal reaction parameters for achieving green synthesis were studied. These MOFs were structurally characterised by powder X-ray diffraction (PXRD). Scanning electron microscopy (SEM) images confirmed the particle nature and size of the synthesised MOFs. All the characterisation techniques strongly supported the conclusion that the synthesised MOFs using PET are like those reported in the literature. These MOFs can also be utilised in various applications, including the adsorption of pollutants and gas storage (Lo et al., 2016).

5.5 CONCLUSION

The synthesis of nanomaterials from waste materials represents a sustainable, cost-effective, and environmentally responsible approach to nanotechnology. Transforming waste into high-value nanomaterials will reduce waste and enable resource recovery. This will also minimise the environmental impact of conventional nanomaterial production, which often requires toxic chemicals, longer synthesis time, high-energy inputs, and expensive raw materials. Additionally, waste-derived nanomaterials have demonstrated competitive performance in applications such as water purification, energy storage, sensing, catalysis, and biomedicine. However, challenges such as inconsistent quality, scaling up production, and standardising synthesis have been encountered. Therefore, in the future, researchers need to develop solutions to these challenges. Finally, advancing research in this area will not only support the circular economy but also open novel ideas for green nanotechnology, ensuring that innovation aligns with global sustainability goals.

REFERENCES

Abdelbasir, S. M., McCourt, K. M., Lee, C. M., & Vanegas, D. C. (2020). Waste-derived nanoparticles: Synthesis approaches, environmental applications, and sustainability considerations. *Frontiers in Chemistry*, *8*, 782.

Adaikalam, K., Hussain, S., Anbu, P., Rajaram, A., Sivanesan, I., & Kim, H.-S. (2024). Eco-friendly facile conversion of waste eggshells into CaO nanoparticles for environmental applications. *Nanomaterials*, *14*(20), 1620.

Afolabi, A. M., Yahaya, M. I., & Jamilu, U. (2025). Sustainable synthesis of carbon nanotubes from potatoes peel for oil spill remediation. *African Journal of Environmental Sciences and Renewable Energy*, *18*(1), 1–29.

Anitha, J., Sherin, P. J., Ranjani, R., & Selvakumar, R. (2024). Recent advances in sugarcane-mediated nanocomposites and its applications. *Value Addition and Product Diversification in Sugarcane*, 213–226.

Arundhathi, K. V., John, N., Abraham, T., Ambady, K. G., & Mathew, B. (2024). Smart nanomaterials for sustainable green energy applications: Environmental technology and sustainability. In *Environmental technology and sustainability* (pp. 259–282). Apple Academic Press.

Çakmak, T. G., Saricaoglu, B., Ozkan, G., Tomas, M., & Capanoglu, E. (2024). Valorization of tea waste: Composition, bioactivity, extraction methods, and utilization. *Food Science & Nutrition*, *12*(5), 3112–3124.

Capanoglu, E., Nemli, E., & Tomas-Barberan, F. (2022). Novel approaches in the valorization of agricultural wastes and their applications. *Journal of Agricultural and Food Chemistry*, *70*(23), 6787–6804.

Chen, D., Yin, L., Wang, H., & He, P. (2014). Pyrolysis technologies for municipal solid waste: A review. *Waste Management*, *34*(12), 2466–2486.

Chinglenthoiba, C., Mahadevan, G., Zuo, J., Prathyumnan, T., & Valiyaveettil, S. (2024). Conversion of PET bottle waste into a terephthalic acid-based metal-organic framework for removing plastic nanoparticles from water. *Nanomaterials*, *14*(3), 257.

Chugh, R., & Kaur, G. (2022). A mini review on green synthesis of nanoparticles by utilization of Musa-balbisiana waste peel extract. *Materials Today: Proceedings.*

Debnath, B., Haldar, D., & Purkait, M. K. (2021). Potential and sustainable utilization of tea waste: A review on present status and future trends. *Journal of Environmental Chemical Engineering*, *9*(5), 106179.

Deepa, Dhanker, R., Kumar, R., Kamble, S. S., Kamakshi, & Goyal, S. (2023). Biosynthesis and characterization of silver nanoparticles generated from peels of Solanum tuberosum (potato) and their antibacterial and wastewater treatment potential. *Frontiers in Nanotechnology*, *5*, 1213160.

Deng, J., You, Y., Sahajwalla, V., & Joshi, R. K. (2016). Transforming waste into carbon-based nanomaterials. *Carbon*, *96*, 105–115.

Dente, S. M. R., Zhang, J., & Hashimoto, S. (2025). Assessing lead waste and secondary resources in major consumer nations: A vanishing resource or a toxic legacy? *Resources*, *14*(4), 52.

Devi, M., Rawat, S., & Sharma, S. (2021). A comprehensive review of the pyrolysis process: From carbon nanomaterial synthesis to waste treatment. *Oxford Open Materials Science*, *1*(1), itab014.

El Essawy, N. A., Ali, S. M., Farag, H. A., Konsowa, A. H., Elnouby, M., & Hamad, H. A. (2017). Green synthesis of graphene from recycled PET bottle wastes for use in the adsorption of dyes in aqueous solution. *Ecotoxicology and Environmental Safety*, *145*, 57–68.

El Essawy, N. A., Konsowa, A. H., Elnouby, M., & Farag, H. A. (2017). A novel one-step synthesis for carbon-based nanomaterials from polyethylene terephthalate (PET) bottles waste. *Journal of the Air & Waste Management Association*, *67*(3), 358–370.

Etxebarria, G., Gomez-Uranga, M., & Barrutia, J. (2012). Tendencies in scientific output on carbon nanotubes and graphene in global centers of excellence for nanotechnology. *Scientometrics*, *91*(1), 253–268.

Falk, G., Shinhe, G. P., Teixeira, L. B., Moraes, E. G., & de Oliveira, A. P. N. (2019). Synthesis of silica nanoparticles from sugarcane bagasse ash and nano-silicon via magnesiothermic reactions. *Ceramics International*, *45*(17), 21618–21624.

Ferronato, N., Rada, E. C., Portillo, M. A. G., Cioca, L. I., Ragazzi, M., & Torretta, V. (2019). Introduction of the circular economy within developing regions: A comparative analysis of advantages and opportunities for waste valorization. *Journal of Environmental Management*, *230*, 366–378.

Gaur, M., Misra, C., Bajpayee, A. K., & Bhardwaj, A. K. (2024). Recent advances in agriculture waste for nanomaterial production. *Green and Sustainable Approaches Using Wastes for the Production of Multifunctional Nanomaterials*, 331–344.

Gomez, C.-Y., Ronald, L.-G., Andry, P.-O., & RF, J. (2025). Literature review: Rubber waste in tire manufacturing. *Journal of Physics: Conference Series*, *3028*(1), 12036.

Gómez-Hernández, R., Panecatl-Bernal, Y., & Méndez-Rojas, M. Á. (2019). High yield and simple one-step production of carbon black nanoparticles from waste tires. *Heliyon*, *5*(7).

Gueboudji, Z. (2024). Waste valorization techniques. In *Generation of energy from municipal solid waste: Circular economy and sustainability* (pp. 29–52). Springer.

Habte, L., Shiferaw, N., Mulatu, D., Thenepalli, T., Chilakala, R., & Ahn, J. W. (2019). Synthesis of nano-calcium oxide from waste eggshell by sol-gel method. *Sustainability*, *11*(11), 3196.

Hamad, B. A., Xu, M., & Liu, W. (2021). Performance of environmentally friendly silica nanoparticles-enhanced drilling mud from sugarcane bagasse. *Particulate Science and Technology*, *39*(2), 168–179.

Hameed, R. S., Fayyad, R. J., Nuaman, R. S., Hamdan, N. T., & Maliki, S. A. J. (2019). Synthesis and characterization of a novel titanium nanoparticals using banana peel extract and investigate its antibacterial and insecticidal activity. *Journal of Pure and Applied Microbiology*, *13*(4), 2241–2249.

Hashmi, A., Singh, A. K., Jain, B., & Singh, A. (2020). Muffle atmosphere promoted fabrication of graphene oxide nanoparticle by agricultural waste. *Fullerenes, Nanotubes and Carbon Nanostructures*, *28*(8), 627–636.

Hou, H., Du, B., Guo, D., Xu, H., & Fan, Y. (2024). Path to the sustainable development of China's secondary lead industry: An overview of the current status of waste lead-acid battery recycling. *Environmental Impact Assessment Review*, *105*, 107389.

Ibrahim, H. M. M. (2015). Green synthesis and characterization of silver nanoparticles using banana peel extract and their antimicrobial activity against representative microorganisms. *Journal of Radiation Research and Applied Sciences*, *8*(3), 265–275.

Ismael, E., Fahim, K. M., Ghorab, S. M. O., Hamouda, R. H., Rady, A. M., Zaki, M. M., & Gamal, A. M. (2024). Sustainable recycling of poultry eggshell waste for the synthesis of calcium oxide nanoparticles and evaluating its antibacterial potency against foodborne pathogens. *Journal of Advanced Veterinary Research*, *14*(1), 130–134.

Issara, U., Chotiko, A., Phothisuwan, S., Kraithong, S., & Rungraeng, N. (2024). Banana peels: Identification, quantification, and characterization. In *Banana peels valorization* (pp. 1–35). Elsevier.

Iwuozor, K. O., Micheal, K. T., Micheal, T. T., Emenike, E. C., Onwuegbusi, C. C., Umoren, L. A., & Adeniyi, A. G. (2024). Sugarcane-based chemicals and its applications. In *Value addition and product diversification in sugarcane* (pp. 125–150). Springer.

Jie, X., Yao, Z., Wang, C., Qiu, D., Chen, Y., Zhang, Y., Ma, B., & Gao, W. (2022). Progress in waste lead paste recycling technology from spent lead-acid battery in China. *Journal of Sustainable Metallurgy*, *8*(3), 978–993.

Jimenez-Champi, D., Romero-Orejon, F. L., Moran-Reyes, A., Muñoz, A. M., & Ramos-Escudero, F. (2023). Bioactive compounds in potato peels, extraction methods, and their applications in the food industry: A review. *Cyta-Journal of Food*, *21*(1), 418–432.

Kulkarni, A. A., & Bhanage, B. M. (2014). Ag@ AgCl nanomaterial synthesis using sugar cane juice and its application in degradation of azo dyes. *ACS Sustainable Chemistry & Engineering*, *2*(4), 1007–1013.

Kumar, R., Kaushal, S., Kumar, P., Kumar, V., Thakur, N., & Kumar, K. (2025). Industrial waste-derived carbon nanomaterials. In *Waste derived carbon nanomaterials* (Vol. 1, pp. 179–201). ACS Publications.

Kumar, V., Bhat, S. A., & Verma, P. (2025). Development of smart nanomaterials from tea waste and their applications: Recent advances, challenges, and perspectives. *Smart Nanomaterials for Environmental Applications*, 677–701.

Li, B., Zhang, Y., Xu, J., Mei, Y., Fan, S., & Xu, H. (2021). Effect of carbonization methods on the properties of tea waste biochars and their application in tetracycline removal from aqueous solutions. *Chemosphere*, *267*, 129283.

Liu, J., Wang, X., Zhao, C., Hao, J., & Fang, G. (2018). Fabrication of porous covalent organic frameworks as selective and advanced adsorbents for the on-line preconcentration of trace elements against the complex sample matrix. *Journal of Hazardous Materials*, *344*, 220–229. https://doi.org/10.1016/j.jhazmat.2017.10.013

Lo, S. H., Raja, D. S., Chen, C.-W., Kang, Y.-H., Chen, J.-J., & Lin, C.-H. (2016). Waste polyethylene terephthalate (PET) materials as sustainable precursors for the synthesis of nanoporous MOFs, MIL-47, MIL-53 (Cr, Al, Ga) and MIL-101 (Cr). *Dalton Transactions*, *45*(23), 9565–9573.

Lv, X. W., Lin, J., Zhang, X. D., Huang, Q. R., Sun, X., Fan, E. S., Chen, R. J., Wu, F., & Li, L. (2024). Selective recycling of spent lithium-ion batteries enables toward aqueous Zn-ion batteries cathode. *Advanced Energy Materials*, *14*(48), 2402560.

Magagula, S. I., Lebelo, K., Motloung, T. M., Mokhena, T. C., & Mochane, M. J. (2023). Recent advances on waste tires: Bibliometric analysis, processes, and waste management approaches. *Environmental Science and Pollution Research*, *30*(56), 118213–118245.

Mahal, E., & Al-Mutlaq, S. (2023). Bio-Based carbon nanomaterials synthesis from waste tea. *Eurasian Chemical Communications*, *5*(3), 264–270.

Malpani, S. K., Meena, R. K., Katara, S., & Goyal, D. (2022). Agro-industrial wastes-derived carbon nanomaterials: Synthesis and multi-faceted applications. In *Innovations in green nanoscience and nanotechnology* (pp. 227–254). CRC Press.

Maroufi, S., Mayyas, M., & Sahajwalla, V. (2017). Nano-carbons from waste tyre rubber: An insight into structure and morphology. *Waste Management*, *69*, 110–116.

Mathura, S., Altmayer, J., Rügamer, T., Donia, N., Cavelius, C., & Kondratiuk, J. (2018). *Nanostructured surfaces through chemical nanotechnology: Opportunities and applications*.

Matović, V., Buha, A., Bulat, Z., & Đukić-Ćosić, D. (2011). Cadmium toxicity revisited: Focus on oxidative stress induction and interactions with zinc and magnesium. *Arhiv Za Higijenu Rada i Toksikologiju*, *62*(1), 65–75.

Meraj, U., Laiq, E., Qurtulen, Bushra, R., Ahmad, M., & Gupta, S. (2024). Enhanced adsorptive removal of methylene blue dye from aqueous media by potato peel-derived carbon quantum dots and their application in cell imaging. *Biomass Conversion and Biorefinery*, 1–18.

Mis-Fernández, R., Rios-Soberanis, C. R., Arenas-Alatorre, J., & Azamar-Barrios, J. A. (2012). Synthesis of carbon nanostructures from residual solids waste tires. *Journal of Applied Polymer Science*, *123*(4), 1960–1967.

Mod, B., Baskar, A. V., Bahadur, R., Tavakkoli, E., Van Zwieten, L., Singh, G., & Vinu, A. (2024). From cane to nano: Advanced nanomaterials derived from sugarcane products with insights into their synthesis and applications. *Science and Technology of Advanced Materials*, *25*(1), 2393568.

Munoz-Sandoval, E. (2014). Trends in nanoscience, nanotechnology, and carbon nanotubes: A bibliometric approach. *Journal of Nanoparticle Research*, *16*(1), 2152.

Mwafy, E. A. (2020). Eco-friendly approach for the synthesis of MWCNTs from waste tires via chemical vapor deposition. *Environmental Nanotechnology, Monitoring & Management*, *14*, 100342.

Nasar, A. (2023). Utilization of tea wastes for the removal of toxic dyes from polluted water—a review. *Biomass Conversion and Biorefinery*, *13*(2), 1399–1415.

Navarrete, A., Herrero, M., Martín, A., Cocero, M. J., & Ibáñez, E. (2011). Valorization of solid wastes from essential oil industry. *Journal of Food Engineering*, *104*(2), 196–201.

Negrão, D. R., & Driemeier, C. (2022). Fate of silica phytoliths in the industrial crushing of sugarcane stalks. *Industrial Crops and Products, 185*, 115132.

Ni'mah, Y. L., Muhaiminah, Z. H., & Suprapto. (2023). Synthesis of silica nanoparticles from sugarcane bagasse by sol-gel method. *AIP Conference Proceedings, 2540*(1), 50011.

Nille, O. S., Patil, A. S., Waghmare, R. D., Naik, V. M., Gunjal, D. B., Kolekar, G. B., & Gore, A. H. (2021). Valorization of tea waste for multifaceted applications: A step toward green and sustainable development. In *Valorization of agri-food wastes and by-products* (pp. 219–236). Elsevier.

Ondiek, J. K. O. (2016). *Synthesis and characterization of iron nanoparticles using banana peels extracts and their application in aptasensor.* University of Nairobi.

Orsuwan, A., Shankar, S., Wang, L.-F., Sothornvit, R., & Rhim, J.-W. (2016). Preparation of antimicrobial agar/banana powder blend films reinforced with silver nanoparticles. *Food Hydrocolloids, 60*, 476–485.

Osman, A. I., Blewitt, J., Abu-Dahrieh, J. K., Farrell, C., Al-Muhtaseb, A. A. H., Harrison, J., & Rooney, D. W. (2019). Production and characterisation of activated carbon and carbon nanotubes from potato peel waste and their application in heavy metal removal. *Environmental Science and Pollution Research, 26*(36), 37228-37241.

Owoeye, S. S., Oji, B., & Aderiye, J. (2015). Effect of temperature, time and atmospheric condition on active silica extraction from corn cob ash. *International Journal of Engineering, Technology and Innovation, 2*, 1–5.

Padhan, S. R., Rathore, S. S., Jat, S. L., Saini, S., Mohanty, A., Panigrahi, K. K., Ranjan, S., Sow, S., Mishra, P., & Baral, K. (2025). Synthesis of carbon nanomaterials from agro-industrial wastes and their extensive applications. In *Waste-derived carbon nanostructures: Synthesis and applications* (pp. 71–106). Springer.

Parkhe, V. S., Patil, T. P., & Tiwari, A. P. (2023). Biowaste-mediated green synthesis of gold nanoparticles using Solanum tuberosum peel extract for antibacterial, antioxidant, and photocatalytic applications. *Nanotechnology for Environmental Engineering, 8*(4), 1067–1081.

Parveen, S., Bukhari, N., Ramzan, N., Musaddiq, S., Mubarak, S., Kalsoom, A., & Qureshi, W. A. (2024). Green synthesis of value-added fluorescent carbon dots from sugarcane bagasse and their antioxidant, reducing and catalytic activities. *Sugar Tech, 26*(1), 33–44.

Pereira, M. A. F., Cesca, K., Poletto, P., & de Oliveira, D. (2021). New perspectives for banana peel polysaccharides and their conversion to oligosaccharides. *Food Research International, 149*, 110706.

Periakaruppan, R., Chen, X., Thangaraj, K., Jeyaraj, A., Nguyen, H. H., Yu, Y., Hu, S., Lu, L., & Li, X. (2021). Utilization of tea resources with the production of superparamagnetic biogenic iron oxide nanoparticles and an assessment of their antioxidant activities. *Journal of Cleaner Production, 278*, 123962.

Pieła, A., Żymańczyk-Duda, E., Brzezińska-Rodak, M., Duda, M., Grzesiak, J., Saeid, A., Mironiuk, M., & Klimek-Ochab, M. (2020). Biogenic synthesis of silica nanoparticles from corn cobs husks: Dependence of the productivity on the method of raw material processing. *Bioorganic Chemistry, 99*, 103773.

Poonsawat, T., Techalertmanee, T., Chumkaeo, P., Yunita, I., Meechai, T., Namkajorn, M., Pornsuwan, S., & Somsook, E. (2019). Facile synthesis of high performance iron oxide/carbon nanocatalysts derived from the calcination of ferrocenium for the decomposition of methylene blue. *Catalysts, 9*(11), 948.

Raigond, P., Raigond, B., Kochhar, T., Sood, A., & Singh, B. (2018). Conversion of potato starch and peel waste to high value nanocrystals. *Potato Research, 61*(4), 341–351.

Rajput, D., Paul, S., & Gupta, A. (2020). Green synthesis of silver nanoparticles using waste tea Leaves. *Advanced Nano Research, 3*(1), 1–14.

Roselló, J., Soriano, L., Santamarina, M. P., Akasaki, J. L., Melges, J. L. P., & Payá, J. (2015). Microscopy characterization of silica-rich agrowastes to be used in cement binders: Bamboo and sugarcane leaves. *Microscopy and Microanalysis, 21*(5), 1314–1326.

Salami, B. A., Oyehan, T. A., Gambo, Y., Badmus, S. O., Tanimu, G., Adamu, S., Lateef, S. A., & Saleh, T. A. (2022). Technological trends in nanosilica synthesis and utilization in advanced treatment of water and wastewater. *Environmental Science and Pollution Research, 29*(28), 42560–42600.

September, L. A., Kheswa, N., Seroka, N. S., & Khotseng, L. (2023). Green synthesis of silica and silicon from agricultural residue sugarcane bagasse ash—a mini review. *RSC Advances, 13*(2), 1370–1380.

Seroka, N. S., Taziwa, R. T., & Khotseng, L. (2022). Extraction and synthesis of silicon nanoparticles (SiNPs) from sugarcane bagasse ash: A mini-review. *Applied Sciences, 12*(5), 2310.

Seth, D., Athparia, M., Singh, A., Rathore, D., Venkatramanan, V., Channashettar, V., Prasad, S., Maddirala, S., Sevda, S., & Kataki, R. (2025). Sustainable environmental practices of tea waste—a comprehensive review. *Environmental Science and Pollution Research, 32*(12), 7449–7467.

Setianingsih, T. (2025). A review: Coffee and tea potential materials for carboneous material synthesis. *Egyptian Journal of Chemistry, 68*(7), 1–16.

Shange, M. G., Mohomane, S. M., & Motaung, T. E. (2025). Effect of synthesized silica on the properties of sugarcane bagasse cellulose/silica nanocomposites via sol-gel processing. *Scientific Reports, 15*(1), 26198.

Sharaf Addin, A. H., Ramli, R. M., Sufian, S., Al-Mahbashi, N., Idris, S. S., Farooqi, A. S., & Shamsu Bakri, S. H. (2025). Sustainable synthesis of carbon nanomaterials from waste tire-derived char using microwave pyrolysis. *Journal of Ecological Engineering, 26*(6).

Sim, L. C., Wong, J. L., Hak, C. H., Tai, J. Y., Leong, K. H., & Saravanan, P. (2018). Sugarcane juice derived carbon dot–graphitic carbon nitride composites for bisphenol A degradation under sunlight irradiation. *Beilstein Journal of Nanotechnology, 9*(1), 353–363.

Sulaiman, F., Adawiyah, Y., Santoso, M. I., & Maryani, Y. (2022). Synthesis and characterization of silica gel from corn cob skin and cob waste. In *2nd International Conference for Smart Agriculture, Food, and Environment (ICSAFE 2021)* (pp. 55–62).

Tawo, O. E., & Mbamalu, M. I. (2025). Advancing waste valorization techniques for sustainable industrial operations and improved environmental safety. *International Journal of Scientific Research and Archives, 14*(02), 127–149.

Tejaswini, M., Pathak, P., & Gupta, D. K. (2022). Sustainable approach for valorization of solid wastes as a secondary resource through urban mining. *Journal of Environmental Management, 319*, 115727.

Thambiraj, S., & Shankaran, R. (2016). Green synthesis of highly fluorescent carbon quantum dots from sugarcane bagasse pulp. *Applied Surface Science, 390*, 435–443.

Thangaraj, B., Mumtaz, F., Abbas, Y., Anjum, D. H., Solomon, P. R., & Hassan, J. (2023). Synthesis of graphene oxide from sugarcane dry leaves by two-stage pyrolysis. *Molecules, 28*(8), 3329.

Thathola, P., Adhikari, P., Dhyani, V., & Chandola, D. (2024). Fundamental scope of nanomaterial synthesis from wastes. In *Green and sustainable approaches using wastes for the production of multifunctional nanomaterials* (pp. 35–48). Elsevier.

Tsang, H.-H. (2013). Uses of scrap rubber tires. *Rubber: Types, Properties and Uses*, 477–491.

Tuck, C. O., Pérez, E., Horváth, I. T., Sheldon, R. A., & Poliakoff, M. (2012). Valorization of biomass: Deriving more value from waste. *Science, 337*(6095), 695–699.

Turgut, P., & Yesilata, B. (2008). Physico-mechanical and thermal performances of newly developed rubber-added bricks. *Energy and Buildings*, *40*(5), 679–688.

Tursunov, O. (2014). A comparison of catalysts zeolite and calcined dolomite for gas production from pyrolysis of municipal solid waste (MSW). *Ecological Engineering*, *69*, 237–243.

Vaibhav, V., Vijayalakshmi, U., & Roopan, S. M. (2015). Agricultural waste as a source for the production of silica nanoparticles. *Spectrochimica Acta Part A: Molecular and Biomolecular Spectroscopy*, *139*, 515–520.

Van der Kuijp, T. J., Huang, L., & Cherry, C. R. (2013). Health hazards of China's lead-acid battery industry: A review of its market drivers, production processes, and health impacts. *Environmental Health*, *12*(1), 61.

Vescovo, D., Manetti, C., Ruggieri, R., Spizzirri, U. G., Aiello, F., Martuscelli, M., & Restuccia, D. (2025). The valorization of potato peels as a functional ingredient in the food industry: A comprehensive review. *Foods*, *14*(8), 1333.

Vieira, I. R. S., da Silva, A. A., da Silva, B. D., Neto, L. T., Tessaro, L., Furtado, C. R. G., de Sousa, A. M. F., Carvalho, N. M. F., & Conte-Junior, C. A. (2024). Eco-friendly synthesis of ZnO nanomaterial from green tea extract: Photocatalytic, antibacterial and antioxidant potential. *Biomass Conversion and Biorefinery*, *14*(19), 24317–24331.

Weerakoon, W., Udawala, R., & Sewvandi, G. A. (2024). Calcium oxide nanoparticles from waste chicken eggshells as a fertilizer in agriculture. In *2024 Moratuwa Engineering Research Conference (MERCon)* (pp. 336–341).

Wong, S., Lee, Y., Ngadi, N., Inuwa, I. M., & Mohamed, N. B. (2018). Synthesis of activated carbon from spent tea leaves for aspirin removal. *Chinese Journal of Chemical Engineering*, *26*(5), 1003–1011.

Wu, L., Xue, D., Wen, R., Jiang, X., Jiao, L., & Chen, H. (2024). Production of activated carbon from tea. In *Recent advances in activated carbon* (pp. 13–26). CRC Press.

Yagmur, E., Ozmak, M., & Aktas, Z. (2008). A novel method for production of activated carbon from waste tea by chemical activation with microwave energy. *Fuel*, *87*(15–16), 3278–3285.

Yang, T., Xie, B., Liu, W., Zhang, D., & Chen, L. (2020). An environment-friendly process of lead recovery from spent lead paste. *Separation and Purification Technology*, *233*, 116035.

Yong, B. C., Abdul Raman, A. A., Buthiyappan, A., & Zainal Abidin, M. I. I. (2020). Synthesis and characterization of sugarcane bagasse cellulose-capped silver nanoparticle using ultrasonic irradiation for the adsorption of heavy metal. *Asia-Pacific Journal of Chemical Engineering*, *15*(3), e2433.

Zahid, M. U., Pervaiz, E., Hussain, A., Shahzad, M. I., & Niazi, M. B. K. (2018). Synthesis of carbon nanomaterials from different pyrolysis techniques: A review. *Materials Research Express*, *5*(5), 52002.

Zhang, Y., Chen, Y., Kang, Z.-W., Gao, X., Zeng, X., Liu, M., & Yang, D.-P. (2021). Waste eggshell membrane-assisted synthesis of magnetic CuFe2O4 nanomaterials with multifunctional properties (adsorptive, catalytic, antibacterial) for water remediation. *Colloids and Surfaces A: Physicochemical and Engineering Aspects*, *612*, 125874.

Zheng, W., Yang, D., Zeng, X., Liu, Q., Wang, C., Wu, X., Li, M., Liu, Y., Chen, L., & Weng, W. (2024). Utilization of the waste aqueous phase from tea residue hydrothermal carbonization for preparing active food packaging films. *Food Chemistry*, *448*, 139141.

Zhou, C., Deng, D., Huang, C., Xie, Y., Dong, H., Wen, Y., Li, B., Zhou, Z., Luo, W., & Zhou, Z. (2024). Catalytic co-pyrolysis of waste tea residue and waste plastics to carbon nanomaterials: Catalyst support, reaction temperature and product application. *Journal of Analytical and Applied Pyrolysis*, *177*, 106323.

Zhou, X., Deng, J., Yang, R., Zhou, D., Fang, C., He, X., Wang, D., Lei, W., Hu, J., & Li, Y. (2020). Facile preparation and characterization of fibrous carbon nanomaterial from waste polyethylene terephthalate. *Waste Management*, *107*, 172–181.

6 Industrial-upscale synthesis of waste-derived nanomaterials

6.1 INTRODUCTION

The synthesis of nanomaterials from waste resources has gained significant momentum in recent years, primarily because it aligns with principles of sustainability, the circular economy, and green chemistry (Abady et al., 2025; Zanbili & Poursattar Marjani, 2025). Traditional nanomaterial synthesis often relies on expensive precursors, energy-intensive processes, and, in some cases, toxic reagents, all of which raise concerns about cost and environmental impact (Zanbili & Poursattar Marjani, 2025). By contrast, waste-derived synthesis offers a double advantage: it provides a sustainable route for managing ever-increasing waste volumes while simultaneously generating high-value nanomaterials with functional properties (Bassey & Bassey, 2024; Lizundia et al., 2022). A wide variety of waste streams have been investigated for this purpose, including agricultural residues rich in biogenic silica and carbon, industrial by-products such as fly ash and slag containing valuable metal oxides and electronic waste that serves as a secondary source of precious and rare earth metals (Bhavan et al., 2023; Branca et al., 2020; Maseko et al., 2021).

At the laboratory-scale, these feedstocks have been successfully converted into a wide range of nanomaterials, including carbon nanostructures, metal nanoparticles, metal oxides, and nanocomposites. These materials exhibit significant potential in applications such as energy storage, catalysis, environmental remediation and biomedicine, as comprehensively reviewed by Abdelbasir et al. (2020). However, only a few synthesis processes have been scaled to industrial levels, as the leap from the lab to large-scale production introduces new complexities in reproducibility, safety, and environmental footprint (Hachhach et al., 2025). The pressing question, therefore, is how to translate these promising results into industrial-scale operations capable of meeting global demand for nanomaterials cost-effectively and sustainably.

Scaling from laboratory to industry requires far more than simply increasing the batch size of an established synthesis process. Industrial translation necessitates a comprehensive re-evaluation of process design that incorporates principles from chemical engineering, materials science, and environmental management. Factors such as reactor configuration, mass and heat transfer, process increase and scalability of input–output ratios must be carefully considered to ensure that upscaling does not compromise product quality. Liu and Meng (2021) highlighted the challenges that scale-up efforts must confront, including mixing, fouling, and maintaining uniform nanoparticle properties across batches. Furthermore, as Hussain (2022) reports, some material properties can degrade or shift when transitioning from

DOI: 10.1201/9781003745846-6

nanoscale control to meso- or macroscale, thereby reducing consistency. Energy efficiency becomes critical at this stage, as methods that are feasible on a gram or milligram scale may become prohibitively costly or unsustainable when extended to kilograms of production. Thus, the industrial production of waste-derived nanomaterials requires not only the optimisation of chemical processes but also the integration of chemical engineering, process intensification, energy and exergy analysis, and LCA to ensure cost-effectiveness, sustainability, and regulatory compliance (Hachhach et al., 2025).

This chapter builds on the foundational discussions presented in earlier sections of the book, particularly the coverage of waste-derived nanomaterials in Chapter 4. While Chapter 4 emphasised the types of nanomaterials obtainable from waste and their laboratory synthesis methods, the present chapter focuses on the next crucial step: industrial upscaling. The aim is not to revisit synthesis routes already described, but rather to critically analyse how these processes can be adapted, optimised and scaled up for larger applications. In doing so, this chapter highlights recent progress, milestone achievements, and key challenges that define the transition of waste-to-nanomaterial technologies from proof-of-concept experiments to industrial reality.

6.2 PRINCIPLES OF UPSCALING IN NANOMATERIAL SYNTHESIS

As outlined in the introduction, scaling nanomaterial synthesis from laboratory to industrial production extends beyond increasing batch size. It requires systematic application of chemical engineering principles to translate optimised small-scale reactions into efficient, reproducible, and sustainable large-scale processes.

6.2.1 Process intensification and reactor design

Process intensification focuses on improving production efficiency by optimising mass and heat transfer, reducing reaction times and minimising energy and resource use (Alcántara Avila et al., 2021; Srivastava et al., 2024). Reactor design is central to achieving these objectives because it dictates how effectively reactants mix, how temperature and concentration gradients are controlled, and how reaction kinetics scale with volume (Hakke et al., 2021). In nanomaterial synthesis, uniform mixing and precise temperature control are crucial for regulating nucleation and growth processes, which, in turn, determine particle size, morphology, and crystallinity (Gimondi et al., 2023).

Continuous stirred-tank reactors (CSTRs) are widely used due to their ability to maintain steady-state conditions and achieve homogeneous mixing, making them suitable for liquid-phase syntheses, such as sol-gel or co-precipitation methods (Cherkasov et al., 2023; Gkogkos et al., 2024), as shown in Figure 6.1. Tubular reactors provide precise control over residence time and are well-suited for continuous flow processes that demand efficient heat and mass transfer, such as hydrothermal and solvothermal syntheses (Li et al., 2022). Fluidised-bed reactors, on the other hand, facilitate excellent gas–solid contact and temperature uniformity, making them advantageous for producing nanostructured catalysts or carbon nanomaterials (Chen et al., 2024). Microreactor systems have also attracted attention for their

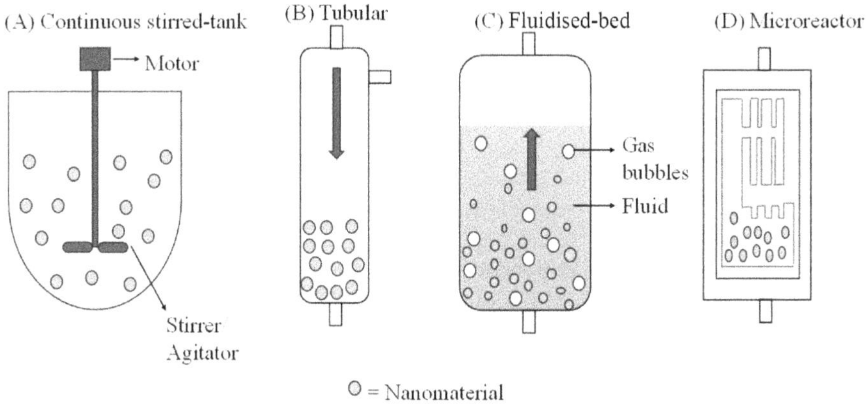

FIGURE 6.1 Comparison between conventional and intensified reactor systems used in nanomaterial synthesis. The conventional (A) continuous stirred-tank reactor (CSTR) relies on mechanical stirring to ensure mixing, while intensified systems such as (B) tubular, (C) fluidised-bed, and (D) microreactors enhance process performance through improved mass and heat transfer, precise control of residence time and superior temperature uniformity.

high surface-to-volume ratios, which enable exceptional heat and mass transfer rates, enabling precise control over nanoparticle formation and consistent product quality (Hakke et al., 2021).

Beyond reactor choice, process intensification can also involve techniques such as microwave- or ultrasound-assisted synthesis, supercritical fluid processing, and plasma-enhanced reactions. These methods shorten reaction times, enhance precursor conversion, and often improve product purity (Gerbec et al., 2005; Low et al., 2021; Tabasum et al., 2025). However, scaling these approaches to industrial levels requires addressing challenges such as optimising energy input, modifying equipment design, and implementing effective safety management. Integrating process analytical technologies (PAT) for real-time monitoring and control further enhances reproducibility and scalability, allowing for dynamic adjustments during synthesis to maintain consistent nanomaterial properties (Gerzon et al., 2022; VandenBerg et al., 2025).

Effective reactor design and process intensification are not only about maximising quantity but also about achieving consistent product quality and sustainability. By combining reaction engineering, thermodynamic modelling, and kinetic optimisation, researchers can develop scalable nanomaterial production systems that maintain the precision of laboratory synthesis while meeting industrial demands for cost efficiency and environmental compliance.

6.2.2 Batch versus continuous flow production

Scaling up the synthesis of waste-derived nanomaterials typically involves either batch or continuous flow production methods. Batch processing, widely used in laboratory protocols, is straightforward to implement and offers flexibility in adjusting reaction conditions, making it suitable for heterogeneous waste feedstocks, such as agricultural residues or electronic waste. For example, sol-gel synthesis of silica nanoparticles from rice husk ash has been demonstrated as an effective route to produce high-purity biogenic silica when feedstock pretreatment and sol-gel parameters are carefully controlled (Shrestha et al., 2023). However, batch methods often face challenges in reproducibility, energy efficiency, and throughput at larger scales. This is due to variations in mixing, heat transfer, or precursor quality that can produce significant batch-to-batch variability in particle size, morphology, and crystallinity, necessitating optimised, reproducible protocols for scale-up (Alvarez, 2023; Andresen et al., 2023).

In contrast, continuous flow production often uses tubular or microreactors, provides superior control over reaction parameters, enhanced heat and mass transfer, and seamless integration into automated production lines. These systems are particularly advantageous for nanomaterials that require precise size and morphology control, such as biomass-derived carbon nanotubes or metal oxide nanoparticles obtained from industrial slag, due to their ability to maintain consistent residence times and reaction conditions (Hakke et al., 2021; Li et al., 2022). Compared to batch methods, continuous hydrothermal flow reactors can deliver a greater than 15% increase in nanoparticle output (Roberts et al., 2017; Li et al., 2023). While continuous hydrothermal systems can enhance efficiency, their cost-effectiveness depends on sustained long-term operation. Economic assessments comparing batch and continuous synthesis of iron oxide nanoparticles indicate that continuous reactors can lower production costs, primarily through reduced labour requirements, as heating and cooling stages become redundant. However, achieving this advantage requires a robust, fouling-resistant system capable of stable, uninterrupted operation (Besenhard et al., 2020).

The choice between batch and continuous flow methods depends on the waste feedstock, desired nanomaterial properties, and economic considerations, such as equipment costs and production scale. Batch processing may be preferred for low-volume, high-value applications, such as specialised nanocomposites, while continuous flow systems are better suited for high-throughput production of nanomaterials (Darr et al., 2017; Besenhard et al., 2020). Figure 6.2 provides a comparative overview of batch and continuous flow synthesis approaches for nanomaterial production. The figure indicates key trade-offs between the two methods in terms of scalability, reproducibility, and resource efficiency. As shown, batch processing generally offers lower scalability and variable reproducibility, making it suitable for small-scale or high-value applications.

6.3 ENERGY, WATER, AND RAW MATERIAL CONSIDERATIONS

Scaling up the synthesis of waste-derived nanomaterials requires careful management of energy, water, and raw material inputs to ensure cost-effectiveness and environmental sustainability. Energy demands often increase significantly at industrial scales, particularly for processes such as pyrolysis, hydrothermal synthesis, and calcination, which convert waste feedstocks into nanomaterials. For example, the hydrothermal or pyrolytic processing of agricultural residues to produce carbon

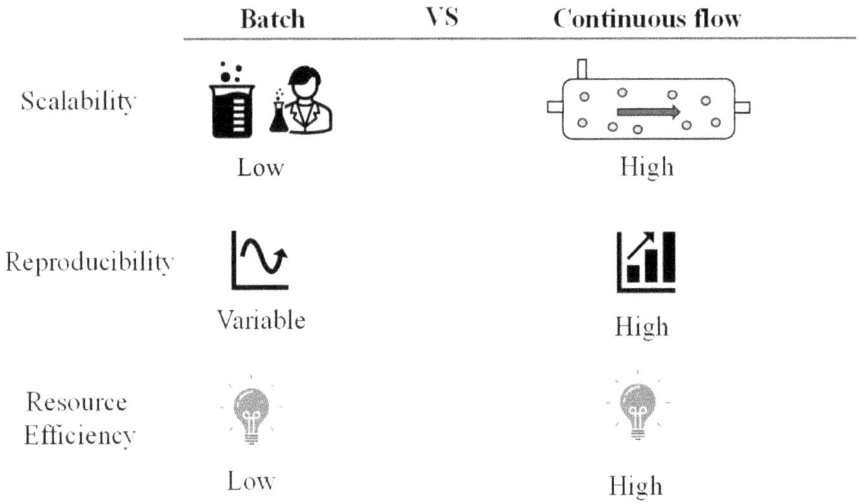

FIGURE 6.2 Comparison of batch and continuous flow synthesis methods for nanomaterial production.

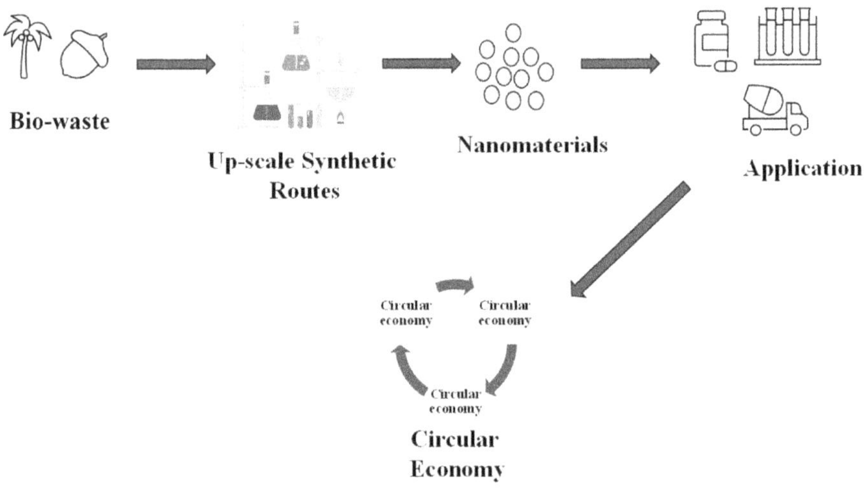

FIGURE 6.3 Schematic illustration of the valorisation of bio-waste into nanomaterials through scalable synthetic routes and their subsequent applications within a circular economy framework.

nanomaterials can consume substantial energy, necessitating strategies such as heat recovery or renewable energy integration to reduce costs and environmental impact (Lesch et al., 2025). Techniques such as microwave-assisted synthesis (MAS) have demonstrated substantially lower energy consumption (80% less) than conventional thermal methods (Sajini & Joseph, 2025).

Water usage is often a critical concern in hydrothermal or precipitation methods for producing metal oxide nanoparticles from industrial by-products, such as fly ash or slag. These processes may require large volumes of water per kilogram of product, particularly for washing or reaction media, though precise values vary between systems. In several studies on colloidal nanoparticle synthesis, solvent recovery (e.g., by distillation) has demonstrably reduced solvent-related resource use (Wrasman et al., 2022). While specific data on water recycling for waste-derived nanoparticle production are scarce or unavailable, these distillation techniques suggest that closed-loop or recycling approaches could significantly reduce water consumption (potentially by tens of percent).

The variability of raw waste feedstocks poses a significant challenge to achieving consistent nanomaterial quality. Feedstocks such as rice husk ash and coal fly ash exhibit compositional fluctuations that can directly affect reaction kinetics, particle morphology, and functional properties. For instance, the silica content in rice husk ash can vary significantly depending on combustion conditions, influencing its reactivity and suitability for nanoparticle synthesis (Rasoul, 2018). Similarly, coal fly ash from different sources shows variable metal oxide content, which can affect the characteristics of nanoparticles derived from it (Singh et al., 2022).

To mitigate these inconsistencies, preprocessing steps such as grinding, chemical purification, and blending are commonly employed. Grinding reduces particle size and increases reactivity, while chemical purification removes impurities that could hinder synthesis. Blending feedstocks from multiple sources may stabilise metal oxide content, ensuring more uniform nanoparticle properties. While this approach may be effective, it also increases operational complexity and costs, underscoring the need for robust quality control and real-time monitoring systems.

6.4 INDUSTRIAL APPROACHES FOR UPSCALING WASTE-DERIVED NANOMATERIALS

The industrial production of waste-derived nanomaterials necessitates adapting laboratory-scale synthesis methods to achieve high yields, consistent quality, and cost-effectiveness, while prioritising environmental sustainability. This involves optimising reaction conditions, equipment and process workflows to accommodate diverse waste feedstocks, such as agricultural residues, industrial by-products, and electronic waste. Several approaches have been developed, including thermal, hydrothermal, chemical, biological, and hybrid methods, each tailored to specific nanomaterials and waste streams.

6.4.1 Thermal and pyrolytic methods

Thermal treatments, such as pyrolysis, calcination, and combustion, are widely used to convert agricultural and industrial waste into carbon-based or metal oxide nanomaterials. Industrial-scale systems employ rotary kilns, fluidised-bed reactors, or furnaces to ensure uniform temperature distribution and high throughput. For example, pyrolysis of rice husk ash in rotary kilns produces silica nanoparticles with yields of up to 75% at scales exceeding 500 kg/day (Subbiah et al., 2025). Process optimisation focuses on controlling residence time, gas flow, and heating profiles to achieve the desired particle size, crystallinity, and surface properties. Energy recovery strategies, such as utilising off-gases for preheating, can reduce energy consumption by 20% (Be-CIS, 2023). However, challenges persist, including managing high-energy demands and ensuring feedstock consistency to maintain uniform nanomaterial properties.

6.4.2 Hydrothermal and solvothermal approaches

Hydrothermal and solvothermal methods are effective for producing metal oxides, sulfides, and composite nanomaterials with controlled morphology from waste streams, such as fly ash or slag (Crouzet et al., 2017; Yadav et al., 2023). Examples include the hydrothermal conversion of fly ash into zeolites and silica/alumina fractions, as well as the hydrothermal or solvothermal production of iron oxide (Fe_2O_3/Fe_3O_4)-based composites from waste precursors (Panek et al., 2021; Besenhard et al., 2020). Industrial-scale setups utilise high-pressure autoclaves or continuous flow reactors to withstand elevated temperatures and pressures. Hydrothermal conversion of fly ash into zeolites has been shown at sub-pilot scale (Wdowin et al., 2014). Reactor designs prioritise enhanced heat transfer and uniform mixing to maintain reproducibility, while solvent recycling systems reduce water and solvent use by up to 50% (Dunne et al., 2015; Freitas et al., 2023). Despite these advancements, scale-up remains a challenge due to the high capital costs of pressure-rated equipment (made of stainless steel) and the need for automated, real-time process control to ensure reproducible product quality (Besenhard et al., 2020).

6.4.3 Chemical precipitation and sol-gel methods

Chemical precipitation and sol-gel techniques are commonly applied to produce metal oxide and composite nanomaterials from waste feedstocks. Reactors such as well-stirred batch reactors with controlled pH or continuous stirred-tank reactors (CSTRs) are often used to improve mixing and pH control, which are critical for achieving uniform particle size distributions. For example, sol-gel synthesis of silica nanoparticles from rice husk ash has yielded average particle sizes of 10–90 nm when appropriate acid washing (e.g., HCl, H_2SO_4) and alkaline extraction steps are used (Shrestha et al., 2023). There are several pilot- and sub-pilot demonstrations of sol-gel and continuous extraction routes that convert waste feedstocks into silica and other nanomaterials. For example, Park et al. (2023) report a pilot-scale continuous one-pot alkali-hydrothermal process for extracting biogenic silica from rice husk with production in the kg/day range and measured energy metrics per kg of silica. Key challenges that remain include removing by-products, recovering reagents, and maintaining homogeneity during scale-up. Table 6.1 shows a few examples of larger-scale or pilot-scale nanomaterial production from waste feedstocks.

TABLE 6.1
Examples of larger-scale or pilot-scale nanomaterial production from waste feedstocks.

Feedstock/\Waste Source	Nanomaterial Produced	Method/Reactor Type	Reported Scale	Key Findings/Features	Reference
Rice husk	Silica (SiO_2)	Continuous one-pot alkali-hydrothermal + ball milling process	Pilot-scale; 20–60 kg/day	Achieved ~86% silica extraction yield; continuous process integration reduces batch steps and energy use	Park et al. (2023)
Rice husk ash	Silica nanoparticles	Sol-gel	Semi-pilot; few kg/day	Produced silica; shows technical feasibility of scaling the sol-gel route for agro-waste feedstock	Hidayat & Setyawan (2023)
Fly ash	Zeolite-type nanomaterials	Hydrothermal (continuous sub-pilot)	10 kg of feed per batch	Scalable to industrial level	Wdowin et al. (2014)
Colloidal silica precursors (synthetic)	Silica nanoparticles (<100 nm)	3D-printed multistage CSTR cascade	Continuous lab/pilot scale	Continuous control of particle size; scalable stirred-tank concept applicable to waste-derived sol-gel feeds	Lignos et al. (2020)

6.4.4 Biological and green synthesis approaches

Biological synthesis, utilising microorganisms, plant extracts, or enzymes, provides an environmentally friendly approach for producing metal and metal oxide nanoparticles. Bioreactors must maintain optimal growth conditions, a sufficient nutrient supply, and efficient oxygen transfer to ensure uniform biological activity. Some studies have demonstrated silver nanoparticle biosynthesis from industrial waste sources (e.g., electroplating effluent, electronic waste leachates) using microbes or plant extracts, with particle sizes ranging from ~50–100 nm and good antimicrobial activity (*Pseudomonas spp., Raoultella planticola*) (Punjabi et al., 2017; Qurbani et al., 2022). Though reported yields are typically in the Gram scale under laboratory or small-pilot conditions, continuous bioreactor setups and immobilised-cell systems are cited as promising ways to scale up production and reduce variation. Like other methods, scaling biological methods requires significant investment in bioreactor design and strict process control to maintain efficiency and reproducibility.

6.4.5 Hybrid and advanced techniques

Hybrid approaches, such as combining microwave-assisted heating with hydrothermal synthesis or plasma treatment with chemical precipitation, enhance control over particle size, morphology, and functionality (Freitas et al., 2023). For example, the microwave-assisted hydrothermal synthesis of CDs from biomass waste was achieved, yielding an average size of approximately 3.3 nm and a fluorescence quantum yield of about 9.56% (George et al., 2024). Industrial adoption requires addressing challenges such as energy optimisation and safety protocols for plasma or microwave systems. These methods show promise for the moderate-scale production of high-value nanomaterials, such as metal nanoparticles that can be utilised in various applications (Darwish et al., 2022), including energy storage or catalysis, but require further development for large-scale implementation.

6.5 ACHIEVEMENTS AND MILESTONES

Over the past decade, the upscaling of waste-derived nanomaterials has progressed beyond laboratory demonstrations, with several notable achievements at pilot and industrial scales. These milestones highlight both the feasibility and the potential of converting waste streams into high-value nanomaterials, while also providing valuable lessons for overcoming scale-up challenges.

6.5.1 Pilot-scale demonstrations

Pilot-scale systems have served as a crucial bridge between laboratory research and industrial adoption of waste-derived nanomaterials. Biomass-based carbon nanomaterials have been produced in fluidised-bed pyrolysis units with capacities ranging from tens to hundreds of kilograms per day, demonstrating scalability and consistent structural properties. Similarly, the hydrothermal carbonisation of agricultural residues in semi-continuous reactors has yielded CDs and mesoporous carbons with properties comparable to those of their lab-scale counterparts.

Recent pilot-scale studies have shown the feasibility of converting agricultural residues, such as rice husk, into high-purity silica nanoparticles. Park et al. (2023) reported a continuous biogenic silica extraction process utilising alkali-hydrothermal treatment and ball milling, achieving production rates of approximately 2–4 kg/day and extraction yields of up to 86% with stable, long-term operation. Comparable results were observed in a pilot plant study in Egypt, where spherical silica nanoparticles with diameters ranging from 18 to 75 nm were synthesised from rice husk ash using sol-gel and chemical precipitation methods (Hafez et al., 2024). These pilot-scale studies demonstrate the potential of integrating waste valorisation technologies into semi-industrial operations.

6.5.2 Industrial implementation and economic viability

The industrial-scale translation of waste-to-nanomaterial processes remains promising, but several emerging cases and programmatic efforts indicate a growing commercial interest. Fly ash, red mud, and other industrial by-products are increasingly targeted as feedstocks for large-scale extraction of aluminosilicate nanoparticles, iron oxides, and functional composite materials (Yadav & Fulekar, 2020; Samal, 2021). Review studies highlight that fly ash and red mud can serve as a reliable source of silica, alumina, and ferrous species convertible to nanoparticulate forms via chemical, biological, or hybrid routes (Yadav & Fulekar, 2020; Samal, 2021)

In the field of geopolymers and aluminosilicate materials, waste-derived systems have already achieved industrial-level deployment in construction and functional material markets. For instance, fly ash–red mud blends are being used in large-scale geopolymer production, where nanocrystalline zeolitic and aluminosilicate gel phases contribute to strength and durability (Koshy et al., 2019; Shi et al., 2020). While these are not direct nanoparticle production plants, they illustrate how nanoscale domains in waste-derived materials are being commercialised.

Although fully industrial facilities producing pure waste-derived nanoparticles (e.g., silica, metal oxides) remain rare or unreported in the open literature, the demonstrated use of waste-based aluminosilicate and geopolymeric technologies provides proof of concept. The economic motivation is compelling: substituting waste feedstocks for virgin materials can reduce raw material costs and enhance the circularity of value chains. Still, a key gap is that many of these industrial implementations produce composite or amorphous materials, for which explicit nanoparticle yields, purities, and performance metrics are not clearly characterised in published reports.

6.5.3 Comparative efficiency and performance

Comparisons between laboratory, pilot, and industrial-scale production of waste-derived nanomaterials reveal valuable insights into efficiency, product quality, and economic performance. At the laboratory-scale, yields and product purities are typically optimised under tightly controlled conditions; however, energy use and batch-to-batch variability often hinder scalability. In contrast, pilot-scale demonstrations, such as continuous silica extraction from rice husk via alkali-hydrothermal treatment, have demonstrated stable operation with extraction yields of up to 86% and daily production rates of 2–4 kg (Park et al., 2023). Similarly, the Egyptian pilot plant for agro-waste valorisation achieved reproducible production of 18–75 nm silica nanoparticles, confirming that process control and waste pretreatment can ensure consistent nanoparticle quality (Hafez et al., 2024).

Economic assessments from both pilot and patent-based studies indicate that continuous systems can reduce production costs by 30%–40% compared to conventional sol-gel or batch-based silica synthesis routes, largely due to shorter cycle times, reduced labour requirements, and partial water and solvent recycling (Mukunda et al., 2003; Nzereogu et al., 2023). Industrial implementations using waste-derived feedstocks, such as fly ash or red mud, further demonstrate cost advantages by replacing virgin silica or alumina precursors (Yadav & Fulekar, 2020). However, these gains may be offset by higher initial equipment costs and the need for advanced process monitoring and automation.

Performance-wise, products synthesised from waste feedstocks have been shown to meet or exceed industrial standards in surface area, crystallinity, and functional performance, particularly in catalysis and adsorbent applications (Park et al., 2023; Yadav & Fulekar, 2020). Nevertheless, maintaining consistency at higher scales requires improved feedstock characterisation, automated quality control, and integrated recovery systems to minimise process waste and energy consumption. Overall, comparative data suggests that while pilot- and industrial-scale systems may exhibit slightly lower yields or purities, their economic and environmental efficiencies provide strong justification for continued industrial upscaling of waste-derived nanomaterials.

6.6 CHALLENGES AND FUTURE PERSPECTIVES

Although waste-derived nanomaterial production has demonstrated considerable progress at pilot and industrial scales, several challenges still hinder its full-scale commercialisation. These challenges span technical, economic, and regulatory domains, emphasising the need for integrated research and policy approaches to achieve sustainable implementation.

6.6.1 Technical and Process Challenges

A major technical limitation to scaling up waste-derived nanomaterial synthesis is feedstock variability. Industrial and agricultural wastes differ significantly in chemical composition and impurity content, leading to inconsistent reaction kinetics and product properties (Yadav & Fulekar, 2020; Nzereogu et al., 2023). This heterogeneity may complicate the standardisation of processes and quality control. Additionally, fouling and clogging in continuous flow and hydrothermal systems at low flow rates remain problematic, particularly when handling high-solid-content feedstocks (Besenhard et al., 2023).

Another significant barrier is energy efficiency. Many thermal or hydrothermal conversion routes still require high-energy inputs for heating, drying, and maintaining pressure. While innovations such as microwave-assisted and supercritical synthesis have shown promise in reducing energy demand, their industrial integration remains limited by scalability and reactor cost (Crouzet et al., 2017; Priecel & Lopez-Sanchez, 2018). Continuous monitoring tools, such as online spectroscopy and AI-based process control, could mitigate these challenges by ensuring real-time optimisation of reaction conditions and minimising batch variability.

6.6.2 Economic and infrastructural constraints

From an economic standpoint, the capital and operational costs of scaling up remain substantial. Pressure-resistant autoclaves, bioreactors, and separation systems represent major financial investments that can be difficult to justify when market demand for waste-derived nanomaterials is still emerging (Hafez et al., 2024). Fragmented integration between waste producers and nanomaterial manufacturers further leads to inconsistent feedstock availability and transportation costs that can offset the sustainability gains. Overcoming these barriers requires reducing costs through economies of scale, improving supply chain coordination, developing harmonised regulatory frameworks, and promoting the dissemination of education and information (Solomon et al., 2024).

6.6.3 Environmental and regulatory considerations

While waste valorisation offers clear environmental benefits, comprehensive LCAs for many emerging waste-derived nanoparticle technologies remain incomplete. Numerous studies on engineered nanomaterials lack full coverage of the use and end-of-life phases, limited inventory data, and sparse quantification of emissions, nanoparticle release, or waste toxicity (Miseljic & Olsen, 2014; Nizam et al., 2021). Furthermore, regulatory frameworks for manufactured nanomaterials are still evolving in many regions. Organisations such as the Organisation for Economic Co-operation and Development (OECD) have published guidance and decision frameworks; however, many jurisdictions lack harmonised safety standards, testing protocols, or transparent reporting mechanisms for nanomaterials derived from waste feedstocks (OECD, 2019). Developing harmonised safety standards, standardised testing protocols, and transparent data-sharing mechanisms among academia, industry, and regulators will be crucial to ensuring the safe deployment of waste-based nanotechnologies.

6.6.4 Future research directions

Future research should prioritise the integration of green synthesis techniques, including microwave-assisted, enzymatic, and electrochemical methods into scalable reactor configurations to reduce energy intensity. In addition, AI-driven process modelling and digital twins can enhance scale-up prediction accuracy by correlating feedstock properties with nanoparticle yield and quality. Lastly, cross-sectoral collaboration between waste management, materials science, and chemical engineering disciplines will be essential to transition from isolated pilot projects to circular, industrial-scale production ecosystems.

6.7 CONCLUSION

The upscaling of waste-derived nanomaterials represents a pivotal step towards integrating circular economy principles within the materials and manufacturing sectors. Over the past decade, significant progress has been made, with several pilot and industrial-scale demonstrations proving that diverse waste streams, ranging from agricultural residues to metallurgical by-products, can serve as viable precursors for functional nanomaterials. These achievements underscore the increasing feasibility of converting environmental liabilities into high-value technological products, thereby simultaneously promoting waste reduction, resource efficiency, and economic sustainability.

However, widespread industrial adoption remains constrained by persistent challenges in process consistency, feedstock heterogeneity, and energy optimisation. Bridging the gap between laboratory innovation and full-scale deployment will require sustained collaboration between academia, industry, and policymakers. Advances in process intensification, automation, and digital control, alongside robust life-cycle and techno-economic assessments, are crucial to ensure that the environmental gains of waste valorisation are not offset by new inefficiencies or emissions.

REFERENCES

Abady, M. M., Mohammed, D. M., Soliman, T. N., Shalaby, R. A., & Sakr, F. A. (2025). Sustainable synthesis of nanomaterials using different renewable sources. *Bulletin of the National Research Centre*, *49*(1), 24.

Abdelbasir, S. M., McCourt, K. M., Lee, C. M., & Vanegas, D. C. (2020). Waste-derived nanoparticles: Synthesis approaches, environmental applications and sustainability considerations. *Frontiers in Chemistry*, *8*, 782.

Alcántara Avila, J. R., Kong, Z. Y., Lee, H. Y., & Sunarso, J. (2021). Advancements in optimisation and control techniques for intensifying processes. *Processes*, *9*(12), 2150.

Alvarez, M. (2023). Challenges in scaling up nanomaterial synthesis for industry: Technical, economic and environmental perspectives. *American Journal of Materials Engineering*, *4*(3), 12–18.

Andresen, E., Islam, F., Prinz, C., Gehrmann, P., Licha, K., Roik, J., Recknagel, S., & Resch-Genger, U. (2023). Assessing the reproducibility and up-scaling of the synthesis of Er, Yb-doped NaYF4-based upconverting nanoparticles and control of size, morphology and optical properties. *Scientific Reports*, *13*(1), 2288.

Bassey, K. E., & Bassey, K. E. (2024). From waste to wonder: Developing engineered nanomaterials for multifaceted applications. *GSC Advanced Research and Reviews*, *20*(3), 109–123.

Be-CIS. (2023). *Top 5 benefits of waste heat recovery for industrial businesses*. Retrieved October 11, 2025, from https://be-cis.com/top-5-benefits-of-waste-heat-recovery-for-industrial-businesses/

Besenhard, M. O., LaGrow, A. P., Famiani, S., Pucciarelli, M., Lettieri, P., Thanh, N. T. K., & Gavriilidis, A. (2020). Continuous production of iron oxide nanoparticles via fast and economical high temperature synthesis. *Reaction Chemistry & Engineering*, *5*(8), 1474–1483.

Besenhard, M. O., Pal, S., Gkogkos, G., & Gavriilidis, A. (2023). Non-fouling flow reactors for nanomaterial synthesis. *Reaction Chemistry & Engineering*, *8*(5), 955–977.

Bhavan, J. S., Joy, J., & Pazhani, A. (2023). Identification and recovery of rare earth elements from electronic waste: Material characterization and recovery strategies. *Materials Today Communications*, *36*, 106921.

Branca, T. A., Colla, V., Algermissen, D., Granbom, H., Martini, U., Morillon, A., Pietruck, R., & Rosendahl, S. (2020). Reuse and recycling of by-products in the steel sector: Recent achievements paving the way to circular economy and industrial symbiosis in Europe. *Metals*, *10*(3), 345.

Chen, S., Jiang, Y., Zhu, Z., Zhang, Q., Zhang, C., Zhang, Q., Qian, W., Zhang, S., & Wei, F. (2024). Fluidization and application of carbon nano agglomerations. *Advanced Science*, *11*(8), 2306355.

Cherkasov, N., Adams, S. J., Bainbridge, E. G., & Thornton, J. A. (2023). Continuous stirred tank reactors in fine chemical synthesis for efficient mixing, solids-handling and rapid scale-up. *Reaction Chemistry & Engineering*, *8*(2), 266–277.

Crouzet, C., Brunet, F., Recham, N., Auzende, A. L., Findling, N., Magnin, V., Ferrasse, J. H., & Goffé, B. (2017). Hydrothermal steel slag valorisation—Part II: Hydrogen and nano-magnetite production. *Frontiers in Earth Science*, *5*, 86.

Darr, J. A., Zhang, J., Makwana, N. M., & Weng, X. (2017). Continuous hydrothermal synthesis of inorganic nanoparticles: Applications and future directions. *Chemical Reviews*, *117*(17), 11125–11238.

Darwish, M., Mafla-Gonzalez, C., Kolenovic, B., Deremer, A., Centeno, D., Liu, T., Kim, D. Y., Cattabiani, T., Drwiega, T. J., Kumar, I., & Li, C. (2022). Rapid synthesis of metal nanoparticles using low-temperature, low-pressure argon plasma chemistry and self-assembly. *Green Chemistry*, *24*(20), 8142–8154.

Dunne, P. W., Munn, A. S., Starkey, C. L., Huddle, T. A., & Lester, E. H. (2015). Continuous-flow hydrothermal synthesis for the production of inorganic nanomaterials. *Philosophical Transactions of the Royal Society A: Mathematical, Physical and Engineering Sciences*, *373*(2057), 20150015.

Freitas, D. C., Mazali, I. O., Sigoli, F. A., da Silva Francischini, D., & Arruda, M. A. Z. (2023). The microwave-assisted synthesis of silica nanoparticles and their applications in a soy plant culture. *RSC Aadvances*, *13*(39), 27648–27656.

George, H. S., Selvaraj, H., Ilangovan, A., Almutairi, B. O., Sheit, M. K., & Sathishkumar, K. (2024). Microwave-assisted one-pot synthesis of carbon dots from biomass: Fe (III) ion sensing and in-situ methylene blue reduction catalytic activity. *Desalination and Water Treatment*, *320*, 100624.

Gerbec, J. A., Magana, D., Washington, A., & Strouse, G. F. (2005). Microwave-enhanced reaction rates for nanoparticle synthesis. *Journal of the American Chemical Society*, *127*(45), 15791–15800.

Gerzon, G., Sheng, Y., & Kirkitadze, M. (2022). Process analytical technologies—advances in bioprocess integration and future perspectives. *Journal of Pharmaceutical and Biomedical Analysis*, *207*, 114379.

Gimondi, S., Ferreira, H., Reis, R. L., & Neves, N. M. (2023). Microfluidic devices: A tool for nanoparticle synthesis and performance evaluation. *ACS Nano*, *17*(15), 14205–14228.

Gkogkos, G., Kahil, E. E., Storozhuk, L., Thanh, N. T. K., & Gavriilidis, A. (2024). Scaling study of miniaturised continuous stirred tank reactors via residence time distribution analysis and application in the production of iron oxide nanoparticles. *Chemical Engineering and Processing-Process Intensification*, *203*, 109880.

Hachhach, M., Bayou, S., El Kasmi, A., Saidi, M. Z., Akram, H., Hanafi, M., Achak, O., El Moujahid, C., & Chafik, T. (2025). Towards sustainable scaling-up of nanomaterials fabrication: Current situation, challenges and future perspectives. *Eng*, *6*(7), 149.

Hafez, A. I., Khedr, M., Ali, H. M., Sabry, R. M., & Gadalla, H. I. (2024). Pilot plant experiments for extraction of inorganic products from agro solid waste and cost estimation of the products. *Egyptian Journal of Chemistry*, *67*(1), 239–253.

Hakke, V., Sonawane, S., Anandan, S., Sonawane, S., & Ashokkumar, M. (2021). Process intensification approach using microreactors for synthesizing nanomaterials—A critical review. *Nanomaterials*, *11*(1), 98.

Hidayat, T., & Setyawan, N. (2023, May). Semi pilot scale study on production of silica from rice husk ash. In *IOP Conference Series: Earth and Environmental Science* (Vol. 1172, No. 1, p. 012049). IOP Publishing.

Hussain, A. (2022, May 18). The challenges behind scaling up nanomaterials. *AZoNano*. Retrieved October 8, 2025, from https://www.azonano.com/article.aspx?ArticleID=6126

Koshy, N., Dondrob, K., Hu, L., Wen, Q., & Meegoda, J. N. (2019). Mechanical properties of geopolymers synthesized from fly ash and red mud under ambient conditions. *Crystals*, *9*(11), 572.

Lesch, R., Visser, E. D., Seroka, N. S., & Khotseng, L. (2025). Biomass-derived carbon-based nanomaterials: Current research, trends and challenges. *Journal of Renewable Materials*, *13*, 1935–1977.

Li, M., Wang, W., & Xu, C. (2023). High-throughput synthesis of nanoparticles using oscillating feedback microreactors: A selective scaling-out strategy. *RSC Advances*, *13*(50), 35435–35444.

Li, Q., Liu, L., Wang, Z., & Wang, X. (2022). Continuous hydrothermal flow synthesis and characterization of ZrO2 nanoparticles doped with CeO2 in supercritical water. *Nanomaterials*, *12*(4), 668.

Lignos, I., Ow, H., Lopez, J. P., McCollum, D. A., Zhang, H., Imbrogno, J., Shen, Y., Chang, S., Wang, W., & Jensen, K. F. (2020). Continuous multistage synthesis and functionalization of sub-100 nm silica nanoparticles in 3D-printed continuous stirred-tank reactor cascades. *ACS Applied Materials & Interfaces*, *12*(5), 6699–6706.

Liu, X., & Meng, H. (2021). Consideration for the scale-up manufacture of nanotherapeutics—A critical step for technology transfer. *View*, *2*(5), 20200190.

Lizundia, E., Luzi, F., & Puglia, D. (2022). Organic waste valorisation towards circular and sustainable biocomposites. *Green Chemistry*, *24*(14), 5429–5459.

Low, D. Y. S., Mahendra, C. K., Supramaniam, J., Tan, L. T. H., Lee, L. H., Manickam, S., Goh, B. H., Tan, K. W., & Tang, S. Y. (2021). Ultrasound-enhanced biosynthesis of uniform ZnO nanorice using Swietenia macrophylla seed extract and its in vitro anticancer activity. *Nanotechnology Reviews*, *10*(1), 572–585.

Maseko, N. N., Schneider, D., Wassersleben, S., Enke, D., Iwarere, S. A., Pocock, J., & Stark, A. (2021). The production of biogenic silica from different South African agricultural residues through a thermo-chemical treatment method. *Sustainability*, *13*(2), 577.

Miseljic, M., & Olsen, S. I. (2014). Life-cycle assessment of engineered nanomaterials: A literature review of assessment status. *Journal of Nanoparticle Research*, *16*(6), 2427.

Mukunda, H. S., Dasappa, S., Paul, P. J., Rajan, N. K. S., & Subbukrishna, D. N. (2003). *A novel process and apparatus for the manufacture of precipitated silica from rice husk ash*. Indian Patent No, 216477. https://patents.google.com/patent/WO2004073600A2/en

Nizam, N. U. M., Hanafiah, M. M., & Woon, K. S. (2021). A content review of life cycle assessment of nanomaterials: Current practices, challenges and future prospects. *Nanomaterials*, *11*(12), 3324.

Nzereogu, P. U., Omah, A. D., Ezema, F. I., Iwuoha, E. I., & Nwanya, A. C. (2023). Silica extraction from rice husk: Comprehensive review and applications. *Hybrid Advances*, *4*, 100111.

OECD. (2019). *Physical-chemical decision framework to inform decisions for risk assessment of manufactured nanomaterials*. Organisation for Economic Co-operation and Development, Paris. Retrieved October 15, 2025, from https://www.oecd.org/en/publications/physical-chemical-decision-framework-to-inform-decisions-for-risk-assessment-of-manufactured-nanomaterials_1fbbaf8c-en.html

Panek, R., Madej, J., Bandura, L., & Słowik, G. (2021). Recycling of waste solution after hydrothermal conversion of fly ash on a semi-technical scale for zeolite synthesis. *Materials*, *14*(6), 1413.

Park, J. Y., Gu, Y. M., Chun, J., Sang, B. I., & Lee, J. H. (2023). Pilot-scale continuous biogenic silica extraction from rice husk by one-pot alkali hydrothermal treatment and ball milling. *Chemical and Biological Technologies in Agriculture*, *10*(1), 102.

Priecel, P., & Lopez-Sanchez, J. A. (2018). Advantages and limitations of microwave reactors: From chemical synthesis to the catalytic valorisation of biobased chemicals. *ACS Sustainable Chemistry & Engineering*, *7*(1), 3–21.

Punjabi, K., Yedurkar, S., Doshi, S., Deshapnde, S., & Vaidya, S. (2017). Biosynthesis of silver nanoparticles by Pseudomonas spp. isolated from effluent of an electroplating industry. *IET Nanobiotechnology*, *11*(5), 584–590.

Qurbani, K., Hussein, S., Hamzah, H., Sulaiman, S., Pirot, R., Motevaseli, E., & Azizi, Z. (2022). Synthesis of silver nanoparticles by Raoultella planticola and their potential antibacterial activity against multidrug-resistant isolates. *Iranian Journal of Biotechnology*, *20*(4), p.e3121.

Rasoul, B. I. (2018). *Effect of incineration temperatures to time on the rice husk ash (RHA) silica structure: A comparative study to the literature with experimental work*. School of Environment & Technology, University of Brighton, UK.

Roberts, E. J., Habas, S. E., Wang, L., Ruddy, D. A., White, E. A., Baddour, F. G., Griffin, M. B., Schaidle, J. A., Malmstadt, N., & Brutchey, R. L. (2017). High-throughput continuous flow synthesis of nickel nanoparticles for the catalytic hydrodeoxygenation of guaiacol. *ACS Sustainable Chemistry & Engineering*, *5*(1), 632–639.

Sajini, T., & Joseph, J. (2025). Microwave-assisted synthesis of nanomaterials: A green chemistry perspective and sustainability assessment. *RSC Sustainability*, *3*(11), 4911–4935.

Samal, S. (2021). Utilization of red mud as a source for metal ions—a review. *Materials*, *14*(9), 2211.

Shi, Y., Zhang, Z., Sang, Z., & Zhao, Q. (2020). Microstructure and composition of red mud-fly ash-based geopolymers incorporating carbide slag. *Frontiers in Materials*, *7*, 563233.

Shrestha, D., Nayaju, T., Kandel, M. R., Pradhananga, R. R., Park, C. H., & Kim, C. S. (2023). Rice husk-derived mesoporous biogenic silica nanoparticles for gravity chromatography. *Heliyon*, *9*(4).

Singh, N. B., Agarwal, A., De, A., & Singh, P. (2022). Coal fly ash: An emerging material for water remediation. *International Journal of Coal Science & Technology*, *9*(1), 44.

Solomon, N. O., Simpa, P., Adenekan, O. A., & Obasi, S. C. (2024). Sustainable nanomaterials' role in green supply chains and environmental sustainability. *Engineering Science & Technology Journal*, *5*(5), 1678–1694.

Srivastava, S., Pandey, V. K., Fatima, A., Pandey, S., Singh, R., Dar, A. H., & Dhillon, B. (2024). A literature review on process intensification: An innovative and sustainable food processing method. *Applied Food Research*, *4*(1), 100363.

Subbiah, G., Singh, R. P., Deepak, K., Nayak, P. P., Tiwari, A., & Sahoo, J. (2025). Continuous pyrolysis of rice husk for sustainable biochar production and carbon sequestration: Recent advances and techno-economic perspectives. *Results in Engineering*, 106991.

Tabasum, N., Saeed, A., Shafiq, R., Khan, B. S., Bashir, M., Yousaf, M., . . . & Siddiqui, M. R. H. (2025). Synthesis of iron oxide nanoparticles via atmospheric pressure microplasma for high-performance energy storage and environmental applications. *Catalysts*, *15*(5), 444.

VandenBerg, M. A., Dong, X., Smith, W. C., Tian, G., Stephens, O., O'Connor, T. F., & Xu, X. (2025). Learning from the future: Towards continuous manufacturing of nanomaterials. *AAPS Open*, *11*(1), 7.

Wdowin, M., Franus, M., Panek, R., Badura, L., & Franus, W. (2014). The conversion technology of fly ash into zeolites. *Clean Technologies and Environmental Policy*, *16*(6), 1217–1223.

Wrasman, C. J., Zhou, C., Aitbekova, A., Goodman, E. D., & Cargnello, M. (2022). Recycling of solvent allows for multiple rounds of reproducible nanoparticle synthesis. *Journal of the American Chemical Society*, *144*(26), 11646–11655.

Yadav, V. K., & Fulekar, M. H. (2020). Advances in methods for recovery of ferrous, alumina and silica nanoparticles from fly ash waste. *Ceramics*, *3*(3), 384–420.

Yadav, V. K., Amari, A., Gacem, A., Elboughdiri, N., Eltayeb, L. B., & Fulekar, M. H. (2023). Treatment of fly-ash-contaminated wastewater loaded with heavy metals by using fly-ash-synthesized iron oxide nanoparticles. *Water*, *15*(5), 908.

Zanbili, F., & Poursattar Marjani, A. (2025). Innovative green and bio-based approaches for photosensitive nanoparticle synthesis: A review on methodologies, characterization and applications. *Micro and Nano Systems Letters*, *13*(1), 3.

7 Preconcentration of potentially toxic metals and organics

7.1 INTRODUCTION

Potential toxic metals and organic compounds are known as emerging pollutants. These chemicals have recently been recognised for their potential to cause significant environmental and health impacts (Bell et al., 2011; Deblonde et al., 2011). They include a variety of compounds, such as antibiotics, drugs, steroids, endocrine disruptors, hormones, industrial additives, chemicals, microbeads, toxic metals, and microplastics (Demnerová et al., 2014; Vasilachi et al., 2021). These pollutants can be found in various sources, including hospitals, industries, agriculture, and households. They pose risks to aquatic life and human health due to their potential toxicity (Arman et al., 2021; Falandysz et al., 2024).

The presence of substantial amounts of these pollutants in environmental compartments has become a major issue and is currently having a serious negative impact on human health and the environment (Wilkinson et al., 2017). To date, several preconcentration methods have been employed for the determination, extraction, and quantification of these pollutants in various water matrices. Generally, in analytical chemistry, preconcentration is conducted with an analytical instrument, as it can indicate the effectiveness of the preconcentration method (Pérez-Fernández et al., 2017; L. Yang et al., 2017). Various techniques, such as GC coupled with different detectors, HPLC, and liquid chromatography-mass spectrometry (LC-MS), enzyme-linked immunosorbent assays (ELISAs), ICP-OES, AAS, and capillary electrophoresis (CE), among others, have been reported in the literature. These techniques have mostly displayed low detection limits, acceptable sensitivity, and selectivity, thereby indicating the effectiveness of the preconcentration methods (Azooz et al., 2024; Bian et al., 2023; Q. Yang et al., 2024). However, another crucial aspect in preconcentration is the adsorbent material. This is the process by which molecules, atoms, or ions adhere to a solid surface, forming a layer known as the adsorbate. This phenomenon occurs when substances from a fluid phase, such as gases or liquids, accumulate on the surface of a solid material, creating a film of the adsorbate on the adsorbent surface (X. Chen et al., 2022). At this stage, the adsorbent of choice traps or adsorbs most of the target analytes using various mechanisms. The adsorbed analytes can then be eluted using various organic solvents. Various adsorbents can be used for adsorption purposes. Researchers can synthesise and tailor adsorbents to specific applications. Some of the adsorbents can be obtained commercially. With the advent of green chemistry, researchers have explored the use of valorised waste

DOI: 10.1201/9781003745846-7

materials to synthesise adsorbents for the preconcentration of potentially toxic metals and organic compounds from various samples.

7.2 SAMPLE PREPARATION METHODS FOR PRECONCENTRATION OF TOXIC METALS AND ORGANIC COMPOUNDS

In analytical chemistry, sample preparation refers to the methods used to treat a sample prior to its analysis using an analytical instrument (Y. Chen et al., 2008; Murray & Cowe, 2004). Preparation is a crucial step in most analytical techniques, as the techniques are often not responsive to the analyte in its in-situ form, or the results are distorted by interfering species (Y. Chen et al., 2008). Sample preparation may involve dissolution, extraction, reaction with specific chemical species, pulverisation, treatment with a chelating agent (e.g., EDTA), masking, filtration, dilution, sub-sampling, or a variety of other techniques (Kole et al., 2011; Ramos, 2012).

The main purposes of the sample preparation step include the removal of interfering species, the extraction of analytes, and increasing the analyte's final signal by increasing its concentration (Nasiri et al., 2020). Extraction of the target analyte from relatively large sample volumes into a few microlitres of eluent results in analyte preconcentration. This process improves the precision and accuracy of results by bringing the analyte concentration into the dynamic range of the specific analytical instrumentation (Fumes et al., 2015). Liquid–liquid extraction (LLE) and solid-phase extraction (SPE), as well as their derivatives, have been used to design and optimise extraction and cleanup (preconcentration) steps. In this study, we will focus primarily on SPE methods. The basic principles and applications of the SPE preparation methods are discussed in the subsequent sections.

7.2.1 CONVENTIONAL SOLID-PHASE EXTRACTION (SPE)

SPE is one of the most widely used sample preparation methods for analysing environmental samples (Backe & Field, 2012; Câmara et al., 2022; Faraji et al., 2019). SPE is one of the preferred and most widely used sample pretreatment techniques over LLE. This is due to its attractive features, such as simplicity, flexibility, and ease of automation (Câmara et al., 2022). During the preconcentration process, an aqueous sample passes through an immobilised phase and is afterwards extracted by suitable organic solvents. However, limited efficiency due to insufficient retention, resulting in low breakthrough volumes, can be observed for highly polar analytes (Kole et al., 2011). Nevertheless, one strategy to overcome breakthrough volume in SPE is to change the adsorbent type, and SPE allows the use of various adsorbents. Some sorbents can be synthesised and functionalised according to their intended usage.

The most used design in SPE is the polypropylene cartridge with a packed sorbent phase. In SPE, the analytes to be extracted are partitioned between a solid phase (sorbent) and a liquid–liquid phase (sample). These analytes must have a greater affinity for the solid phase than for the sample matrix. Generally, SPE consists of four steps: column preparation, sample loading, column post-wash, and sample desorption. The pre-wash step conditions the stationary phase. Additionally, the wash step removes undesirable substances. The target analytes are retained on the appropriate sorbent bed after washing out interfering compounds. The relevant elution solvents are then used to achieve full recovery of the target analyte.

The selection of an appropriate SPE extraction sorbent depends on understanding the mechanism(s) of interaction between the sorbent and the analyte of interest. Understanding depends on knowledge of the hydrophobic, polar, and ionogenic properties of both the solute and the sorbent (Murtada & Pawliszyn, 2023). The most common retention mechanisms in SPE are van der Waals forces ("non-polar interactions"), hydrogen bonding, dipole–dipole forces ("polar interactions"), and cation–anion interactions ("ionic interactions"). Each sorbent offers a unique combination of these properties, which can be applied to a wide variety of extractions (Augusto et al., 2013). Three general theory interactions exist: (i) Reversed phase involves a polar or moderately polar sample matrix (mobile phase) and a nonpolar stationary phase (Płotka-Wasylka et al., 2016). The analyte of interest is typically mid- to nonpolar. Retention of organic analytes from polar solutions (e.g., water) onto these SPE materials is due primarily to the attractive forces between the carbon–hydrogen bonds in the analyte and the functional groups on the sorbent surface. These nonpolar–nonpolar attractive forces are commonly referred to as London dispersion forces or van der Waals forces. The nonpolar solvent, which can disrupt the forces between the sorbent and compound, is used to elute an adsorbed compound from a reversed-phase SPE tube or disk (Spietelun et al., 2013). The following materials are used as reversed-phase media: carbon-based media, polymer-based media, and polymer-coated and bonded silica media. Carbon-based media consist of graphitic, non-porous carbon with a high affinity for organic polar and nonpolar compounds from both polar and nonpolar matrices (Murtada & Pawliszyn, 2023). Retention of analytes is based primarily on the analyte's structure, rather than on interactions of functional groups on the analyte with the sorbent surface. Polymer-based sorbents are styrene/divinylbenzene materials. It is used to retain hydrophobic compounds that contain some hydrophilic functionality, especially aromatics (Souza-Silva et al., 2015). Elution steps can be performed with mid- and nonpolar solvents because the polymeric packing is stable across almost all matrices. Polymer-coated and bonded silica media is a hydrophobic-bonded silica that is coated with a hydrophilic polymer. The pores in the polymer allow small, hydrophobic organic compounds of interest (e.g., drugs) to reach the bonded silica surface, while large interfering compounds (e.g., proteins) are shielded from the bonded silica by the polymer and flushed through the SPE tube (Abd Rahim et al., 2016).

(ii) Normal phase involves a polar analyte, a mid- to nonpolar matrix (e.g., acetone, chlorinated solvents, and hexane), and a polar stationary phase. Retention of an analyte under normal-phase conditions is primarily due to interactions between the analyte's polar functional groups and polar groups on the sorbent surface (Murtada & Pawliszyn, 2023; Souza-Silva et al., 2015). A compound adsorbed by these mechanisms is eluted by passing a solvent that disrupts the binding mechanism, usually a solvent that is more polar than the sample's matrix. The bonded silicas have short alkyl chains with polar functional groups bonded to the surface. These silicas, due to their polar functional groups, are significantly more hydrophilic compared to bonded reversed-phase silicas (Abd Rahim et al., 2016; Zulkifli et al., 2018). As with typical normal-phase silicas, these sorbents can be used to adsorb polar compounds from nonpolar matrices. The polar adsorption material is modified silica gel, commonly used as the base for all bonded phases. The functional groups that are involved in the adsorption of compounds from nonpolar matrices are the free hydroxyl groups on the surface of the silica particles. This method may be used to adsorb polar compounds from nonpolar matrices, with subsequent elution of the compounds in an organic solvent more polar than the original sample matrix (Wu et al., 2021).

(iii) Ion exchange SPE can be used for compounds that are in a solution. Anionic (negatively charged) compounds can be isolated on an aliphatic quaternary amine group that is bonded to the silica surface. Cationic (positively charged) compounds are isolated by using silica with aliphatic sulfonic acid groups that are bonded to the surface (Jagirani & Soylak, 2024). The primary retention mechanism of the compound is based mainly on the electrostatic attraction of the charged functional group in the compound to the charged group that is bonded to the silica surface. The significance of pH in SPE is crucial. Solutions used in SPE procedures have a very broad pH range. Silica-based packings typically have a stable pH range of 2–7.5. At pH levels above and below this range, the bonded phase can be hydrolysed and cleaved from the silica surface, or the silica itself can be dissolved. In SPE, however, the solutions are typically in contact with the sorbent for a short period. The fact that SPE cartridges are disposable and intended for single-use allows the use of any pH to optimise analyte retention or elution (Chromatographica et al., 2021; Tim et al., 2024).

Several studies from the literature have reported on the use of solid-phase extraction for the preconcentration of toxic metals and organic pollutants using valorised materials. These studies encompass both traditional and current solid-phase extraction methods. The primary difference between these methods is that traditional methods require the sorbent to be packed into a column.

In a study conducted by Leepipatpiboon, a combination of chelating heavy metals such as Ag(I), Cd(II), Cu(II), Ni(II), and Pb(II) with 4–(2-pyridylazo) resorcinol (PAR) and extraction on C18 cartridges prior to atomic absorption spectrometry was developed for trace metal analysis. Various factors, including pH, PAR concentration, C18 amount, and flow rate, were investigated to optimise the enrichment procedure. Water samples with an enrichment factor of 20 were analysed. The results obtained showed a precision of 1%–5% and a recovery of 81%–100%. The errors were in the range of 0.5%–0.7% at the $\mu g\ L^{-1}$ level (Leepipatpiboon, 1995).

In another study reported by Tuzen and co-workers, a multi-element preconcentration procedure for SPE using Chromosorb 108 as bathocuproinedisulfonic acid chelates, followed by flame atomic absorption spectrometric determinations of some heavy metal ions in environmental samples, is proposed in the present work. The influences of analytical parameters, including the pH of the aqueous solution, amounts of reagents, flow rates of sample and eluent solutions, sample volume, and others, on the quantitative recoveries of copper, cadmium, lead, zinc, manganese, iron, chromium, nickel, and cobalt ions were investigated. The effects of concomitant ions on analyte retention were also examined. The presented preconcentration procedure was applied to the determination of metal ions in reference standard materials (NRCC-SLRS 4 Riverine Water, SRM 1515 Apple leaves, and GBW 07605 Tea) and some real samples, including tap and river water, red wine, rice, black tea, and honey (Tuzen et al., 2005).

Jiao and Gao conducted on-site SPE, consisting of sorption, separation, and elution units, designed for in-situ preconcentration of heavy metal ions. The D401 resin powder was employed as a sorbent to capture Pb^{2+}, Cu^{2+}, Zn^{2+}, Cd^{2+}, Co^{2+}, and Ni^{2+}, and the sorbed species were desorbed with 2 mol L^{-1} nitric acid as the eluent. Under optimised conditions, these heavy metal ions in West Lake, Taihu Lake, and the Yangtze River of China were captured and then determined by ICP-OES, with recoveries ranging from 92.5% to 111.5%. The on-site SPE achieved rapid preconcentration of heavy metals, thereby eliminating the need to transport and store large volumes of water samples. It is suitable for monitoring of water quality in mountainous areas, tablelands, and other remote locations (Jiao & Gao, 2013).

Gómez also reported a study where a multi-residue analytical method has been developed and validated for determining a selection of 16 pharmaceuticals: the anti-epileptic carbamazepine, seven analgesic/anti-inflammatory drugs (mefenamic acid, indomethacine, ibuprofen, naproxen, diclofenac, ketorolac, and acetaminophen), the analgesic opiate codeine, two antidepressants (fluoxetine and paroxetine), β-blockers (atenolol and propranolol), antibiotic (trimethoprim, metronidazole, and erythromycin), and the anti-ulcer ranitidine in hospital effluent wastewaters. The method enables simultaneous extraction of pharmaceutical compounds using SPE with Waters Oasis HLB at pH 7. The analytes were then identified and quantitatively determined by liquid chromatography–tandem mass spectrometry (LC–MS–MS) using multiple reaction monitoring (MRM). Recovery rates for the pharmaceuticals were above 75%, except for ranitidine (45%), and the overall method variability was below 9%. The instrumental detection limit (IDL) ranged from 2 to 31 pg of injected material, and the method detection limit (LOD) ranged from 7 to 47 ng L^{-1} in spiked hospital effluent. The method precision, calculated as the relative standard deviation (RSD), ranged from 0.3% to 4.9%. A detailed study of the matrix effect was included in this work, examining signal suppression in effluent wastewater from a hospital complex. The developed analytical method was applied to preliminary data from hospital effluent wastewater samples (Gómez et al., 2006). In another report by Babić and colleagues (Babić et al., 2006), SPE-HPLC analysis of important veterinary pharmaceuticals from different classes in a highly complex wastewater matrix was conducted. The pharmaceuticals investigated included three sulfonamides (sulfamethazine, sulfadiazine, and sulfaguanidine), a sulfonamide synergist (trimethoprim), a tetracycline (oxytetracycline), a fluoroquinolone (enrofloxacin), and a β-lactam (penicillin G/procaine). The reported method involved preconcentration and cleanup by SPE using Oasis HLB extraction cartridges. The final analysis of the selected pharmaceutical compounds was performed using high-performance liquid chromatography (HPLC) coupled with a diode array detector (DAD). Recoveries ranged from 68.3% to 97.9% with a relative standard deviation of less than 8.4%. Only for sulfaguanidine was a low recovery obtained. The limits of quantification ranged from 1.5 to 100 μg L^{-1}, depending on the pharmaceutical. The described method was applied to the determination of pharmaceuticals in wastewater samples from the pharmaceutical industry.

7.2.2 Dispersive solid-phase extraction (DSPE) and magnetic solid-phase extraction (MSPE)

Dispersive solid-phase extraction (DSPE) and magnetic solid-phase extraction (MSPE) are innovative SPE technologies in which adsorbents of choice are added directly to the sample solution (Selahle et al., 2021). The column packing and conditioning steps are avoided. These methods result in simpler, rapid, and more efficient sample preparation procedures (Rasc et al., 2018). Furthermore, the attractive feature of DSPE and MSPE is the enhanced contact area between the adsorbent and target analytes (Selahle et al., 2021). DSPE has been widely used since its development around 2000 (Faraji et al., 2019) and has been successfully applied for extraction, isolation, and purification in the analytical analysis of a wide variety of veterinary drugs used in the livestock industry. DSPE simplifies SPE cleanup, allows more samples to be analysed at once, is quite rapid, and requires low solvent consumption (Badawy et al., 2022). DSPE involves adding a solid sorbent, usually silica or polymer-based, directly to the sample solution (Farajzadeh et al., 2021; Islas et al., 2017).

Waste materials are also used to synthesise various adsorbents. The dispersion process increases the contact area between the sorbent and the analyte. The sorbents employed in DSPE for the preconcentration of heavy metal and organic compound residues are solids chemically modified by the addition of several chemical compounds that modify their affinities (Büyüktiryaki et al., 2020). Valorised waste materials can also enhance selectivity for the analytes of interest, thereby maximising retention and minimising interferences in the analytical matrix (Ścigalski & Kosobucki, 2020). After dispersion, the sorbent is isolated by centrifugation or filtration. Once the solid phase is isolated, the analytes or interferences adsorbed on the sorbent surface can be easily eluted or removed by adding appropriate organic solvents. Table 7.1 summarises some reported dispersive SPE methods for the preconcentration of organic compounds and heavy metals from various matrices using valorised waste materials. As shown in the table, various adsorbents, including MOFs, activated carbons, and nanoparticles, can be synthesised from waste materials and used for the preconcentration of both heavy metals and organic compounds. These adsorbents yielded high percentage recoveries of the target analytes, ranging from 72% to 116%. Various analytical techniques were also reported, including AAS and FAAS, which are known to be the least sensitive analytical instruments.

To improve the performance of DSPE, magnetic or magnetically modified adsorbents have been introduced (Dimpe & Nomngongo, 2016; Selahle et al., 2021). The procedure is based on the use of magnetic or magnetisable adsorbents placed directly into a solution or suspension containing analytes, and the application of an external magnetic field arranged outside the extraction vessel for rapid separation of the sorbent with the adsorbed analytes from the crude sample matrix (Herrero-Latorre et al., 2015). According to many authors, this technique highly simplifies the SPE procedure, as it reduces the stage of centrifugation and enhances the extraction efficiency (Capriotti et al., 2019; Vasconcelos & Fernandes, 2017).

Magnetic separation, based on DSPE, is a convenient miniaturised preconcentration technique developed for the analysis of contaminants. MSPE is a noteworthy improvement of the DSPE technique, in which the extraction occurs when a more

TABLE 7.1
Some of the reported dispersive solid-phase extraction methods for the preconcentration of organic compounds and heavy metals from various matrices using waste-derived adsorbents.

Waste Material	Analytes	Matrix	%Recoveries	Detection method	Reference
Banana peelings	Cu and Pb	Wastewater treatment	-	ICP-OES	Seng (2018)
Microporous activated carbon from brewery waste	Sulfamethoxazole, trimethoprim, and ciprofloxacin	Wastewater	82–94	Micellar electrokinetic chromatography	Pouretedal and Sadegh (2014)
Sawdust	Cu, Ni, Fe, and Mn	Mine wastewater	98.31	FAAS	Stevancovic et al. (2011)
Avocado seed	Parabens	Environmental	>100	HPLC-DAD	Nephiphidi et al. (2025)
Activated carbon from lignocellulosic biomass	Fluoroquinolines antibiotics	River water	96.12–98.13	UV-Visible spectrophotometer	Ahmed and Theydan (2014)
Eggshells	Cr and Cd	Water samples	>99	FAAS	Park et al. (2007)
Activated carbon nanoparticles from vine wood	Amoxicillin, tetracyclines, and Cephalexin	Aqueous solutions	74–88	UV-Visible spectrophotometer	Pouretedal and Sadegh (2014)
Waste PET Plastic-derived UIO-66(Zr)-MOF	Penicillin antibiotics	Water samples	89.9–101	HPLC-DAD	Skans et al. (2025)

uniform sorbent dispersion is formed in a sample solution (Azzouz et al., 2018). Compared to SPE, MSPE provides a larger contact area between analytes and sorbents, is easier to perform, saves time, and reduces waste. Thanks to various synthetic strategies available, functionalised magnetic materials with high adsorption capabilities have enabled the development of efficient methods for determining heavy metals and organic compounds in environmental samples. The magnetic material is added to the sample, adsorbed, and then separated from the mixture using a magnet while the sample is decanted. A higher-affinity solvent will then be added to the vial to separate the analytes from the magnetic material. Magnetic materials can also be synthesised from waste materials.

Some researchers hypothesised that a particle of ferromagnetic material below a critical particle size (<15 nm) would consist of a single magnetic domain, that is, a particle in a state of uniform magnetisation at any field. Direct contact between the magnetic sorbent and the analytes is believed to facilitate selective adsorption of the compounds onto the solid surface (Mi et al., 2023). Due to the small particle size of magnetite nanoparticles (MNPs), they exhibit a high specific surface area, high sorption capacity, and rapid adsorption rates. As a consequence, smaller amounts of sorbent and shorter equilibrium times are required to extract analytes from large sample volumes (Wang et al., 2023). Moreover, high selectivity is expected through a retention mechanism that primarily relies on hydrogen bonding, dipole–dipole interactions, and π–π interactions (Mi et al., 2023). However, a size of several nanometres is often insufficient to isolate analytes from large volumes of samples, because pure iron oxides are prone to aggregation, resulting in a loss of magnetism (Pathak, 2024). In order to overcome these difficulties, the surface of the magnetic core (mainly iron, nickel, cobalt, and their oxides such as magnetite (Fe_3O_4) and maghemite (γ-Fe_3O_4)) (Mohapatra et al., 2019) is coated with suitable inorganic substances (e.g., silica, alumina, manganese oxide (IV). A suitable coating for the core increases the durability of the chemical sorbent and prevents its oxidation. Moreover, further modification of the inorganic or organic coating of the core by suitable functional groups enhances the sorption properties of the sorbent (Ta, 2012), particularly when inorganic substances such as silica are, directly or via a second step, functionalised with different molecules such as antibodies, specific receptors, metals, enzymes, and protein alkyl chains (Andrade-Eiroa et al., 2016).

One of the main advantages of magnetic solid separations is their speed (Pathak, 2024). As stated earlier, MNPs can efficiently extract analytes from large sample volumes after a short equilibrium time. In addition, coating MNPs with specific functionalities endows them with high selectivity towards target analytes. The main advantages of MNPs are that they can be recycled, usually after appropriate rinsing, and that the adsorbed analytes can be desorbed by procedures such as sonication (Tatarchuk et al., 2023). Other advantages of this procedure compared to other SPE methods include easier preparation, simpler sample handling, reduced solvent consumption, and increased extraction recovery (Mehta et al., 2015).

Although MSPE is considered the best preconcentration method, it has its own drawbacks. One disadvantage of magnetic sorbents is that the stationary phase and thermally unstable components can be partially or completely degraded during desorption at high temperatures. This leads to deterioration in the accuracy and precision of analysis (Shen et al., 2023). In fact, at pH < 4.0, degradation of the magnetite present in the magnetic adsorbent can take place, allowing the formation of chelates between Fe(III) and target analytes such as sulfonamides (Shen et al., 2023). On the other hand, at pH > 9.0, magnetite particles acquired a negative charge because of the binding of hydroxide groups, causing electrostatic repulsion between the adsorbent and anionic forms of target analytes (Mohapatra et al., 2019). This is slightly confusing: it is supposed that magnetite is both a magnet and an anion in motion. It is well known that an ion is a charged particle in motion, which behaves like a magnet when moving in the bulk of a stirred solution, creating a small magnetic field in its vicinity. Although the motion of ions in solutions is similar to the thermal motion of electrons in a conductor with no net current, the net magnetic field is cancelled out, leaving a very small magnetic field near each moving ion. Furthermore, interactions between magnetic fields and electrons or charged moving particles are only possible under certain circumstances ($v \neq 0$ and not parallel or antiparallel to the magnetic field created by the MNP) (Tatarchuk et al., 2023).

The only difference between the DSPE and MSPE is the fact that MSPE uses a magnetic sorbent and DSPE does not use a magnetic sorbent. The remaining steps in these preconcentration methods are identical.

Various applications of the MSPE method using waste materials have been reported in the literature for the preconcentration of both organic pollutants and heavy metals. This includes a study reported by Oliveira and co-workers, in which AMD was used to synthesise magnetic sorbents. AMD is one of the main environmental problems associated with mining activity, whether the mine is operational or abandoned. In their work, several precipitates from this mine drainage, generated by the oxidation of sulfide minerals, were utilised as magnetic adsorbents upon exposure to weathering. Such AMD precipitates from abandoned Portuguese mines (AGO, AGO-1, CF, and V9) were compared with two Moroccan raw materials (ClayMA and pyrophyllite) regarding their efficiency in wastewater treatment. The adsorption properties were studied by optimising various experimental factors, including adsorbent type, adsorbent mass, and dye concentration, using the Box–Behnken Design, with methylene blue (MB) and crystal violet (CV) as organic pollutants. The obtained kinetic data were analysed using the pseudo-first-order and pseudo-second-order equations, and the equilibrium adsorption data were examined using the Freundlich and Langmuir models. The adsorption behaviour of the different adsorbents was well-fitted by the pseudo-second-order kinetic model and the Langmuir isotherm. The most efficient adsorbent for both dyes was AGO-1, due to its cellulose content, with Q_{max} values of 40.5 and 16.0 mg g^{-1} for CV and MB, respectively. This study confirmed the feasibility of using AMD precipitates to adsorb organic pollutants from water, providing valuable information for developing future, affordable solutions to reduce waste associated with mining activities (Oliveira et al., 2024).

In another study, Etale and co-authors investigated the adsorptive removal of Cu(II), Mn(II), and U(VI) by maghemite nanoparticles (NPs) under AMD conditions to assess the potential of NPs for remediating AMD-contaminated water. This maghemite was found to be magnetic. The effects of time, NP, and metal concentration, as well as manganese and sulfate ions, were quantified at pH 3. Adsorption of all three ions was rapid, and equilibrium was attained in 5 min or less. 56% of

Cu, 53% of Mn, and 49% of U were adsorbed. In addition, adsorption efficiencies were enhanced by ≥10% in the presence of manganese and sulfate ions, although Cu sorption was reduced in 1:2 Cu-to-Mn solutions. Adsorption also increased with pH: 86% Cu, 62% Mn, and 77% U were removed from solution at pH 9 and higher initial metal concentrations. Increasing NP concentrations, however, did not always result in increased metal removal. Kinetics data were best described by a pseudo-second-order model, suggesting chemisorption, whereas isotherm data were better fit by the Freundlich model. Metal removal by NPs was then tested on an AMD-contaminated surface and in groundwater. Removal efficiencies of up to 46% for Cu and 54% for Mn in surface water and 8% for Cu, and 50% for Mn in ground water were achieved, confirming that maghemite NPs can be applied for the removal of these ions from AMD-contaminated waters (Etale et al., 2016).

A new preconcentration method based on magnetic solid-phase extraction using palmitate-coated magnetite nanoparticles, coupled with gas chromatography-tandem mass spectrometry, was also reported by Ana et al. (2014). This method was developed for the extraction of estrone, androstenedione, and progesterone hormones from water. The amount of nanoparticles and extraction time were optimised using an experimental design to estimate their combined influence on recoveries. Results showed that the extraction efficiency was highly affected by the addition of salt, and acetonitrile was used as the solvent to elute the analytes from the nanoparticles. Derivatisation via enolisation of the ketone group was performed on the extract to simultaneously derivatise the hydroxyl and ketone groups of steroid hormones, prior to their chromatographic analysis. The extraction efficiencies of the nanoparticles were studied by spiking at four concentration levels, and recoveries ranged from 94.6% to 109.7%. The limits of detection ranged from 4 to 8 ng L^{-1}. The developed method provided a preconcentration factor of 100. The present work demonstrates that palmitate-coated magnetite nanoparticles are well-suited for extracting natural oestrogenic hormones that may be pollutants in real water samples.

Mvala and co-workers also published a report stating that plastic waste poses a serious environmental risk, but it can be recycled to produce a variety of nanomaterials for water treatment. In this study, PET waste and AMD were used to prepare magnetic mesoporous carbon (MMC) nanocomposites for the adsorptive removal of pharmaceuticals and personal care products (PPCPs) from water samples. The latter were then characterised using Fourier transform infrared (FTIR) spectroscopy, X-ray diffraction (XRD), scanning electron microscopy with energy-dispersive X-ray spectroscopy (SEM-EDX), transmission electron microscopy (TEM), Brunauer–Emmett–Teller (BET) analysis, and ζ-potential measurements. The results of BET isotherms revealed high specific surface areas of 404, 664, and 936 m^2/g with corresponding pore sizes 2.51, 2.28, and 2.26 nm for MMC, MMAC-25%, and MMAC-50% adsorbents, respectively. Under optimised conditions, the equilibrium studies were best described by the Langmuir and the Freundlich models, and the kinetics were best described by the pseudo-second-order model. The maximum monolayer adsorption capacities predicted by the Langmuir model were 112, 102, and 106 mg g^{-1} for acetaminophen, caffeine, and carbamazepine, respectively. The composites could be reused for up to six cycles without losing their adsorption efficiency. Furthermore, prepared adsorbents were used to remove acetaminophen, caffeine, and carbamazepine from wastewater samples, achieving removal efficiencies of up to 95% (Mvala et al., 2024).

Do and co-workers described a study in which magnetic Fe_3O_4 nanoparticles (Fe_3O_4 NPs) were synthesised via a green method using *Camellia sinensis* leaf extract as both a reducing and stabilising agent. The obtained superparamagnetic nanoparticles exhibited a saturation magnetisation of ~50 emu/g and an average crystallite size of ~20 nm, formed at 180°C over 8 h. These nanoparticles were used as adsorbents for the preconcentration and extraction of polyacrylamide (PAM) from aqueous solutions. Adsorption followed a pseudo-second-order kinetic model and was fitted to the Langmuir isotherm, with optimal conditions at pH 8, a contact time of 50 min, and a dosage of 1.5 g L^{-1}, achieving 99% removal efficiency. The Fe_3O_4 NPs were easily recovered using an external magnetic field, highlighting their potential as low-cost, eco-friendly, and reusable adsorbents for water treatment applications (Do et al., 2011).

In another study by Lunge and colleagues, magnetic iron oxide nanoparticles were successfully synthesised using a tea waste template. The nanoparticles from tea waste exhibited supermagnetic properties under an external magnetic field with a saturation magnetisation value of 6.9 emu/g at room temperature. The TEM images revealed a particle size range of 5–25 nm. Magnetic nanoparticles were tested for the extraction of As(III) and As(V) from aqueous solution. The adsorption studies were conducted, and the data were found to obey the Langmuir equation, with high adsorption capacities of 188.69 mg g^{-1} for As(III) and 153.8 mg g^{-1} for As(V). The mean sorption energy (E) calculated from the D-R model indicated a physicochemical sorption process. A pseudo-second-order kinetic model provided the best fit for As(III) adsorption on tea nanoparticles, yielding an activation energy of 64.27 kJ/mol. Thermodynamics showed that the adsorption process is endothermic. The tea-derived nanoparticles were low-cost and could be reused for up to 5 adsorption cycles, after which they were regenerated with NaOH (Lunge et al., 2014).

Zhang and co-workers have also immobilised iminodiacetic acid on mesoporous $Fe_3O_4@SiO_2$ microspheres and used this material for an efficient, cost-effective magnetic SPE method for the extraction of trace levels of Cd, Mn, and Pb. The microspheres were characterised by infrared spectroscopy, scanning electron microscopy, and transmission electron microscopy. The loaded microspheres can be easily separated from the aqueous sample solution by applying an external magnetic field. The method had detection limits of 0.16, 0.26, and 0.26 ng L^{-1} for Cd, Mn, and Pb, respectively. The RSDs at a concentration of 1 μg L^{-1} ($n = 7$) were 4.8%, 4.6%, and 7.4%, respectively. The method was successfully applied to the determination of these metals in biological and environmental samples using ICP-MS (Zhang et al., 2011).

In a study conducted by Belibağlı & Uysal, Black Tea-Fe_3O_4 (BT-Fe_3O_4) magnetite nanocomposites were synthesised using an eco-friendly method from the extract of black tea (BT) leaves, derived from *Camellia sinensis*, and employed to investigate the extraction of nonylphenol ethoxylate from water. The results of the characterisation determination studies performed on the adsorbent revealed that the herbal extract obtained from black tea leaves successfully covered the surface of magnetite (Fe_3O_4) nanoparticles. Batch tests were conducted to determine the effect of BT-Fe_3O_4 dose, initial NPEO concentration, pH, and contact time on the adsorption behaviour of BT-Fe_3O_4. Adsorption isotherms and kinetics were described by a Freundlich isotherm model and a pseudo-second-order kinetic model. The experimental results showed that the maximum adsorption efficiency of nonylphenol ethoxylate (73.45%) occurred at an optimal pH of 7.0, a nonylphenol ethoxylate concentration of 10 mg L^{-1}, a BT-Fe_3O_4 dose of 1 g L^{-1}, and a contact time of 60 min. The results of the study have shown that BT-Fe_3O_4 particles have promising applicability in the extraction of nonylphenol ethoxylate from aqueous media by the adsorption method as an environmentally friendly and low-cost alternative adsorbent with rapid separation ability for wastewater treatment (Belibağlı & Uysal, 2022).

In another study by Bassim and co-workers, extracted *Citrus aurantium* (CA) was used as a green approach for the preparation of Fe_3O_4 nanoparticles. The green Fe_3O_4 (Fe_3O_4/CA) was characterised by X-ray diffraction (XRD), scanning electron microscopy (SEM), energy-dispersive X-ray spectroscopy analysis (EDX), FTIR spectroscopy, BET surface area measurement, and vibrating sample magnetometry (VSM). The synthesised Fe_3O_4/CA was used to preconcentrate the methylene blue (MB) dye from an aqueous solution. A four-factor central composite design (CCD) combined with Response Surface Methodology (RSM) was used to maximise MB dye removal. The four independent variables, which were initial dye concentration (10–50 mg L^{-1}), solution pH (3–9), adsorbent dose (ranging from 200 to 1000 mg L^{-1}), and contact time (30–90 min), were used as inputs to the model for the percentage dye removal. The results of an analysis of variance (ANOVA) confirmed the regression model's high significance. The predicted MB dye removal values agreed with the corresponding experimental values. The validity of the quadratic model was examined, and good agreement between the experimental and predicted values was observed. Their findings demonstrated that green Fe_3O_4 NPs are a suitable adsorbent for extracting MB from aqueous samples (Bassim et al., 2022).

From the above-discussed applications, it is evident that most magnetic sorbents can be synthesised from waste materials derived from AMD and plants. Given the availability of various plant wastes, especially from the agricultural sector, this creates an opportunity for researchers to explore other plant wastes for the synthesis of magnetic sorbents.

7.3 HYBRID SAMPLE PREPARATION METHOD

Hybrid sample preparation is the practice of combining two or more techniques to prepare samples for analysis (Selahle et al., 2021). The goal is to leverage the strengths of different methods to improve accuracy, efficiency, or compatibility with downstream processes. MSPE and liquid-phase extraction methods are commonly combined for enhanced enrichment and reliable results. Therefore, various components of the SPE and LLE obtained from waste materials can be used in designing a hybrid method for the preconcentration of heavy metals and organic compounds. The combination of these techniques will yield notable advantages, including reduced analysis time, lower solvent requirements, and enhanced instrument sensitivity However, combining two preconcentration methods also has its own drawbacks. This includes the fact that combining two extraction techniques in sequence adds operational complexity (Badawy et al., 2022). It requires more steps, careful optimisation, and handling, which can increase the risk of human error. Furthermore, multiple transfer steps between phases and sorbents increase the chances of analyte loss, especially for trace-level compounds (Abd Rahim et al., 2016). However, no studies have been reported in the literature that combine different preconcentration methods using valorised waste materials. Therefore, this also leaves an opportunity for researchers to explore such novel preconcentration techniques.

7.4 CONCLUSIONS

The use of waste-derived materials for the preconcentration of potentially toxic metals and organic pollutants offers a promising and sustainable alternative to conventional adsorbents. By transforming into effective adsorbents, this approach addresses two major environmental challenges: waste management and pollutant remediation. These low-cost, abundantly available materials have demonstrated strong adsorption efficiency and compatibility with various analytical techniques. Further research is required to enhance material stability, improve regeneration and stability, and standardise preparation methods. Sustained research and innovation in the design of hyphenated methods for the preconcentration of potentially toxic metals and organic pollutants is needed. Overall, employing waste materials for preconcentration of toxic metals and organic pollutants supports circular economy principles and promotes greener, more sustainable approaches to pollutant monitoring and remediation.

REFERENCES

Abd Rahim, M., Ibrahim, W. A. W., Ramli, Z., Sanagi, M. M., Aboul-Enein, H. Y., Ahmad, W., Al-Sibaai, A. A., Bashammakh, A. S., Alwael, H., El-Shahawi, M. S., Al-Rubaye, A. F., Hameed, I. H., Kadhim, M. J., Amelin, V. G., Bol'Shakov, D. S., Tretiakov, A. V., Tretyakov, A. V., Amiri, A., Tayebee, R., . . . Praveena, S. M. (2016). Combination of solid-phase extraction with dispersive liquid–liquid microextraction followed by GC-MS for determination of pesticide residues from water, milk, honey and fruit juice. *Journal of Chromatography A*, *9*(1), 288–294. https://doi.org/10.1007/s11418-017-1144-z

Ahmed, M. J., & Theydan, S. K. (2014). Fluoroquinolones antibiotics adsorption onto microporous activated carbon from lignocellulosic biomass by microwave pyrolysis. *Journal of the Taiwan Institute of Chemical Engineers*, *45*(1), 219–226.

Ana, R., Beatriz, P., Jose, A., Tadeo, L., Molero, E., & Sánchez, C. (2014). Analysis of steroid hormones in water using palmitate—Coated magnetite nanoparticles solid—phase extraction and gas chromatography—tandem mass spectrometry. *Chromatographia*, 837–843. https://doi.org/10.1007/s10337-014-2688-7

Andrade-Eiroa, A., Canle, M., Leroy-Cancellieri, V., & Cerdà, V. (2016). Trends in analytical chemistry solid-phase extraction of organic compounds: A critical review (Part I). *Trends in Analytical Chemistry*, *80*, 641–654. https://doi.org/10.1016/j.trac.2015.08.015

Arman, N. Z., Salmiati, S., Aris, A., Salim, M. R., Nazifa, T. H., Muhamad, M. S., & Marpongahtun, M. (2021). A review on emerging pollutants in the water environment: Existences, health effects and treatment processes. *Water, 13*(22), 3258.

Augusto, F., Hantao, L. W., Mogollo, N. G. S., & Braga, S. C. G. N. (2013). New materials and trends in sorbents for solid-phase extraction. *TrAC Trends in Analytical Chemistry*, *43*, 14–23. https://doi.org/10.1016/j.trac.2012.08.012

Azooz, E. A., Tuzen, M., Mortada, W. I., & Ullah, N. (2024). A critical review of selected preconcentration techniques used for selenium determination in analytical samples. *Critical Reviews in Analytical Chemistry*, *54*(7), 2072–2086.

Azzouz, A., Kailasa, S. K., Lee, S. S., J. Rascón, A., Ballesteros, E., Zhang, M., & Kim, K. H. (2018). Review of nanomaterials as sorbents in solid-phase extraction for environmental samples. *TrAC—Trends in Analytical Chemistry*, *108*, 347–369. https://doi.org/10.1016/j.trac.2018.08.009

Babić, S., Ašperger, D., Mutavdžić, D., Horvat, A. J. M., & Kaštelan-Macan, M. (2006). Solid phase extraction and HPLC determination of veterinary pharmaceuticals in wastewater. *Talanta*, *70*(4), 732–738.

Backe, W. J., & Field, J. A. (2012). Is SPE necessary for environmental analysis? A quantitative comparison of matrix effects from large-volume injection and solid-phase extraction based methods. *Environmental Science & Technology*, *46*(12), 6750–6758.

Badawy, M. E. I., El-Nouby, M. A. M., Kimani, P. K., Lim, L. W., & Rabea, E. I. (2022). A review of the modern principles and applications of solid-phase extraction techniques in chromatographic analysis. *Analytical Sciences*, *38*(12), 1457–1487.

Bassim, S., Mageed, A. K., AbdulRazak, A. A., & Majdi, H. S. (2022). Green synthesis of Fe3O4 nanoparticles and its applications in wastewater treatment. *Inorganics*, *10*(12), 260.

Belibağlı, P., & Uysal, Y. (2022). Green synthesis of BT-Fe3O4 nanocomposite using Camellia sinensis leaves extract: Characterization, NPEO adsorption experiments, isotherms, and kinetics. *Sigma Journal of Engineering and Natural Sciences*, *40*(1), 132–143.

Bell, K. Y., Wells, M. J. M., Traexler, K. A., Pellegrin, M., Morse, A., & Bandy, J. (2011). Emerging pollutants. *Water Environment Research*, *83*(10), 1906–1984.

Bian, Y., Zhang, Y., Zhou, Y., Wei, B., & Feng, X. (2023). Recent insights into sample pretreatment methods for mycotoxins in different food matrices: A critical review on novel materials. *Toxins*, *15*(3), 215.

Büyüktiryaki, S., Keçili, R., & Hussain, C. M. (2020). Functionalized nanomaterials in dispersive solid phase extraction: Advances & prospects. *TrAC Trends in Analytical Chemistry*, *127*, 115893.

Câmara, J. S., Perestrelo, R., Berenguer, C. V., Andrade, C. F. P., Gomes, T. M., Olayanju, B., Kabir, A., MR Rocha, C., Teixeira, J. A., & Pereira, J. A. M. (2022). Green extraction techniques as advanced sample preparation approaches in biological, food, and environmental matrices: A review. *Molecules*, *27*(9), 2953.

Capriotti, A. L., Cavaliere, C., La Barbera, G., Montone, C. M., Piovesana, S., & Laganà, A. (2019). Recent applications of magnetic solid-phase extraction for sample preparation. *Chromatographia*, *82*(8), 1251–1274.

Chen, X., Hossain, M. F., Duan, C., Lu, J., Tsang, Y. F., Islam, M. S., & Zhou, Y. (2022). Isotherm models for adsorption of heavy metals from water-A review. *Chemosphere*, *307*, 135545.

Chen, Y., Guo, Z., Wang, X., & Qiu, C. (2008). Sample preparation. *Journal of Chromatography A*, *1184*(1–2), 191–219.

Chromatographica, A., Dugheri, S., Marrubini, G., Mucci, N., Cappelli, G., Bonari, A., Pompilio, I., Trevisani, L., & Arcangeli, G. (2021). *A review of micro-solid-phase extraction techniques and devices applied in sample pretreatment coupled with chromatographic analysis*. https://doi.org/10.1556/1326.2020.00790

Deblonde, T., Cossu-Leguille, C., & Hartemann, P. (2011). Emerging pollutants in wastewater: A review of the literature. *International Journal of Hygiene and Environmental Health*, *214*(6), 442–448.

Demnerová, K., Gavrilescu, M., Aamand, J., Fava, F., & Agathos, S. (2014). Emerging pollutants in the environment: Present and future challenges in biomonitoring, ecological risks and bioremediation. *New Biotechnology*, *32*(1), 147–156. https://doi.org/10.1016/j.nbt.2014.01.001

Dimpe, K. M., & Nomngongo, P. N. (2016). Current sample preparation methodologies for analysis of emerging pollutants in different environmental matrices. *TrAC Trends in Analytical Chemistry*, *82*, 199–207.

Do, M. H., Phan, N. H., Nguyen, T. D., Pham, T. T. S., Vu, T. T. T., & Nguyen, T. K. P. (2011). Activated carbon/Fe3O4 nanoparticle composite: Fabrication, methyl orange removal and regeneration by hydrogen peroxide. *Chemosphere*, *85*(8), 1269–1276.

Etale, A., Tutu, H., & Drake, D. C. (2016). Application of maghemite nanoparticles as sorbents for the removal of Cu (II), Mn (II) and U (VI) ions from aqueous solution in acid mine drainage conditions. *Applied Water Science*, *6*(2), 187–197.

Falandysz, J., Liu, G., & Rutkowska, M. (2024). Analytical progress on emerging pollutants in the environment: An overview of the topics. *TrAC Trends in Analytical Chemistry*, *175*, 117719.

Faraji, M., Yamini, Y., & Gholami, M. (2019). Recent advances and trends in applications of solid-phase extraction techniques in food and environmental analysis. *Chromatographia*, *82*(8), 1207–1249.

Farajzadeh, M. A., Pezhhanfar, S., & Mohebbi, A. (2021). Development of a dispersive solid phase extraction procedure using a natural adsorbent as an efficient and costless sorbent followed by dispersive liquid–liquid microextraction. *International Journal of Environmental Analytical Chemistry*, *101*(11), 1499–1512.

Fumes, B. H., Silva, M. R., Andrade, F. N., Nazario, C. E. D., & Lanças, F. M. (2015). Recent advances and future trends in new materials for sample preparation. *TrAC—Trends in Analytical Chemistry*, *71*, 9–25. https://doi.org/10.1016/j.trac.2015.04.011

Gómez, M. J., Petrović, M., Fernández-Alba, A. R., & Barceló, D. (2006). Determination of pharmaceuticals of various therapeutic classes by solid-phase extraction and liquid chromatography–tandem mass spectrometry analysis in hospital effluent wastewaters. *Journal of Chromatography A*, *1114*(2), 224–233.

Herrero-Latorre, C., Barciela-García, J., García-Martín, S., Peña-Crecente, R. M., & Otárola-Jiménez, J. (2015). Magnetic solid-phase extraction using carbon nanotubes as sorbents: A review. *Analytica Chimica Acta*, *892*, 10–26.

Islas, G., Ibarra, I. S., Hernandez, P., Miranda, J. M., & Cepeda, A. (2017). Dispersive solid phase extraction for the analysis of veterinary drugs applied to food samples: A review. *International Journal of Analytical Chemistry*, *2017*(1), 8215271.

Jagirani, M. S., & Soylak, M. (2024). ScienceDirect Green sorbents for the solid phase extraction of trace species. *Current Opinion in Green and Sustainable Chemistry*, *47*, 100899. https://doi.org/10.1016/j.cogsc.2024.100899

Jiao, F., & Gao, H.-W. (2013). On-site solid-phase extraction and application to in situ preconcentration of heavy metals in surface water. *Environmental Monitoring and Assessment*, *185*(1), 39–44.

Kole, P. L., Venkatesh, G., Kotecha, J., & Sheshala, R. (2011). Recent advances in sample preparation techniques for effective bioanalytical methods. *Biomedical Chromatography*, *25*(1–2), 199–217.

Chromatographia, *82*(8), 1251–1274. https://doi.org/10.1007/s10337-019-03721-0

Leepipatpiboon, V. (1995). Trace enrichment by solid-phase extraction for the analysis of heavy metals in water. *Journal of Chromatography A*, *697*(1–2), 137–143.

Lunge, S., Singh, S., & Sinha, A. (2014). Magnetic iron oxide (Fe3O4) nanoparticles from tea waste for arsenic removal. *Journal of Magnetism and Magnetic Materials*, *356*, 21–31.

Mehta, D., Mazumdar, S., & Singh, S. K. (2015). Magnetic adsorbents for the treatment of water/wastewater—A review. *Journal of Water Process Engineering*, *7*, 244–265. https://doi.org/10.1016/j.jwpe.2015.07.001

Mi, Y., Ma, C., Zheng, W., Li, Z., & Zhang, M. (2023). Magnetic losses in single-domain magnetic particles. *The European Physical Journal Special Topics*, *232*(8), 1353–1368. https://doi.org/10.1140/epjs/s11734-022-00763-8

Mohapatra, J., Xing, M., & Liu, J. P. (2019). Inductive thermal effect of ferrite magnetic nanoparticles. *Materials*, 12(19), 3208.

Murray, I., & Cowe, I. (2004). Sample preparation. *Near-Infrared Spectroscopy in Agriculture*, *44*, 75–112.

Murtada, K., & Pawliszyn, J. (2023). Trends in analytical chemistry evolution in understandings of the design, optimization, and application of sorbent-based extraction devices. *Trends in Analytical Chemistry*, *169*(October), 117396. https://doi.org/10.1016/j.trac.2023.117396

Mvala, B. A., Munonde, T. S., Mpupa, A., Bambo, M. F., Matabola, K. P., & Nomngongo, P. N. (2024). Valorization and upcycling of acid mine drainage and plastic waste via the preparation of magnetic sorbents for adsorption of emerging contaminants. *ACS Omega*, *9*(32), 34700–34718.

Nasiri, M., Ahmadzadeh, H., & Amiri, A. (2020). Sample preparation and extraction methods for pesticides in aquatic environments: A review. *TrAC—Trends in Analytical Chemistry*, *123*, 115772. https://doi.org/10.1016/j.trac.2019.115772

Nephiphidi, A. S., Mudzielwana, R., & Selahle, S. K. (2025). Preparation of biosorbent from avocado seeds for preconcentration and simultaneous extraction of trace parabens in environmental wastewater samples. *Frontiers in Chemistry*, *13*, 1688983.

Oliveira, M. S. F., Assila, O., Fonseca, A. M., Parpot, P., Valente, T., Rombi, E., & Neves, I. C. (2024). Acid mine drainage precipitates from mining effluents as adsorbents of organic pollutants for water treatment. *Molecules*, *29*(15), 3521.

Park, H. J., Jeong, S. W., Yang, J. K., Kim, B. G., & Lee, S. M. (2007). Removal of heavy metals using waste eggshell. *Journal of Environmental Sciences*, *19*(12), 1436–1441.

Pathak, S. (2024). *Optimization of magneto-viscoelasticity of magnetic fluids and development of its applications in thermal and mechanical systems* (Doctoral dissertation, RMIT University).

Pérez-Fernández, V., Mainero Rocca, L., Tomai, P., Fanali, S., & Gentili, A. (2017). Recent advancements and future trends in environmental analysis: Sample preparation, liquid chromatography and mass spectrometry. *Analytica Chimica Acta*, *983*, 9–41. https://doi.org/10.1016/j.aca.2017.06.029

Płotka-Wasylka, J., Szczepańska, N., de la Guardia, M., & Namieśnik, J. (2016). Modern trends in solid phase extraction: New sorbent media. *TrAC—Trends in Analytical Chemistry*, *77*, 23–43. https://doi.org/10.1016/j.trac.2015.10.010

Pouretedal, H. R., & Sadegh, N. (2014). Effective removal of amoxicillin, cephalexin, tetracycline and penicillin G from aqueous solutions using activated carbon nanoparticles prepared from vine wood. *Journal of Water Process Engineering*, *1*, 64–73.

Ramos, L. (2012). Critical overview of selected contemporary sample preparation techniques. *Journal of Chromatography A*, *1221*, 84–98.

Rasc, J., Azzouz, A., Kumar, S., Soo, S., Ballesteros, E., Zhang, M., & Kim, K. (2018). Trends in Analytical Chemistry Review of nanomaterials as sorbents in solid-phase extraction for environmental samples. *Trends in Analytical Chemistry*, *108*. https://doi.org/10.1016/j.trac.2018.08.009

Ścigalski, P., & Kosobucki, P. (2020). Recent materials developed for dispersive solid phase extraction. *Molecules*, *25*(21), 4869.

Selahle, S. K., Mpupa, A., & Nomngongo, P. N. (2021). A review of extraction, analytical, and advanced methods for the determination of neonicotinoid insecticides in environmental water matrices. *Reviews in Analytical Chemistry*, *40*(1), 187–203.

Seng, L. K. (2018). Adsorption of heavy metals using banana peels in wastewater treatment. *The Eurasia Proceedings of Science Technology Engineering and Mathematics*, (2), 312-317.

Shen, Z., Kuang, Y., Zhou, S., Zheng, J., & Ouyang, G. (2023). Trends in analytical chemistry preparation of magnetic adsorbent and its adsorption removal of pollutants: An overview. *Trends in Analytical Chemistry*, *167*(April), 117241. https://doi.org/10.1016/j.trac.2023.117241

Skans, I., Mnguni, M., & Nomngongo, P. N. (2025). Waste PET plastic-derived UIO-66 (Zr)-MOF adsorbent for dispersive solid phase extraction and removal of penicillin antibiotics water systems. *Analytical Science Advances*, *6*(1), e70023.

Souza-Silva, É. A., Gionfriddo, E., & Pawliszyn, J. (2015). A critical review of the state of the art of solid-phase microextraction of complex matrices II. Food analysis. *TrAC—Trends in Analytical Chemistry*, *71*, 236–248. https://doi.org/10.1016/j.trac.2015.04.018

Spietelun, A., Kloskowski, A., Chrzanowski, W., & Namieśnik, J. (2013). Understanding solid-phase microextraction: key factors influencing the extraction process and trends in improving the technique. *Chemical reviews*, 113(3), 1667-1685.

Stevanovic, J., Stevanovic, Z., Bugarin, M., & Grujic, A. (2011). Using the low-cost waste materials for heavy metals removal from the mine wastewater. *Materials Transactions*, *52*(10), 10–13. https://doi.org/10.2320/matertrans.M2011191

Ta, C. (2012). *Synthesis of superparamagnetic nanoparticles dispersed in spherically shaped carbon nanoballs*. https://doi.org/10.1007/s11051-012-1118-8

Tatarchuk, T., Soltys, L., & Macyk, W. (2023). Magnetic adsorbents for removal of pharmaceuticals: A review of adsorption properties. *Journal of Molecular Liquids*, *384*(April), 122174. https://doi.org/10.1016/j.molliq.2023.122174

Tuzen, M., Saygi, K. O., & Soylak, M. (2008). Solid phase extraction of heavy metal ions in environmental samples on multiwalled carbon nanotubes. *Journal of Hazardous Materials*, *152*(2), 632–639.

Vasconcelos, I., & Fernandes, C. (2017). Magnetic solid phase extraction for determination of drugs in biological matrices. *TrAC Trends in Analytical Chemistry*, *89*, 41–52.

Vasilachi, I. C., Asiminicesei, D. M., Fertu, D. I., & Gavrilescu, M. (2021). Occurrence and fate of emerging pollutants in water environment and options for their removal. *Water*, *13*(2), 181.

Wang, J., Han, Q., Wang, K., Li, S., Luo, W., Liang, Q., Zhong, J., & Ding, M. (2023). Talanta Recent advances in development of functional magnetic adsorbents for selective separation of proteins/peptides. *Talanta*, *253*(August 2022), 123919. https://doi.org/10.1016/j.talanta.2022.123919

Wilkinson, J., Hooda, P. S., Barker, J., Barton, S., & Swinden, J. (2017). Occurrence, fate and transformation of emerging contaminants in water: An overarching review of the field. *Environmental Pollution*, *231*, 954–970. https://doi.org/10.1016/j.envpol.2017.08.032

Wu, A., Zhao, X., Wang, J., Tang, Z., Zhao, T., Niu, L., Yu, W., Yang, C., Fang, M., Lv, H., & Liu, S. (2021). Technology Application of solid-phase extraction based on magnetic nanoparticle adsorbents for the analysis of selected persistent organic pollutants in environmental water: A review of recent advances. *Critical Reviews in Environmental Science and Technology*, *51*(1), 44–112. https://doi.org/10.1080/10643389.2020.1720493

Yang, L., Said, R., & Abdel-Rehim, M. (2017). Sorbent, device, matrix and application in microextraction by packed sorbent (MEPS): A review. *Journal of Chromatography B*, *1043*, 33–43. https://doi.org/10.1016/j.jchromb.2016.10.044

Yang, Q., Zhao, Y., Yu, H., Xiong, X., & Huang, K. (2024). Recent advances of sample separation and preconcentration techniques for analytical atomic spectrometry. *Applied Spectroscopy Reviews*, *59*(5), 652–677.

Zhang, N., Peng, H., Wang, S., & Hu, B. (2011). *Fast and selective magnetic solid phase extraction of trace Cd, Mn and Pb in environmental and biological samples and their determination by ICP-MS*. 121–128. https://doi.org/10.1007/s00604-011-0659-3

Zulkifli, S. N., Rahim, H. A., & Lau, W. J. (2018). Detection of contaminants in water supply: A review on state-of-the-art monitoring technologies and their applications. *Sensors and Actuators, B: Chemical*, *255*, 2657–2689. https://doi.org/10.1016/j.snb.2017.09.078

8 Electrochemical and optical sensing of pollutants

8.1 INTRODUCTION

Our water systems, including groundwater, wastewater, and surface water, are polluted by metals and trace organic compounds. The contaminants are linked to various negative environmental and health effects on living organisms. Therefore, it is necessary to monitor the presence and concentration of these pollutants in our water systems to develop effective removal techniques. Over the years, metals have been detected and quantified in water using atomic absorption spectroscopy (AAS), inductively coupled plasma optical emission spectroscopy (ICP-OES), and inductively coupled plasma-mass spectrometry (ICP-MS). On the other hand, the analysis of organic contaminants has been enabled by gas chromatography-mass spectrometry (GC-MS), liquid chromatography-mass spectrometry (LC-MS), and high-performance liquid chromatography diode array detection (HPLC-DAD). Although these analytical techniques are sensitive, selective, and accurate, they are costly, time-consuming, and complex to operate. Furthermore, they are not suitable for in-situ monitoring of pollutants, as they require complex sample pretreatment to ensure the sample is compatible with the intended analytical detection technique. Electrochemical and optical sensors have recently emerged as a suitable solution to address the shortcomings of traditional analytical techniques, as they offer rapid analysis, ease of operation, high sensitivity, small size and portability, low cost, and no complex pre-analysis sample preparation steps. As a result, electrochemical and optical sensors have received significant attention for monitoring metals and trace organic compounds in water.

The working electrode plays a pivotal role in the performance and the cost of the entire electrochemical sensor. This has triggered extensive research to develop and produce chemically modified electrodes (CMEs) to shift away from non-renewable, expensive, heavy-metal-based electrodes. Nanomaterials are receiving increased attention as excellent materials for modifying electrodes as they enhance their surface area, electron diffusion, and binding affinity for targeted pollutants. These qualities are linked to the excellent physicochemical properties of nanomaterials, namely, limited internal diffusion resistance, high binding energy, and large surface area (Cui et al., 2006; Khajeh et al., 2013; Khajeh & Sanchooli, 2011; Liu et al., 2005). Various nanomaterials, such as carbon nanotubes (Fozing Mekeuo et al., 2022), reduced graphene oxide (Lee et al., 2022), CQDs (Nazari et al., 2023), zinc oxide (Pabel et al., 2025), and silver nanoparticles (Shivakumar et al., 2020), have been applied in electrochemical sensing of different pollutants. Currently, there is an

DOI: 10.1201/9781003745846-8

upsurge in interest in producing nanomaterials from waste, due to the relatively high cost of synthesising nanomaterials and the uncontrollable generation of waste from various sectors. Nanomaterials are produced from waste sources such as plastics, electronic and electric waste, wastewater sludge, rubber tyres, batteries, and biomass. The waste-generated nanomaterials will further reduce the cost of electrochemical sensors, as they are produced from waste, and further enhance their performance, as they can be easily functionalised and modified during the preparation stage.

8.2 ELECTROCHEMICAL AND OPTICAL SENSORS

Optical and electrochemical detection have garnered significant interest for environmental detection of heavy metal(loid)s and trace organics owing to their cost-effectiveness, high sensitivity, rapidity, and simplicity. Electrochemical analysis takes advantage of most environmental contaminants exhibiting a redox potential; therefore, when voltage is applied, the contaminants gain or lose electrons via a redox reaction, thus generating a current signal (Li & Zhu, 2020). This signal is displayed on the electrochemical workstation and is proportional to the concentration of the electroactive contaminants, helping ensure they are electrochemically detected. The most employed techniques in the detection and quantification of environmental pollutants include voltammetry and amperometry. Voltammetric techniques are subdivided into cyclic voltammetry (CV), differential pulse voltammetry (DPV), square wave voltammetry (SWV), anodic stripping voltammetry (ASV), and linear sweep voltammetry (LSV).

In cyclic voltammetry, a linearly varying potential is applied to the working electrode with respect to the reference electrode at a constant scan rate (Elgrishi et al., 2018; Li & Zhu, 2020). As the potential is scanned, a current signal is generated by redox reactions at the working electrode's surface. When the potential reaches the terminating potential (set vertex), it reverses direction to the set starting potential (Li & Zhu, 2020). CV may be applied for various factors, including (i) to determine if the electrochemical process occurring at the working electrode is reversible or not; (ii) to determine the electrochemically active surface area of the CME; (iii) to determine the control process of the working electrode, whether it is adsorption- or diffusion-controlled; (iv) to determine the reaction mechanism occurring at the electrode; (v) to qualitatively analyse electroactive substances. However, CV exhibits low sensitivity due to high background (capacitive) current associated with the method; therefore, it is not commonly used for quantitative analysis.

DPV is one of the most widely used techniques for both qualitative and quantitative analysis of environmental samples, owing to its high sensitivity because of reduced background (charging) current (Li & Zhu, 2020). In this technique, a series of potential pulses is applied to the working electrode, which is charged, generating a charging current. At the same time, the applied pulse renders the target analyte to produce an electrode signal, which generates a Faradaic current during the electrochemical reaction. Fortunately, the charging current quickly decays to nearly zero, whereas the Faradaic current decays slowly. As a result, current is sampled at a delayed stage of each applied potential pulse, and the measured current will be almost the same as the Faradaic current. This phenomenon makes DPV significantly more sensitive than CV.

SWV is widely used in the electrochemical analysis of environmental samples due to its short analysis time, excellent sensitivity, and high selectivity. Unlike in DPV, the potential waveform in SWV consists of both the forward pulse and the reverse pulse within each square wave cycle (Li & Zhu, 2020; Osteryoung & Osteryoung, 1985; Ramaley & Krause, 1969). The current is measured at the end of the forward and reverse pulse. The reverse pulse will ensure the product formed during the forward pulse undergoes a reverse (reduction or oxidation) reaction. The concentration of the target analytes is directly proportional to the peak current. Since SWV can regulate frequency, this allows the method to control response time and shorten analysis time.

Amperometric current–time (i–t) curves are usually linked with other electrochemical techniques, since they are not suitable for qualitative analysis because of their poor selectivity. However, they boast qualities such as high sensitivity and rapid analysis time, making them attractive for quantitative analysis. In amperometric i–t curves, a constant potential is applied to the working electrode to drive the electrochemical reaction of the target analyte, generating a Faradaic current (Li & Zhu, 2020; Yi et al., 2006). The concentration of the electroactive target analyte is directly proportional to the generated Faradaic current.

Heavy metal(loid)s are usually analysed using stripping voltammetric techniques, mainly anodic stripping voltammetry (ASV). ASV is an electrochemical method that is mainly used in analysing metal ions that can form amalgam (Ding et al., 2021). In this technique, the targeted metal ions in solution are preconcentrated via oxidation–reduction deposition on a thin layer of a zero-valent metal (usually mercury) on a working electrode (Borrill et al., 2019). Thereafter, the mercury (zero-valent metal) is re-oxidised back to its corresponding metal ion by linearly applying sweeping voltage in the anode direction, so that the analyte deposited on the electrode is stripped off at a certain voltage. The specific oxidation potential of the metal plays a central role in ensuring the selective detection of metal ions.

The fluorescence technique is one of the most widely used optical sensors for detecting heavy metals in water. This optical sensor is based on the variation in the photoluminescence of fluorescent materials upon interaction with target analytes (heavy metals) (Garg et al., 2024). During the interaction between the heavy metal and the fluorescence molecule, the electrons in the molecule's outer shell are excited and transferred to higher energy levels. When excited electrons return to the ground state, emitting a photon, luminescence occurs (Garg et al., 2024). The luminescence is either enhanced, quenched, or results in a shift in emission wavelength. This phenomenon is employed in quantitative and qualitative analysis of targeted analytes from the sample solution.

8.3 SYNTHESIS AND CHARACTERISATION OF WASTE-GENERATED NANOMATERIALS

8.3.1 Carbon dots

Over the years, carbon-based materials such as carbon nanotubes, graphene, carbon nanofibres, and fullerenes have limitations in biological and aquatic studies owing to their poor water solubility and biocompatibility. Notwithstanding, CQDs have emerged as one of the most studied materials in the sensing of contaminants in water, owing to their simple synthesis method, photoluminescence characteristics, water solubility, biocompatibility, low toxicity, controllable surface functionality, and photostability (Keerthana & Cherian, 2025). The small size of CQDs ensures enhanced surface contact with targeted analytes and thus increases electrochemical sensing (Bressi et al., 2023). Various waste sources have been instrumental in designing CQDs to reduce waste generation and negative environmental impact, cut production costs, and produce sustainable materials. The waste-derived CQDs have been produced utilising various synthetic methods, including hydrothermal treatment (Fernandes et al., 2025; Surendran et al., 2021; S. Zhang et al., 2023; W. Zhang et al., 2023), microwave (Zhao et al., 2021), and chemical oxidation (Wang et al., 2015). The physicochemical properties and optical characteristics of CQDs highly depend on the nature of the carbon source and the synthetic method (W. Zhang et al., 2023). Therefore, it is not surprising that researchers have explored multiple carbon sources and several preparation methods to generate CQDs with excellent properties for various applications.

The hydrothermal method is one of the most studied synthetic procedures for CQDs from waste materials, owing to its simplicity. Fernandes et al. prepared CQDs from various waste materials, including spent coffee grounds, sawdust, corn stover, and cork powder, in combination with citric acid as a co-precursor (Fernandes et al., 2025). This study demonstrated a high fluorescence quantum yield up to 40%. Moreover, the production of CDs from waste showed a lower environmental impact than those prepared from commercial materials. This study highlighted that waste materials not only increase the percentage production yield but also reduce environmental impact. In another study, Surendran et al. prepared CQDs from pineapple (*Ananas comosus*) waste peels via hydrothermal processing (Surendran et al., 2021). The peels were cut into small pieces, rinsed with water, and dried in an oven before being transferred to an autoclave and heated at 200 °C for 6 h. The produced CQDs exhibit particle sizes of 2.4 ± 0.5 nm and strong blue emission, as measured by HR-TEM and fluorescence, respectively (Surendran et al., 2021).

The hydrothermal method is compatible with various waste materials, including surgical masks, fruit peels, and traditional medicines. Arrigo et al. prepared CQDs from surgical facemasks, which became a global environmental concern, especially during the height of the COVID-19 pandemic (Arrigo et al., 2023). This material produced CQDs with sizes ranging from 4 nm to 13 nm, as confirmed by SEM; they also showed excellent photostability and luminescence, as confirmed by UV-Vis and fluorescence spectroscopy. The hydrothermal method also makes it easy to dope CQDs, thereby enhancing their physicochemical properties. Zhang et al. prepared N- and Cl-doped CQDs using *Ligusticum wallichii* and *Corydalis* waste materials. The products produced a unique optical triple channel and particle size of 1.9 ± 0.26 nm, which were measured via fluorescence spectrum and HR-SEM, respectively.

The microwave-assisted method is another preparation method that improves the fluorescence performance of CQDs and reduces energy consumption during the synthesis. Zhao et al. generated *Forsythia*@carbon quantum dots from the Chinese herbal medicine *Forsythia* (Zhao et al., 2021). This method has a short synthesis time (5 min), which aids in saving energy and cutting production costs as compared to the hydrothermal method, which normally lasts up to 6 h at high temperatures. HR-TEM indicated that the CQDs have a diameter of approximately 2.5 nm; the fluorescence spectrum showed that the CQDs exhibit excellent blue fluorescence (Zhao et al., 2021). Moreover, Forsythia-based CQDs contain abundant functional groups and heteroatoms, including C=C, C–C, C=O, C–O, –NH, –N–, C–N, and C=N, as confirmed by XPS. This phenomenon is due to the nature of plant waste material, which contains thousands of compounds with different functional groups. The presence of heteroatoms enhances the solubility and optical properties of CQDs.

Chemical oxidation is a synthetic route commonly used to produce CQDs with abundant oxygen-containing functional groups, thereby increasing their hydrophilicity, physicochemical, and optical properties. This method utilises strong acids or oxidising agents. Wang et al. utilised ultrasonic-assisted chemical oxidation to engineer yellow-fluorescence CQDs from petroleum coke (Wang et al., 2015). This method used sulphuric acid and nitric acid as activation agents. The CQDs contain –OH, C=O, and C–O functional groups, as determined by FTIR spectroscopy. HR-TEM confirmed that CQDs have an average particle size of 2.8 nm. The generated material displayed excellent UV-Vis absorption and fluorescence properties, which were attributed to the presence of C=C, C–OH, and C=O functional groups.

8.3.2 Graphene oxide and reduced graphene oxide

Graphene oxide (GO) is a nanomaterial that can be derived from waste materials and has shown great potential for sensing and removing contaminants from water. Upoma et al. produced GO utilising graphite from waste dry cell batteries as an adsorbent in the removal of azithromycin from water (Upoma et al., 2022). This method was cost-effective and environmentally friendly as it used waste batteries to prevent the disposal of and leaching of chemicals to the environment. The sheet-like morphology was confirmed by SEM and TEM images, whereas the crystallinity was determined by XRD. In a separate study, Pabel et al. synthesised GO from graphite derived from waste dry cell batteries to produce rGO/Ag nanocomposites (Pabel et al., 2024). GO was synthesised using a modified Hummers' method; GO was reduced to rGO using $NaBH_4$ as a reducing agent; and Ag(I) was reduced to Ag nanoparticles and combined with rGO to form rGO/Ag. HR-TEM clearly revealed an rGO sheet with Ag nanoparticles (2.5 nm) embedded on its surface (Pabel et al., 2024).

8.4 APPLICATION OF WASTE-DERIVED NANOMATERIALS FOR ELECTROCHEMICAL SENSING OF EMERGING TRACE ORGANIC POLLUTANTS

Waste-derived nanomaterials and nanocomposites have played a huge role in the electrochemical detection of emerging trace organic compounds. Table 8.1 summarises some waste-derived nanomaterials and nanocomposites that have been applied for sensing emerging trace organics. This is expected for biowaste-derived nanomaterials, since they are composed of numerous organic compounds that

TABLE 8.1
Waste-derived utilised in electrochemical sensing of emerging trace organic pollutants.

Material	Source	Modified Electrode	Analyte	Detection Method	LOD (μM)	Reference
rGO	Red dragon fruit peels	SPE	Sulfamethoxazole	DPV	0.456	Lee et al. (2022)
Ag nanoparticles	*Eucalyptus* bark	GCE	Nitrobenzene	DPV	0.027	Shivakumar et al. (2020)
Au/$CaCO_3$ nanocomposite	Eggshell	GCE	4-nitrobenzene	CV	0.00054	Ding et al. (2020)
Ayous Sawdust-carbon nanotubes composite	*Ayous* sawdust	GCE	Carbendazim	CV	0.04	Fozing Mekeuo et al. (2022)
OMCNs	Crab shell	-	Malachite green	DPV	0.05	Yang et al. (2020)
Au/rGO/AC nanocomposite	Date seeds	GCE	4-nitrophenol	Amperometry	0.36	Harraz et al. (2022)
CDs	Orange peels	SPCE	Nitrobenzene	DPV	0.013	Bressi et al. (2023)
RuO_2@PC	*Citrus limetta*	ITO	Hexestrol	CV, chronoamperometry, and DPV	0.0232	Chaudhary et al. (2020)
$NiCo_2S_4$-wastechar	Discarded food packaging, fibre bags, used face masks and corrugated paper	GCE	Dopamine	CV, LSV, DPV, and chronoamperometry	0.0095	Zelai (2025)
NcS-CDs	Eggshells	GCE	Metronidazole	CV, DPV, EIS	9.0×10^{-10}	Ghiasi et al. (2025)
ZnO–porous graphene	Cork oak (*Quercus suber*)	-	Paraquet	SWASV	0.09	Chaudhary et al. (2025)
ZnO	Zinc shells of recycled batteries	GCE	Dopamine	DPV	0.00047	Pabel et al. (2025)
BPAC/$NiFe_2O_4$/MnCoFe-LDH nanocomposite	Banana peel	GCE	Palbociclib	DPV	0.0035	Erk et al. (2024)
CAs-Fe	Seaweed residue	GCE	Dopamine	CV	0.0033	Wu et al. (2024)
BH-PS	Black liquor—a pulp industry by-product	-	Caffeic acid, chlorogenic acid, oleuropein, and catechin	DPV	0.03–0.55	Fiori et al. (2025)
Zn-BTC	Battery waste	GCE	Uric acid	CV	1.43	Hossain et al. (2025)

contain functional groups (such as the –C=C– network and nitrogen- and oxygen-containing groups) with high affinity for targeted organic compounds. For instance, Bressi et al. utilised orange peel waste to engineer CDs for electrochemical detection of nitrobenzene in wastewater (Bressi et al., 2023). The abundance of aromatic moieties in the CDs and CDs/SPCE facilitated π–π interactions with the aromatic nitrobenzene, thereby attracting it to the electrode. Moreover, the amide and carboxyl groups facilitated interaction with the nitro group of nitrobenzene via a donor–acceptor electron-transfer mechanism. The small size of CDs and their attractive functional groups, owing to the biowaste source, ensured the CDs/SPCE sensor had high sensitivity to nitrobenzene, yielding a low LOD of 0.013 μM.

rGO, one of the carbon-based nanomaterials, exhibits an abundant –C=C– network that is attractive to trace organics in sensing. Lee et al. prepared rGO employing dragon fruit (*Hylocereus polyrhizus*) peel extracts as reducing agents to reduce GO (Lee et al., 2022). rGO was coated onto SPE to develop a sensor for sulfamethoxazole in the aquatic environment. The rGO/SPE exhibited a low LOD of 0.456 μM for sulfamethoxazole. The sensor's detection and sensitivity were largely attributed to the deoxygenation of the oxygen-containing group in GO, thereby generating a large –C=C– network that facilitates a π–π interaction mechanism with sulfamethoxazole and attracts it to the electrode surface. Moreover, a flat-like sheet of rGO ensured there were plenty of active sites on the electrode, which enhanced sensing.

There is growing evidence for the use of waste extracts as reducing/stabilising agents in the synthesis of nanomaterials for sensing, thereby rendering the technique cost-effective and environmentally friendly. In their study, Shivakumar et al. extracted reducing and stabilising agents from *Eucalyptus* bark to generate Ag nanoparticles for Ag/GCE electrode in electrochemical detection of nitrobenzene in tap and lake water (Shivakumar et al., 2020). Ag nanoparticles are attractive for electrochemical sensing owing to their large surface area, good conductivity, and high sensitivity. The Ag/GCE sensor showed high sensitivity, selectivity, and a low LOD of 0.027 μM, owing to the excellent electrocatalytic activity of Ag nanoparticles in reducing nitrobenzene.

Metal oxides are another group of nanomaterials that have contributed immensely to the development of electrochemical sensors for the detection of emerging organic compounds, owing to their nanostructured nature, good conductivity, and large surface area. Therefore, it is important to develop waste-derived alternatives to prepare these nanomaterials, thereby reducing costs and ensuring sustainable, environmentally friendly synthesis methods for metal oxides. To explore this, Pabel et al. collected discarded zinc–carbon batteries and used zinc hulls to generate ZnO nanoparticles for electrochemical detection of dopamine (Pabel et al., 2025). The ZnO/GCE electrode exhibited high electrocatalytic activity for the oxidation of dopamine, enabling sensitive detection. As a result, the electrode demonstrated high selectivity and a very low LOD of 0.00047 μM.

The formation of a nanocomposite also plays a central role in reducing the cost of the electrode, which may be a combination of two or more materials, with one or more components derived from waste materials. Normally, metals and metal oxides are combined with waste derived from organic materials to enhance the adsorption of target organic compounds onto the electrode, as they exhibit poor adsorptive interactions with organic compounds. For instance, Chaudhary et al. prepared porous carbon (PC) from *Citrus lamella* waste pulp, which was then hybridised with RuO_2 nanoparticles to generate RuO_2@PC nanocomposite (K. Chaudhary et al., 2020). The RuO_2@PC nanocomposite was coated on indium tin oxide (ITO)-coated glass to form a RuO_2@PC/ITO electrode for electrochemical sensing of hexestrol. The electrode demonstrated high electrochemical sensitivity and the detection of hexestrol, with an LOD of 0.0232 μM, owing to the synergistic effect of metal oxide nanoparticles and PCB on the adsorption and oxidation of hexestrol.

8.5 APPLICATION OF WASTE-DERIVED NANOMATERIALS FOR OPTICAL AND ELECTROCHEMICAL SENSING OF TRACE METALS

Fluorescence-based sensors are abundantly used for the quantitative and qualitative analysis of trace metals from water. Keerthana et al. employed the fluorescence sensor for Au^{3+} utilising CQDs derived from vitamin B12 waste, which was synthesised via microwave method (Keerthana & Cherian, 2025). CQDs exhibit various functional groups such as N–H, O–H, C–N, C=O, and C=C, indicating the presence of heteroatoms and surface functionalisation. This phenomenon is attributed to the nature of vitamin B12 which has abundant carbon content and functional groups such as –OH, $–NH_2$, –C–N, and –COOH. The graphitic content and surface functional groups rendered CQDs to have excellent optical properties as evident by strong emission maximum at 400 nm, 320 nm, and 237 nm. Moreover, the CQDs produced about 30% quantum yield. As a result of these physicochemical and optical properties, vitamin B12-derived CQDs selectively and sensitively detected Au^{3+} with a limit of detection of 0.041 μM, Table 8.2. Furthermore, the recoveries in river water, lake water, and tap water collected near the industrial area were in a range of 99.3%–102.2%. It is of paramount interest to select waste material that has functional groups suitable to interact with a targeted analyte. In this case, Au^{3+} ions have a high affinity for nitrogen- and oxygen-containing functional groups, which ensured high fluorescence sensing performance (Keerthana & Cherian, 2025). This study developed a cheap, environmentally friendly, and fast analysis compared to traditional analysis methods.

Fluorescence and electrochemical sensors still represented the best detection techniques for trace metals owing partly to their fast response time. Bressi et al. employed beer bagasse waste-derived carbon nanodots for fluorometric and electrochemical detection of heavy metals from water (Bressi et al., 2024). The fluorescence probes demonstrated successful detection of Hg(II) and Pb(II) than other heavy metals (Cd(II) and Ni(II)) due to their high affinity for oxygen- and nitrogen-containing functional groups on the surface of CNDs. The CNDs fluorescence sensor produced a low LOD of 0.0113 μM and 0.0788 μM for Hg(II) and Pb(II), respectively, Table 8.2. The disadvantage of the designed fluorescence probe is its inability to simultaneously detect all four heavy metals. On the other hand, CNDs/SPE sensors successfully detect all heavy metals simultaneously with distinct peaks. However, the LODs generated by CNDs/SPE sensors were in a range

TABLE 8.2
Waste-derived nanomaterials used in electrochemical and optical sensing of trace elements.

Material	Source	Modified Electrode	Analyte	Detection Method	LOD (μM)	Reference
CQDs	*Ligusticum wallichii* and *Corydalis*	-	Cr(VI)	Fluorescence	0.28	S. Zhang et al. (2023)
CQDs	Coke	-	Cu(II)	Fluorescence	0.0295	Wang et al. (2015)
CQDs	Coal washery rejects	-	Hg(II) and Cr(VII)	Fluorescence	0.240 and 0.145	Boruah and Saikia (2022)
Carbon nanoparticles	Waste polystyrene	-	Heavy metals	Fluorescence	26.0	Teo et al. (2022)
rGO/Ag	Graphite from waste dry cells	GCE	Arsenic	DPASV	2.40×10^{-7}	Pabel et al. (2024)
CQDs	Expired vitamin B12 tablets	-	Au(III)	Fluorescence	0.041	Keerthana and Cherian (2025)
CQD	Discarded chewing gum	-	Fe(III)	Fluorescence	0.028	Hu and Gao (2019)
MWCNT-MES-CPE	Eggshell	Carbon paste electrode	Cd(II)	DPV	0.00240	Mohammadi et al. (2019)
AgNP-MO	*Moringa oleifera*'s bark extract	Platinum electrode	Cu(II)	CV	0.530	Sebastian et al. (2019)
Carbon nanodots	Beer bagasse	SPCE	Ni(II), Cd(II), Pb(II), and Hg(II)	SWASV	0.000124–0.000608	Bressi et al. (2024)
Graphene quantum dots	*Citrus limetta* waste pulp	-	Fe(III)	Fluorescence	0.0000193	Thakur et al. (2019)
$NiCo_2S_4$-wastechar	Discarded food packaging, fibre bags, used face masks, and corrugated paper	GCE	Pb(II)	CV, LSV, DPV, and chronoamperometry	0.0054	Zelai (2025)
PLCDs	Plastic waste	-	Uranium	Fluorescence	$1{,}05 \times 10^{-5}$	Thakkar et al. (2025)
ZnO–porous graphene	Waste cork	-	Cd(II) and Pb(II)	SWASV	0.002 and 0.001	Chaudhary et al. (2025)
N-rich nanocarbon	Coconut waste	GCE	Pb(II)	DPASV	0.022	Sharma et al. (2024)
HA/GEL nanocomposites	Talang queenfish bone wastes	CPE	Pb(II)	DPV	0.056	Ghaneh et al. (2024)
Nano-biochar	Tea residue	GCE	Pb(II)	DPASV	0.490	Su et al. (2024)
CQDs	Watermelon rind	-	Fe(III) and Cu(II)	Fluorescence	1,75 and 0.43	Zhang et al. (2024)

of 0.124–0.608 μM, which is higher than those for CNDs fluorescence sensor for Hg(II) and Pb(II), thus indicating that the latter is more sensitive for these two heavy metals. This study indicates that for simple matrices that may contain only Hg(II) or Pb(II), a CNDs fluorescence sensor may be a more suitable detection technique due to its sensitivity, whereas for complex matrices containing multiple heavy metals, the CNDs/SPE sensor is the most suitable.

Sharma et al. produced nitrogen and sulfur coconut husk-derived nanocarbons to modify the GCE in Pb(II) sensing (Sharma et al., 2024). The N,S-rich nanocarbon/GCE sensor showed high sensitivity (with a low LOD of 0.022 μM) and selectivity towards Pb(II) than other trace metals such as Hg(II), Cd(II), and Cu(II), Table 8.2. This excellent electrochemical performance owes to the nanostructured nature of the coating material, which exhibited a large surface area (1672 $m^2\ g^{-1}$) and pore volume (0.476 $cm^3\ g^{-1}$) which aided in increasing conductivity and thus enhanced the sensitivity towards the targeted analyte. The nitrogen- and sulfur-containing functional groups facilitated selectivity and sensitivity due to a high affinity towards Pb(II) than other competing trace metals. This study also illustrates that in addition to selecting the best material to coat the electrode, selecting the most suitable detection techniques also plays a significant role in increasing sensitivity. This was achieved by comparing DPASV and DPNV in sensing Pb(II) where the latter generated a high LOD of 0.0507 μM showing about two times lower sensitivity than the former detection method.

Hybrid nanomaterials also play a significant role in electrochemical sensing of trace metals by combining the electrochemical and physicochemical properties of two or more materials. For instance, Chaudhary et al. utilised a direct laser writing method to produce ZnO–porous graphene from waste cork (Chaudhary et al., 2025). The porous nanostructure nature of graphene provides a large surface area and active sites to accommodate metal ions, while the porosity ensures fast diffusion of ions. On the other hand, ZnO enhances the kinetics of electron transfer from the electrode to the electrolyte, thus enhancing electrochemical sensing of Cd(II) and Pb(II). The synergy of the two materials in tandem with square wave anodic stripping voltammetry (SWASV) detection techniques produced low LDOs which were 0.001 μM and 0.002 μM for Pb(II) and Cd(II), respectively. The ZnO–porous graphene nanostructured electrode showed high electrochemical sensing for trace metals even in a complex seawater matrix. This study illustrates the importance of combining the properties of waste-derived materials with metal oxide to produce cost-effective and sustainable electrodes with high selectivity and sensitivity for trace metals.

This study by Zelai et al. also highlighted the synergist effect between metal-based nanomaterials and waste-based char (Zelai, 2025). Zelai et al. anchored $NiCo_2S_4$ nanomaterials on the surface of waste-derived char (generated from discarded food packaging, fibre bags, used face masks, and corrugated paper) to produce $NiCo_2S_4$-waste char nanocomposite. The nanocomposite was then coated on the GCE for electrochemical sensing of Pb(II). Waste-char, $NiCo_2S_4$, and $NiCo_2S_4$-waste char nanocomposite were applied on the GCE tested for electrochemical sensing of Pb(II); it was demonstrated that the nanocomposite has excellent sensitivity for Pb(II) than separate materials. This emphasises the synergistic effect of waste-char and $NiCo_2S_4$.

8.6 CONCLUSION AND FUTURE PERSPECTIVE

This study demonstrated that waste-derived nanomaterials, such as CQDs, carbon nanotubes, metal oxides, and reduced graphene oxide, have played a significant role in lowering costs and enhancing the sensitivity of electrochemical and optical sensors for trace metals and emerging organic pollutants. Among these materials, CQD is the most widely synthesised and employed owing to its ease of synthesis and enhanced photoluminescence and electrochemical properties. The LOD for heavy metal(loid)s ranged from 2.40×10^{-7} μM to 1.75 μM obtained from various electrochemical detection techniques and fluorescence. On the other hand, trace organic pollutants exhibited LODs in the range of 9.0×10^{-10} μM to 1.43 μM using electrochemical detection techniques only. This area of research still requires more attention, as converting waste into value-added nanomaterials helps reduce the negative environmental impact of waste disposal and lowers the cost of qualitative and quantitative analysis of inorganic and organic pollutants in water. More focus should be placed on expanding the range of nanomaterials derived from waste for electrochemical and optical sensing, so that this area is not skewed towards CQDs.

REFERENCES

Arrigo, A., Cancelliere, A. M., Galletta, M., Burtone, A., Lanteri, G., Nastasi, F., & Puntoriero, F. (2023). From waste to energy: Luminescent solar concentrators based on carbon dots derived from surgical facemasks. *Materials Advances*, *4*(21), 5200–5205.

Borrill, A. J., Reily, N. E., & Macpherson, J. V. (2019). Addressing the practicalities of anodic stripping voltammetry for heavy metal detection: A tutorial review. In *Analyst* (Vol. 144, Issue 23, pp. 6834–6849). Royal Society of Chemistry.

Boruah, A., & Saikia, B. K. (2022). Chemical fabrication of efficient blue-luminescent carbon quantum dots from coal washery rejects (waste) for detection of Hg2+ and Cr6+ ions in water. *ChemistrySelect*, *7*(14).

Bressi, V., Celesti, C., Ferlazzo, A., Len, T., Moulaee, K., Neri, G., Luque, R., & Espro, C. (2024). Waste-derived carbon nanodots for fluorimetric and simultaneous electrochemical detection of heavy metals in water. *Environmental Science: Nano*, *11*(3), 1245–1258.

Bressi, V., Chiarotto, I., Ferlazzo, A., Celesti, C., Michenzi, C., Len, T., Iannazzo, D., Neri, G., & Espro, C. (2023). Voltammetric sensor based on waste-derived carbon nanodots for enhanced detection of nitrobenzene. *ChemElectroChem*, *10*(13).

Chaudhary, K., Mogha, N. K., Lalwani, S., Sharma, R. K., & Masram, D. T. (2020). Ruthenium oxide nanoparticles immobilized over: *Citrus limetta* waste derived carbon material for electrochemical detection of hexestrol. *Journal of Materials Chemistry B*, *8*(35), 7956–7965.

Chaudhary, Y., Rao, M. S., Rakesh, B., & Sankaran, K. J. (2025). Hybrid ZnO–porous graphene nanostructured electrode for the electrochemical detection of cadmium, lead and paraquat. *Journal of Materials Chemistry A*, *13*, 27446–27460.

Cui, Y., Chang, X., Zhai, Y., Zhu, X., Zheng, H., & Lian, N. (2006). ICP-AES determination of trace elements after preconcentrated with p-dimethylaminobenzaldehyde-modified nanometer SiO_2 from sample solution. *Microchemical Journal*, *83*(1), 35–41.

Ding, Q., Kang, Z., Cao, L., Lin, M., Lin, H., & Yang, D. P. (2020). Conversion of waste eggshell into difunctional $Au/CaCO_3$ nanocomposite for 4-Nitrophenol electrochemical detection and catalytic reduction. *Applied Surface Science*, *510*, 145526.

Ding, Q., Li, C., Wang, H., Xu, C., & Kuang, H. (2021). Electrochemical detection of heavy metal ions in water. *Chemical Communications*, *57*(59), 7215–7231.

Elgrishi, N., Rountree, K. J., McCarthy, B. D., Rountree, E. S., Eisenhart, T. T., & Dempsey, J. L. (2018). A practical beginner's guide to cyclic voltammetry. *Journal of Chemical Education*, *95*(2), 197–206.

Erk, N., Bouali, W., Genc, A. A., Salamat, Q., & Soylak, M. (2024). An ultrasensitive electrochemical sensor using banana peel activated carbon/$NiFe_2O_4$/MnCoFe-LDH nanocomposites for anticancer drug determination. *ACS Omega*, *9*(25), 27446–27457.

Fernandes, S., Algarra, M., Gil, A., Esteves da Silva, J., & Pinto da Silva, L. (2025). Development of a facile and green synthesis strategy for brightly fluorescent carbon dots from various waste materials. *ChemSusChem*, *18*(2).

Fiori, S., Scroccarello, A., Della Pelle, F., Del Carlo, M., Cozzoni, E., & Compagnone, D. (2025). Integrating electrochemical sensors in circular economy: Biochar-film sensors based on paper industry waste for agri-food by-product valorization. *Green Analytical*, *13*, 100277.

Fozing Mekeuo, G. A., Despas, C., Péguy Nanseu-Njiki, C., Walcarius, A., & Ngameni, E. (2022). Preparation of functionalized ayous sawdust-carbon nanotubes composite for the electrochemical determination of carbendazim pesticide. *Electroanalysis*, *34*(4), 667–676.

Garg, N., Deep, A., & Sharma, A. L. (2024). Recent trends and advances in porous metal-organic framework nanostructures for the electrochemical and optical sensing of heavy metals in water. In *Critical reviews in analytical chemistry* (Vol. 54, Issue 5, pp. 1121–1145). Taylor and Francis Ltd.

Ghaneh, M. A., Hashemi, M., Hashemnia, S., & Solati, Z. (2024). Proposing a selective electrochemical sensor for Pb2+ prepared from fish-waste-derived hydroxyapatite/gelatin nanocomposite. *ChemistrySelect*, *9*(5).

Ghiasi, M., Ghanbarzadeh, M., Ghaffarinejad, A., & Shahdost-Fard, F. (2025). Green nitrogen and sulfur co-doped carbon dots derived from eggshell as a high performance aptasensing interface for non-invasive detection of metronidazole. *Talanta*, *285*.

Harraz, F. A., Rashed, M. A., Faisal, M., Alsaiari, M., & Alsareii, S. A. (2022). Sensitive and selective electrochemical sensor for detecting 4-nitrophenole using novel gold nanoparticles/reduced graphene oxide/activated carbon nanocomposite. *Colloids and Surfaces A: Physicochemical and Engineering Aspects*, *654*, 130068.

Hossain, M. S., Aumi, S. A., Tarannum, N., Chowdhury, F., Sahadat Hossain, M., Ahmed, M. F., Tanvir, N. I., Akhtar, U. S., Ahmed, S., & Mobarak, M. Bin. (2025). Upcycling battery waste into Zn-BTC metal organic framework for dual applications in electrochemical sensing of uric acid and antibacterial applications. *RSC Advances*, *15*(37), 30564–30575.

Hu, Y., & Gao, Z. (2019). Highly photoluminescent carbon dots derived from discarded chewing gum: Toward multiple sensing of pH, ferric ion, and adenosine triphosphate. *ChemistrySelect*, *4*(43), 12807–12814.

Keerthana, P., & Cherian, A. R. (2025). Pharmaceutical waste-derived carbon quantum dots via microwave method for selective Au3+ ion detection. *ChemistrySelect*, *10*(30).

Khajeh, M., Laurent, S., & Dastafkan, K. (2013). Nanoadsorbents: Classification, preparation, and applications (with emphasis on aqueous media). *Chemical Reviews*, *113*(10), 7728–7768.

Khajeh, M., & Sanchooli, E. (2011). Synthesis and evaluation of silver nanoparticles material for solid phase extraction of cobalt from water samples. *Applied Nanoscience*, *1*, 205–209.

Lee, T. W., Tsai, I. C., Liu, Y. F., & Chen, C. (2022). Upcycling fruit peel waste into a green reductant to reduce graphene oxide for fabricating an electrochemical sensing platform for sulfamethoxazole determination in aquatic environments. *Science of the Total Environment*, *812*, 152273.

Li, Z., & Zhu, M. (2020). Detection of pollutants in water bodies: Electrochemical detection or photo-electrochemical detection? *Chemical Communications*, *56*(93), 14541–14552.

Liu, Y., Liang, P., & Guo, L. (2005). Nanometer titanium dioxide immobilized on silica gel as sorbent for preconcentration of metal ions prior to their determination by inductively coupled plasma atomic emission spectrometry. *Talanta*, *68*(1), 25–30.

Mohammadi, S., Taher, M. A., Beitollahi, H., & Naghizadeh, M. (2019). Sensitive voltammetric determination of cadmium at a carbon nanotubes/Fe_3O_4/eggshell composites modified carbon paste electrode. *Environmental Nanotechnology, Monitoring & Management*, *12*, 100241.

Nazari, Z., Hadi Nematollahi, M., Zareh, F., Pouramiri, B., & Mehrabani, M. (2023). An electrochemical sensor based on carbon quantum dots and ionic liquids for selective detection of dopamine. *ChemistrySelect*, *8*(3).

Osteryoung, J. G., & Osteryoung, R. A. (1985). Square wave voltammetry. *Analytical Chemistry*, *57(1)*, 101–109.

Pabel, M. Y., Kabir, M. H., Hossain, M. S., Mojumder, F., Datta, S., Bashar, M. S., & Yasmin, S. (2025). Dopamine detection using leaf-shaped ZnO synthesized from zinc shells of recycled batteries. *Materials Advances*, *6*(7), 2243–2252.

Ramaley, L., & Krause, M. S. (1969). Theory of square wave voltammetry. *Analytical* Chemistry, *41*(11), 1362–1365.

Sebastian, M., Aravind, A., & Mathew, B. (2019). Green silver nanoparticles based multitechnique sensor for environmental hazardous Cu(II) ion. *BioNanoScience*, *9*(2), 373–385.

Sharma, R., Rana, D. S., Awasthi, A., Singh, D., Ibrahim, A. A., Umar, A., & Baskoutas, S. (2024). Nitrogen functionalized biomass derived mesoporous carbon nanomaterials for electrochemical detection of lead (II) ions. *Heliyon*, *10*(20).

Shivakumar, M., Dharmaprakash, M. S., Manjappa, S., & Nagashree, K. L. (2020). Green synthesis of silver nanoparticles (SNPs)-modified electrode for electrochemical detection of nitrobenzene. *Journal of the Iranian Chemical Society*, *17*(4), 893–900.

Su, Z., Wang, J., Hu, S., Cheng, Y., Yang, Y., Zhou, S., Chen, M., Cao, Q., Zhang, S., Yang, L., Liu, Z., & Tu, X. (2024). In-situ reshaping nano-biochar on electrode surface for machine learning assisted selective sensing of Pb2+ in real water samples. *Applied Surface Science*, *665*, 160294.

Surendran, P., Lakshmanan, A., Priya, S. S., Geetha, P., Rameshkumar, P., Kannan, K., Hegde, T. A., & Vinitha, G. (2021). Fluorescent carbon quantum dots from Ananas comosus waste peels: A promising material for NLO behaviour, antibacterial, and antioxidant activities. *Inorganic Chemistry Communications*, *124*, 108397.

Teo, J. Y. Q., Zheng, X. T., Seng, D. H. L., Hui, H. K., Chee, P. L., Su, X., Loh, X. J., & Lim, J. Y. C. (2022). Waste polystyrene-derived sulfonated fluorescent carbon nanoparticles for cation sensing. *ChemistrySelect*, *7*(36).

Thakkar, H., Vaswani, K., & Thakore, S. (2025). Bifunctional composite films embedded with plastic waste-derived carbon quantum dots for synergistic uranium detection and enrichment. *ACS Applied Nano Materials*, *8*(27), 13650–13662.

Thakur, A., Devi, P., Saini, S., Jain, R., Sinha, R. K., & Kumar, P. (2019). *Citrus limetta* organic waste recycled carbon nanolights: Photoelectro catalytic, sensing, and biomedical applications. *ACS Sustainable Chemistry and Engineering*, *7*(1), 502–512.

Upoma, B. P., Yasmin, S., Ali Shaikh, M. A., Jahan, T., Haque, M. A., Moniruzzaman, M., & Kabir, M. H. (2022). A fast adsorption of azithromycin on waste-product-derived graphene oxide induced by h-bonding and electrostatic interactions. *ACS Omega*, *7*(34), 29655–29665.

Wang, Y., Wu, W. T., Wu, M. B., Sun, H. Di, Xie, H., Hu, C., Wu, X. Y., & Qiu, J. S. (2015). Yellow-visual fluorescent carbon quantum dots from petroleum coke for the efficient detection of Cu2+ ions. *New Carbon Materials*, *30*(6), 550–559.

Wu, H., Wen, Q., Luan, X., Yang, W., Guo, L., & Wei, G. (2024). Facile synthesis of Fe-doped, algae residue-derived carbon aerogels for electrochemical dopamine biosensors. *Sensors*, *24*(9).

Yang, B., Bin, D., Tian, T., Liu, Y., & Liu, B. (2020). An ordered mesoporous carbon nanofiber array for the sensitive electrochemical detection of malachite green. *ChemElectroChem*, *7*(3), 659–664.

Yeasin Pabel, M., Yasmin, S., Shaikh, M. A. A., & Kabir, M. H. (2024). Electronic waste derived reduced graphene oxide supported silver nanoparticles for the electrochemical sensing of trace level arsenite in aqueous medium. *Sensors and Actuators A: Physical*, *366*, 115028.

Yi, C., Zhang, Q., Li, C. W., Yang, J., Zhao, J., & Yang, M. (2006). Optical and electrochemical detection techniques for cell-based microfluidic systems. *Analytical and Bioanalytical Chemistry*, *384*(6), 1259–1268.

Zelai, T. (2025). Integrating $NiCo_2S_4$ with waste-derived carbonaceous material: A highly efficient and selective electrochemical sensor for simultaneous dopamine and Pb (II) detection. *Surfaces and Interfaces*, *72*, 107409.

Zhang, J., Chen, Y., He, Y., Bai, Y., Wang, W., Yang, G., Kong, C., Cao, X., & Gu, L. (2024). Fruit waste-derived carbon dots with rhodamine B for the ratiometric detection of Fe3+ and Cu2+. *Analytical Methods*, *16*(43), 7413–7423.

Zhang, S., Mao, Y., Sun, J., Song, T., Song, Z., Zhao, X., & Wang, W. (2023). One-pot solvothermal preparation of triple-emission N, Cl doped carbon quantum dots from waste traditional Chinese medicines as a fluorescent sensor for sensing water and Cr (VI). *Colloids and Surfaces A: Physicochemical and Engineering Aspects*, *669*, 131471.

Zhang, W., Li, L., Yan, M., Ma, J., Wang, J., Liu, C., Bao, Y., Jin, H., & Fan, Q. (2023). Turning waste into treasure: Multicolor carbon dots synthesized from waste leather scrap and their application in anti-counterfeiting. *ACS Sustainable Chemistry and Engineering*, *11*(13), 5082–5092.

Zhao, X., Wang, L., Ren, S., Hu, Z., & Wang, Y. (2021). One-pot synthesis of Forsythia@ carbon quantum dots with natural anti-wood rot fungus activity. *Materials & Design*, *206*, 109800.

9 Chemical sensors for metals and trace organics

9.1 INTRODUCTION

A requirement in the discussion of chemical sensors from waste-derived nanomaterials will require, to a certain degree, definitions of the important terms used in the statement. For the purpose of this chapter, the chemical sensors will be described first. According to the International Union of Pure and Applied Chemistry (IUPAC), a chemical sensor is a device that converts chemical information into an analytically useful signal (Hulanicki et al., 1991). The chemical information may originate from a chemical reaction or a physical property of the investigated system (IUPAC). Chemical sensors consist of two functional units: a receptor and a transducer (Gouma, 2010; Hulanicki et al., 1991).

The receptor transforms chemical information into a form that the transducer can measure. The transducer converts energy from the receptor to a useful analytical signal (Hulanicki et al., 1991). Chemical sensors are classified according to the transducer principle. These include (i) optical sensors, which transform optical phenomena resulting from the chemical interaction of the analyte and receptor (Brecht & Gauglitz, 1997; Hulanicki et al., 1991; Seitz & Sepaniak, 1988). (ii) Electrochemical sensors transform an electrochemical interaction into a useful analytical signal (Saputra, 2023). (iii) Electrical sensors, unlike their electrochemical counterparts, have a signal that is a result of the change in electrical properties rather than an electrochemical interaction. (iv) Mass-sensitive sensors transform the mass change caused by the accumulation of the analyte at a modified surface. (vi) Lastly, thermometric sensors are based on the measurement of the heat effects of a specific chemical reaction that involves the analyte (Hulanicki et al., 1991).

Generally, chemical sensor receptors may be based on the following principles. Physical, where no chemical reaction is taking place. Chemical, where there is a chemical reaction giving rise to a signal, and biochemical, in which the source of the analytical signal is a biochemical process. This chapter focuses on receptors prepared from waste-derived nanomaterials.

The term "waste" in brief refers to something that the holder intends to dispose (Thürer et al., 2017). Though not necessarily useless, waste is often considered useless by the holder (Sirach & Dave, 2024; Thürer et al., 2017). There are different categories of waste, as described by Aslam et al. (2024). Nanomaterials in brief can be defined as materials with at least one dimension within the range of

DOI: 10.1201/9781003745846-9

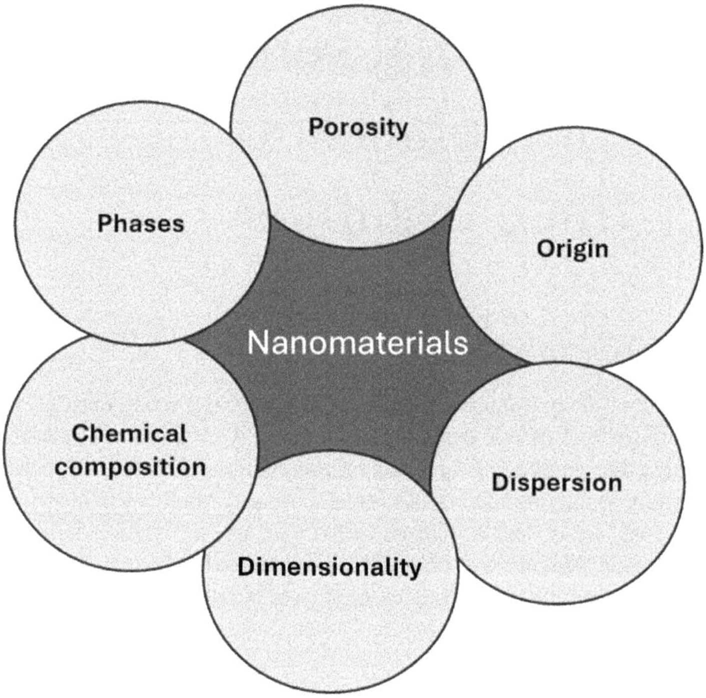

FIGURE 9.1 Classification of materials at the nanoscale.

1–100 nm (nanoscale). Figure 9.1 shows the different classifications of nanomaterials (Barhoum et al., 2022; Harish et al., 2022; Sirach & Dave, 2024). There are two primary methods for preparing waste-derived nanomaterials. This includes using waste as a source material or using waste to assist in the preparation of nanomaterials (Sirach & Dave, 2024).

In recent years, there has been significant research done on the application of nanomaterials in sensor devices (Munonde & Nomngongo, 2020). These are mainly focused on the nanomaterials' ability to serve as transducers with high selectivity and sensitivity (Gouma, 2010). This is due to their size-dependent properties, such as large surface-to-volume ratios, surface chemistry, and their tuneable physical properties, which make them ideal for target analyte interfacing (Malode et al., 2023; Munonde & Nomngongo, 2020). Different classes of nanomaterials have been studied for the fabrication of chemical sensors. Transduction in chemical sensors with nanomaterials has been shown to have superior performance in terms of sensitivity and detection limits (Samaddar et al., 2018)

Thus, the aim of this book is to review the different types of waste-derived nanomaterials and their use in chemical sensing. This chapter focuses on the synthesis, characterisation, and application of waste-derived nanomaterials on the chemical sensing of trace metals and emerging trace organic compounds.

9.2 CHEMICAL SENSORS

9.2.1 Synthesis and characterisation

Before one considers preparing nanomaterials from waste materials, the initial stage before synthesis is often pretreatment (Sharma et al., 2024). This process can be achieved by either of the two ways, a combination of both chemical and physical pretreatment. The most common chemical pretreatment methods are acid or base treatments. Their physical counterparts include grinding and milling (Razzak et al., 2025). Various procedures have been explored for the synthesis of waste-derived nanomaterials. These can be divided into (i) top-down and (ii) bottom-up approaches, (iii) physical methods, (iv and v) physical and chemical methods (Aslam et al., 2024). Chapter 5 of this book covers in detail the different waste materials that can be used for the synthesis of nanomaterials. The review by Chand et al. also details different synthesis routes for biowaste-derived carbon nanomaterials (Chand et al., 2025).

In the preparation of chemical sensors, the common characterisation techniques that are often used include Fourier transform infrared spectroscopy (FTIR) to study the key surface functional groups and chemical bonds of the prepared materials (Mandal et al., 2025) X-Ray Diffraction (XRD) (Mishra et al., 2025), X-ray photo-electron (XPS) to obtain information on the chemical state of elements and relative composition by atomic layers present in CDs and UV-Vis spectroscopy to study the quality of the carbon dots, which is essential for the fluorescence emission (Mandal et al., 2025; Mishra et al., 2025; Ngu et al., 2016; Sharma et al., 2017). Electron microscopy is also used in the form of a transmission electron microscope to determine the shapes and study the dispersion of the nanomaterials (Yahaya Pudza et al., 2020; Zheng et al., 2018a).

9.2.2 Application of waste-derived nanomaterials for the chemical sensing of trace metals

The release of metals into natural water from wastewater is a concern. Anthropogenic processes, such as agricultural and chemical practices and energy conversion, are among the causes of high levels of water pollution by trace metals (Munonde et al., 2017). Excessive ingestion of trace metals has been linked to adverse health effects by the World Health Organization (WHO) (Cotruvo, 2017). Traditionally, trace metal detection is often performed using spectroscopic techniques (Munonde & Nomngongo, 2020; Ullah et al., 2018; Willner & Vikesland, 2018). However, the use of these analytical techniques is tedious due to their bulkiness, limiting their use to laboratories.

Chemical sensors offer an alternative, providing good selectivity and sensitivity while reducing analysis costs. Furthermore, the incorporation of nanomaterials into the sensors not only reduces the size of the analytical device but also enhances the sensor performance owing to the unique properties of nanomaterials (Munonde & Nomngongo, 2020). In the subsequent section, the application of waste-derived chemical sensors for trace metal sensing is discussed. In the discussion, the use of electrochemical sensors has been omitted, as this is covered in Chapter 8.

Though there is extensive literature reports on the chemical sensing of trace metals using nanomaterials in a variety of samples, as reported in Dutta & Das (2012) and Ullah et al. (2018). There is limited evidence of the use of waste-derived nanomaterials in chemical sensing of trace metals. Most studies, as reviewed in Chapter 8 of this book, and reviews such as Dutta & Das (2012), Malode et al. (2023), Munonde & Nomngongo (2020), Ullah et al. (2018), and Willner & Vikesland, 2018) report the use of waste-derived nanomaterials in electrochemical sensing.

Though scarce, fluorescent detection of metal ions has been reported by some researchers. The detection of Au^{3+} in aqueous solutions was reported by Ramanan et al. (2018) using plastic waste-derived carbon dots. The nanomaterial was prepared via solvothermal synthesis. The authors described a selective and sensitive fluorescent sensor with a limit of detection of 53 nM. The prepared sensor demonstrated translatable performance in real water samples. The described mechanism of action was likely to induce aggregation (Ramanan et al., 2018).

Other studies also reported the colourimetric detection of metal ions using plastic waste-based carbon nanomaterials (Cruz et al., 2019; Miao et al., 2024; Kumari et al., 2018; Chaudhary et al., 2021). Plastic waste-derived carbon nanomaterials showed great potential for detecting metal ions in water samples. These studies showed good selectivity of the sensors in real samples, even in the presence of non-target ions. This highlighted the benefit of valorising waste plastic into usable materials.

Various other types of waste have also been used to prepare carbon nanomaterials for the sensing of metal ions. In 2015, Ngu et al. used carbon nanoparticles from waste rice husks to prepare an optical sensor for metal ions in water (Ngu et al., 2016). Sharma et al. detected chromium (IV) using waste pineapple juice-derived carbon nanoglobules (Sharma et al., 2017). Much later, studies by Roshni et al. (Roshni et al., 2019), Mohammadzadeh et al. (Mohammadzadeh et al., 2022), Mandal et al. (Mandal et al., 2025), and Mirsha et al. (Mishra et al., 2025) also explored various starting materials for the preparation of chemical sensors for metal-ion detection.

9.2.3 Application of waste-derived nanomaterials for the chemical sensing of emerging trace organics

The detection of emerging contaminants (ECs) in the environment is an issue of concern (La Farre et al., 2008; Poynton & Robinson, 2018). Emerging organic compounds are used in daily products such as plasticisers, pesticides, surfactants, personal care products, and pharmaceuticals. These compounds are reported to be difficult to remove in the short term (Wilkinson et al., 2017). Different types of emerging contaminants enter the environment via intentional or unintentional disposal (Khan et al., 2022). As such, the detection of these contaminants is of importance. In recent times, the detection and quantification of emerging contaminants have been achieved primarily through chromatography-based methods. There are very few studies currently employing waste-derived nanomaterials.

One such study reported by Zheng and others used *Fructus lycii* to prepare carbon dots (CDs), which were further capped with silver nanoparticles to form CDs-AgNPs. The prepared CDs-AgNPs were used to detect phoxim in water. For real-world applications, they collected tap water, river water, and an apple spiked with known concentrations of phoxim. They obtained good recoveries of 87%–110.0% and LOD of 0.04 μM. The sensor worked by the coordination of amino and carboxylate groups via electrostatic interactions. This created a cross-linkage between the nanoparticles and phoxim (Zheng et al., 2018b).

Murugesan and others used orange peels to prepare a carbon-dot silver nanohybrid (Ag@CDs) for the detection of melamine. Ag@CDs demonstrated good sensitivity and selectivity for melamine sensing in aqueous solution, with an LOD of ~0.85 μM. For real applicability of the prepared carbon dot, milk with melamine was used, and it showed a recovery of 98.03%, which shows that the material prepared from orange peels and conjugated with silver is efficient in sensing applications (Murugesan et al., 2023)

9.3 CONCLUSION AND FUTURE PERSPECTIVES

This chapter aims to review waste-derived nanomaterials and their use in chemical sensing of metal ions and emerging trace organic compounds. The synthesis and characterisation techniques for waste-derived nanomaterials were briefly covered. The techniques included spectroscopy and microscopy. However, in achieving all the aims, this chapter has shown that more studies are needed in the field of chemical sensing of trace metal ions and emerging trace organic compounds. Although studies on waste-derived nanomaterials for heavy metal remediation exist, there are few on chemical sensing of trace metal ions. Similarly, this was a challenge in studies of emerging trace organic compounds. Most studies on chemical sensors focus only on electrochemical sensors. Waste-derived nanomaterials show strong potential for chemical sensing, and further studies are needed for the analysis of trace metals and emerging organic compounds.

REFERENCES

Aslam, J., Aslam, R., & Hussain, C. M. (2024). *Waste-derived nanoparticles: Synthesis, applications and sustainability*. Elsevier.

Barhoum, A., García-Betancourt, M. L., Jeevanandam, J., Hussien, E. A., Mekkawy, S. A., Mostafa, M., Omran, M. M., S. Abdalla, M., & Bechelany, M. (2022). Review on natural, incidental, bioinspired, and engineered nanomaterials: History, definitions, classifications, synthesis, properties, market, toxicities, risks, and regulations. *Nanomaterials, 12*(2), 177.

Brecht, A., & Gauglitz, G. (1997). Recent developments in optical transducers for chemical or biochemical applications. *Sensors and Actuators B: Chemical, 38*(1–3), 1–7.

Chand, S., Kumar, R., Thakur, N., Kumar, K., Umar, A., Almas, T., & Baskoutas, S. (2025). Sustainable synthesis and multifunctional applications of biowaste-derived carbon nanomaterials and metal oxide composites: A review. *Chemosphere, 385*, 144540.

Chaudhary, S., Kumari, M., Chauhan, P., & Chaudhary, G. R. (2021). Upcycling of plastic waste into fluorescent carbon dots: An environmentally viable transformation to biocompatible C-dots with potential prospective in analytical applications. *Waste Management, 120*, 675–686.

Cotruvo, J. A. (2017). 2017 WHO guidelines for drinking water quality: First addendum to the fourth edition. *Journal-American Water Works Association, 109*(7), 44–51.

Cruz, M. I. S. Dela, Thongsai, N., de Luna, M. D. G., In, I., & Paoprasert, P. (2019). Preparation of highly photoluminescent carbon dots from polyurethane: Optimization using response surface methodology and selective detection of silver (I) ion. *Colloids and Surfaces A: Physicochemical and Engineering Aspects*, *568*, 184–194.

Dutta, M., & Das, D. (2012). Recent developments in fluorescent sensors for trace-level determination of toxic-metal ions. *TrAC Trends in Analytical Chemistry*, *32*, 113–132.

Gouma, P.-I. (2010). *Nanomaterials for chemical sensors and biotechnology*. Pan Stanford Publishing.

Harish, V., Ansari, M. M., Tewari, D., Gaur, M., Yadav, A. B., García-Betancourt, M.-L., Abdel-Haleem, F. M., Bechelany, M., & Barhoum, A. (2022). Nanoparticle and nanostructure synthesis and controlled growth methods. *Nanomaterials*, *12*(18), 3226.

Hulanicki, A., Glab, S., & Ingman, F. (1991). Chemical sensors: Definitions and classification. *Pure and Applied Chemistry*, *63*(9), 1247–1250.

Khan, S., Naushad, M., Govarthanan, M., Iqbal, J., & Alfadul, S. M. (2022). Emerging contaminants of high concern for the environment: Current trends and future research. *Environmental Research*, *207*, 112609.

Kumari, A., Kumar, A., Sahu, S. K., & Kumar, S. (2018). Synthesis of green fluorescent carbon quantum dots using waste polyolefins residue for Cu2+ ion sensing and live cell imaging. *Sensors and Actuators B: Chemical*, *254*, 197–205.

La Farre, M., Pérez, S., Kantiani, L., & Barceló, D. (2008). Fate and toxicity of emerging pollutants, their metabolites and transformation products in the aquatic environment. *TrAC Trends in Analytical Chemistry*, *27*(11), 991–1007.

Malode, S. J., Shanbhag, M. M., Kumari, R., Dkhar, D. S., Chandra, P., & Shetti, N. P. (2023). Biomass-derived carbon nanomaterials for sensor applications. *Journal of Pharmaceutical and Biomedical Analysis*, *222*, 115102.

Mandal, T., Senapati, R. N., Ghosh, A. K., Masto, R. E., & Singh, V. (2025). Kitchen waste-derived oxygen-doped fluorescent carbon quantum dots for ultra-trace detection of chromium (VI) in environmental water matrices. *ACS Sustainable Resource Management*, *2*(9), 1786–1796.

Miao, C., Wang, Q., Yang, S., Tang, Y., Liu, X., & Lu, S. (2024). Hydrothermal route upcycling surgical masks into dual-emitting carbon dots as ratiometric fluorescent probe for Cr (VI) and corrosion inhibitor in saline solution. *Talanta*, *275*, 126070.

Mishra, A., Kushwaha, A., & Verma, R. (2025). Detection of Pb2+ and Fe3+ ions in wastewater using a low cost portable optical sensor derived from green synthesized metal oxide nanoparticle on a paper substrate. *Analytica Chimica Acta*, 344477.

Mohammadzadeh, S. E., Faghiri, F., & Ghorbani, F. (2022). Green synthesis of phenolic capping Ag NPs by green walnut husk extract and its application for colorimetric detection of Cd2+ and Ni2+ ions in environmental samples. *Microchemical Journal*, *179*, 107475.

Munonde, T. S., Maxakato, N. W., & Nomngongo, P. N. (2017). Preconcentration and speciation of chromium species using ICP-OES after ultrasound-assisted magnetic solid phase extraction with an amino-modified magnetic nanocomposite prepared from Fe3O4, MnO2 and Al2O3. *Microchimica Acta*, *184*(4), 1223–1232.

Munonde, T. S., & Nomngongo, P. N. (2020). Nanocomposites for electrochemical sensors and their applications on the detection of trace metals in environmental water samples. *Sensors*, *21*(1), 131.

Murugesan, P., Libiya, N., Moses, J. A., & Anandharamakrishnan, C. (2023). Fluorescence resonance energy transfer-based sensor with silver-conjugated orange peel waste-derived carbon dots for melamine detection. *Carbon Letters*, *33*(7), 2335–2348.

Ngu, P. Z. Z., Chia, S. P. P., Fong, J. F. Y., & Ng, S. M. (2016). Synthesis of carbon nanoparticles from waste rice husk used for the optical sensing of metal ions. *New Carbon Materials*, *31*(2), 135–143.

Poynton, H. C., & Robinson, W. E. (2018). Contaminants of emerging concern, with an emphasis on nanomaterials and pharmaceuticals. In *Green chemistry* (pp. 291–315). Elsevier.

Ramanan, V., Siddaiah, B., Raji, K., & Ramamurthy, P. (2018). Green synthesis of multi-functionalized, nitrogen-doped, highly fluorescent carbon dots from waste expanded polystyrene and its application in the fluorimetric detection of Au3+ ions in aqueous media. *ACS Sustainable Chemistry & Engineering*, *6*(2), 1627–1638.

Razzak, S. A., Uddin, S., Hasan, A., Nawaz, A., Siddiquee, M. N., Al Bari, M. A., & Hossain, M. M. (2025). Biomass-derived carbon materials as sustainable platforms for advanced biomedical applications. *Environmental Surfaces and Interfaces*, *4*, 45–70.

Roshni, V., Misra, S., Santra, M. K., & Ottoor, D. (2019). One pot green synthesis of C-dots from groundnuts and its application as Cr (VI) sensor and in vitro bioimaging agent. *Journal of Photochemistry and Photobiology A: Chemistry*, *373*, 28–36.

Samaddar, P., Ok, Y. S., Kim, K.-H., Kwon, E. E., & Tsang, D. C. W. (2018). Synthesis of nanomaterials from various wastes and their new age applications. *Journal of Cleaner Production*, *197*, 1190–1209.

Saputra, H. A. (2023). Electrochemical sensors: Basic principles, engineering, and state of the art. *Monatshefte Für Chemie-Chemical Monthly*, *154*(10), 1083–1100.

Seitz, W. R., & Sepaniak, M. J. (1988). Chemical sensors based on immobilized indicators and fiber optics. *C R C Critical Reviews in Analytical Chemistry*, *19*(2), 135–173.

Sharma, S., Kumar, R., & Kumar, K. (2024). Synthesis of biowaste derived carbon nanomaterials: Efficiency, quality, and environmental impact. *Indian Chemical Engineer*, 1–20.

Sharma, S., Umar, A., Mehta, S. K., & Kansal, S. K. (2017). Fluorescent spongy carbon nanoglobules derived from pineapple juice: A potential sensing probe for specific and selective detection of chromium (VI) ions. *Ceramics International*, *43*(9), 7011–7019.

Sirach, R., & Dave, P. N. (2024). Introduction of waste-derived nanoparticles. In *Waste-derived nanoparticles* (pp. 3–42). Elsevier.

Thürer, M., Tomašević, I., & Stevenson, M. (2017). On the meaning of 'waste': Review and definition. *Production Planning & Control*, *28*(3), 244–255.

Ullah, N., Mansha, M., Khan, I., & Qurashi, A. (2018). Nanomaterial-based optical chemical sensors for the detection of heavy metals in water: Recent advances and challenges. *TrAC Trends in Analytical Chemistry*, *100*, 155–166.

Wilkinson, J. L., Hooda, P. S., Swinden, J., Barker, J., & Barton, S. (2017). Spatial distribution of organic contaminants in three rivers of Southern England bound to suspended particulate material and dissolved in water. *Science of the Total Environment*, *593*, 487–497.

Willner, M. R., & Vikesland, P. J. (2018). Nanomaterial enabled sensors for environmental contaminants. *Journal of Nanobiotechnology*, *16*(1), 95.

Yahaya Pudza, M., Zainal Abidin, Z., Abdul Rashid, S., Md Yasin, F., Noor, A. S. M., & Issa, M. A. (2020). Eco-friendly sustainable fluorescent carbon dots for the adsorption of heavy metal ions in aqueous environment. *Nanomaterials*, *10*(2), 315.

Zheng, M., Wang, C., Wang, Y., Wei, W., Ma, S., Sun, X., & He, J. (2018a). Green synthesis of carbon dots functionalized silver nanoparticles for the colorimetric detection of phoxim. *Talanta*, *185*, 309–315.

Zheng, M., Wang, C., Wang, Y., Wei, W., Ma, S., Sun, X., & He, J. (2018b). Green synthesis of carbon dots functionalized silver nanoparticles for the colorimetric detection of phoxim. *Talanta*, *185*, 309–315.

10 Remediation of trace metals and organics using nanomaterials derived from waste

10.1 INTRODUCTION

Water pollution is enormously threatening ecosystems and posing a serious risk to biodiversity. Rapid urbanisation and industrialisation have led to the release of huge amounts of both inorganic and organic pollutants into the environment, causing severe adverse effects to the aquatic and terrestrial ecosystems (Dehkordi et al., 2024). Therefore, designing innovative, effective and sustainable methods for wastewater remediation is urgent. A considerable number of wastewater remediation methods have been developed, including photocatalysis, adsorption, membrane filtration, advanced oxidation processes (AOPs), and electrochemical treatment, among others (Ejraei et al., 2019; Guo et al., 2024). Although these methods differ in their operational processes and mechanisms, they all rely on functional nanomaterials (FNs) to attain efficient pollutant removal. Various conventionally synthesised functional nanomaterials (CSFNs) have been applied as adsorbents, electrocatalysts and photocatalysts for the removal of metals and organic pollutants (Priya et al., 2024; Gao et al., 2025). However, preparing CSFNs requires expensive precursor materials, which undermines economic and environmental sustainability.

Recently, waste-derived nanomaterials (WDNs) have attracted significant attention across various fields, including gas storage, catalysis, food safety, information technology, and wastewater remediation (Mecha et al., 2025). To date, the synthesis of WDNs using effective approaches continues to attract growing attention. Thus far, a considerable number of metal-based nanoparticles and carbon-based nanomaterials have been successfully made from waste (Patel, Kulwant, et al., 2024; Ndagijimana et al., 2025). These nanoparticles demonstrated high performance in wastewater remediation. Various synthetic strategies have been employed to convert environmental solid waste into functional WDNs, including ball milling, chemical co-precipitation, pyrolysis, and hydrothermal methods (Chen et al., 2022). Solid wastes that are generally converted to WDNs (photocatalysts, adsorbents, and electrocatalysts) for wastewater remediation include industrial waste (e.g, red mud, brick waste, sewage sludge, and coal mining waste) (Judith & Vasudevan, 2022; Kumar et al., 2025), biomass waste (e.g, tea waste, lotus seedpod, soya waste, and animal faeces) (Kshirsagar et al., 2023), and electronic waste or e-waste (spent batteries, waste printed circuit boards, electronic

DOI: 10.1201/9781003745846-10

packaging waste, and waste liquid crystal displays) (Pineda-Vásquez et al., 2023). Industrial and electronic wastes are considered special wastes due to their composition, origin, and environmental risks, whereas biomass waste is considered sustainable waste because it produces fewer harmful by-products during processing (Chen et al., 2022).

According to the World Bank, agricultural and industrial wastes are among the special wastes that contribute 3/4 of the total waste (Chen et al., 2022). Therefore, converting industrial and agricultural waste into WDNs solves the entire environmental pollution while significantly mitigating the cost of nanomaterials synthesis. The WDNs produced from waste vary in chemical composition, size, morphology, and moisture content. A considerable number of studies have been conducted employing WDNs for the removal of microorganisms, organic and inorganic pollutants (J. Singh et al., 2025; Aththanayaka et al., 2025). For example, Boruah and co-workers reported photocatalytic degradation of three dyes (MB, RhB, and CR) using magnetic mixed metal oxide nanomaterials derived from industrial waste, and the nanomaterials achieved over 95% degradation efficiency for all dyes (Boruah et al., 2020). In a different study, Panneerselvam and co-workers employed a nanocomposite constituting silver nanoparticles incorporated into activated carbon derived from cigarette butt waste for anti-larval and antibacterial activity in wastewater remediation (Panneerselvam et al., 2022). Apart from the application of WDNs for wastewater remediation, controlling their shapes, sizes, and tunability while enhancing performance can be challenging. Therefore, designing effective methods for converting waste into useful nanomaterials with desirable properties (high surface area, tunable moieties, and high stability) is indispensable.

Physicochemical properties of WDNs differ depending on the type of the synthetic strategy (ball milling, pyrolysis, hydrothermal, chemical co-precipitation, etc.) and the features of waste precursors (Aslam et al., 2024). For example, metal oxide nanoparticles are successfully synthesised using sol-gel and combustion methods, while pyrolysis is used to prepare carbon materials (Dunnigan et al., 2018). Thus far, there is a considerable number of studies reporting biochar derived from biomass waste in wastewater remediation (Yunus & Fu, 2024; da Silva Araújo et al., 2024; Soundari & Prasanna, 2025). The application of FNs derived from industrial, electronic, and biomass wastes for the removal of metals and organic pollutants, employing various methods such as electrochemical treatment, AOPs, adsorption, and photodegradation, remains limited. Therefore, there is a need to review the advances in FNs derived from various solid wastes for the removal of both organic and metal pollutants.

This review gives an overview of the recent advances in WDNs and their nanocomposites for the removal of trace metals and organic pollutants from wastewater. First, FNs derived from various solid wastes are discussed. Second, different methods for preparing WDNs and their nanocomposites are summarised, and significant factors in synthetic strategies are discussed. Third, recent advances in remediation technologies employing WDNs for the removal of trace metals and organic pollutants are also discussed. Lastly, challenges and prospects of remediation techniques are also summarised.

10.2 NANOMATERIALS DERIVED FROM BIOMASS, INDUSTRIAL AND ELECTRONIC WASTE

Recently, various waste-derived nanomaterials and nanocomposites have been reported for wastewater remediation. Depending on the source of precursor materials and design strategies, various WDNs have demonstrated remarkable performance as adsorbents, photocatalysts, and electroactive materials.

10.2.1 Functional nanomaterials derived from biomass waste

Biomass waste has gained enormous attention as a precursor for functional carbon materials (FCMs). Plants account for over 82% of biomass, with the rest coming from animal leftovers or biological processes (Ullah et al., 2024). FNs can be obtained from abundant biomass resources such as (i) animal wastes from livestock, (ii) agricultural crops and their wastes, (iii) residues from forestry, (iv) wood wastes, (v) wastewater sludge and solid municipal waste, and (vi) algae from agriculture and marine plants (H. He et al., 2023). Various carbon-based nanomaterials (CNMs) with diverse shapes have been prepared from biomass waste and used for wastewater remediation. These include carbon nanotubes (CNTs), carbon nanocages (CNs), carbon nanospheres (CNSs), graphene, fullerenes, activated carbon (AC), hydrochar and biochar (Khandaker et al., 2025; A. Feng et al., 2024). These CNMs offer a promising solution, including sustainable, economical materials for wastewater remediation. Among CNMs derived from biomass, biochar prepared through pyrolysis has been broadly used and has exhibited high performance for wastewater remediation (Tanveer et al., 2025; E. Singh et al., 2021). Attractive characteristics such as high surface area, microporosity and stable structure have earned biochar diverse applications in the removal of heavy metals, organic pollutants, and dyes from wastewater (Liang et al., 2021; Praveen et al., 2022). As mentioned, these CBMs have shown outstanding performance as adsorbents, photocatalysts, and electroactive materials.

10.2.2 Functional nanomaterials derived from industrial waste

Industrialisation has led to the release of hazardous metals, including chromium, nickel, arsenic, copper, zinc, cadmium, mercury, and lead, into the environment. The environmental ubiquity of these metals arises from various sectors, including metal processing, mining, paint manufacturing, battery manufacturing, textile manufacturing, electroplating, and the pulp and paper industry (R. Sharma et al., 2024; Mo et al., 2018). Synthesis of nanomaterials from industrial waste presents an effective solution for wastewater remediation, while promoting economic and environmental sustainability through waste valorisation. In addition, converting industrial waste into value-added products reduces secondary pollution generated during conventional treatment. FNs derived from industrial waste are attracting attention in wastewater remediation. Among industrial waste contributors, electroplating waste serves as a precursor for the preparation of WDNs. This is because electroplating waste is enriched in metals (Cr, Fe, Ag, Sn, Cu, Ni, etc.) and organic matter, typically available in various forms such as hydroxides, metal oxides, silicates, phosphates, and sulfates (Bayrakdar et al., 2025). These are used as potential precursors for the preparation of advanced nanomaterials for wastewater remediation. Thus far, magnetic chitosan, magnetic biochar, zeolite, TiO_2-doped fly ash, MFe_2O_4 (M=Cu, Mn, Ag, Ni, Fe, Zn), and metals (Ni, Al, Cr, Fe, Cu) have been recovered from industrial waste for wastewater remediation (Mo et al., 2018; R. Sharma et al., 2024; Judith & Vasudevan, 2022). These were recovered from various waste sources, such as steel industry waste, fly ash, paper mill sludge, electroplating waste, tannery waste, pickling waste, and phosphogypsum waste (Kaithwas et al., 2012).

10.2.3 Functional nanomaterials derived from e-waste

Electronic waste is increasingly becoming a societal problem due to high amounts of electronic devices being discarded into the environment. According to the world record on environmental sustainability and waste management, approximately 54 million tonnes of e-waste were generated in 2019, with projections suggesting an increase to ≈75 million tonnes by 2030 (Jain et al., 2023). Therefore, converting e-waste to value-added products is urgent. Various e-waste sources reported include waste printed circuit boards (WPCBs), spent batteries, electronic packaging waste, and waste liquid crystal displays (Seif et al., 2023). Metals (e.g., Li, Mn, Pb, Hg, Pt, Rh, Cu, Ni, Co, Pd, Au, rare earth metals, etc.) and nanomaterials such Zn–Mn magnetite ferrite materials, ZnO NPs, MnO_2/graphene nanocomposites, $ZnMn_2O_4$NPs, PbI_2 perovskite nanocrystals, CuO/TiO_2 NPs, Cu_2O NPs, carbon nanotubes, nanozero-valent Cu particles, polymers, $CoFe_2O_4$ nanocrystals and Zn–Mn NPs, have been successfully recovered from e-waste and used in wastewater remediation processes (Chen et al., 2022; Seif et al., 2023; Pineda-Vásquez et al., 2023). Spent batteries are the most prevalent category A e-waste materials owing to their wide use as portable power sources. Therefore, most metals, metal oxides, and nanomaterials are recovered from battery waste. Conversely, used machines and instruments such as mobile phones, supercapacitors, automobiles, personal computer circuit boards and laptops are classified as category B e-wastes (Robinson, 2009). Polymers and most metals recovered from battery waste have also been extracted from used machines and instruments (Achilias et al., 2009).

10.3 METHODS FOR PREPARATION OF WASTE-DERIVED NANOMATERIALS

Studies reporting the direct use of untreated solid waste for wastewater remediation are scarce, underscoring the need for deliberate strategies to effectively convert such waste into FNs. Various methods are employed to prepare WDNs with controlled sizes, stability, chemical composition, density, and morphology. Mainstream methods widely used to convert solid waste into FNs include sol-gel, hydrothermal, co-precipitation, pyrolysis and ball milling, all of which are discussed in this section.

10.3.1 Hydrothermal method

Preparation of WDNs can be achieved through the hydrothermal method, which converts solid waste into FNs by heating at high temperatures and pressures in a sealed reaction vessel (Katika et al., 2025). Transforming solid waste into FNs can be achieved using two hydrothermal synthesis routes: (i) First, hydrothermal synthesis can be used to convert waste into metal oxides and metal-based composites, such as metal oxides, metal sulfides, metal hydroxide/carbon composites, red mud/biomass waste composites, metal sulfide/biochar composites, metal oxide, and sulfide composites. Second, high-moisture biomass waste, such as animal excreta and sewage sludge, can be converted to FNs via hydrothermal processing, eliminating the need for a separate drying step (Y. Zhang et al., 2025). To obtain well-defined, nanostructured materials, hydrothermal process parameters, such as solution pH, reaction temperature, reaction time, and liquid-to-solid ratio, must be carefully optimised (Y. Zhao et al., 2025).

Many studies have reported the use of a single hydrothermal process to convert solid waste into FNs (Y. Zhao et al., 2025; Y. Zhang et al., 2025). However, other studies have attempted to integrate the hydrothermal process with other techniques to obtain efficient FNs from waste. For example, Lestari and co-workers integrated pyrolysis (300°C)—hydrothermal synthesis (180°C) for the preparation of CQDs from plastic bag waste (Lestari et al., 2025). Similarly, Li and co-workers applied a hydrothermal–pyrolytic integrated method for the preparation of hierarchical porous Al_2O_3 and AlOOH microspheres from red mud (Li et al., 2017). It is worth noting that integrating these methods can conveniently and efficiently convert various solid wastes into FNs while reducing resource utilisation and minimising secondary pollution.

10.3.2 Pyrolysis

Pyrolysis is a powerful method for efficiently transforming waste into FNs with tailored physicochemical characteristics. Pyrolysis is conducted in an inert environment that promotes the formation of hydrocarbons, biochar, and volatile gases (Devi et al., 2021). By contrast, calcination involves heating solid waste to form FNs at high temperatures in an oxygenated atmosphere, yielding metal oxide-based nanomaterials (Cheng et al., 2018). Pyrolysis is primarily used to convert biomass waste into FNs, and the physicochemical properties of the resulting biomass-derived FNs depend on parameters such as heating rate, pyrolysis temperature, and incubation time (Papari et al., 2021). It is imperative to optimise pyrolysis parameters to attain well-tailored FNs that meet application-specific performance criteria. For example, biochar prepared by pyrolysis contains specific functional groups (e.g., –COOH, –C=O, and –OH) that are beneficial for wastewater remediation (Ganesan et al., 2025). However, changing pyrolysis parameters can significantly alter surface chemistry and chemical compositions of WDNs. This is observed when oxygen-containing moieties on biochar are fully destroyed at pyrolysis temperatures exceeding 800°C, resulting in hydrophobic aromatic biochar that exhibits exceptional performance as an adsorbent for the removal of a wide range of organic pollutants (Huang et al., 2024). However, excessively high temperatures can destroy the biochar structure, thereby attenuating its adsorption performance. At low temperatures below 400°C, biochar contains more functional moieties than those prepared at high temperatures. Obviously, these functional moieties offer significant advantages, as they contribute significantly to the removal of organic and inorganic pollutants through adsorption and photodegradation (Huang et al., 2024). Nonetheless, a different study revealed that, compared to oxygen-rich biochar obtained at low temperature (200°C), biochar prepared at a higher temperature (≈700°C) demonstrated remarkable persulfate activation, facilitating efficient charge separation for the photocatalytic degradation of acid orange 7 (Han et al., 2021). Therefore, it is encouraged to optimise the surface properties of pyrolysed biochar to meet application-specific requirements, thereby improving performance by altering pH, functional groups, surface chemistry, and porosity. Normally, biochar obtained from biomass pyrolysis exhibits unfavourable properties, such as a small surface area, low porosity, and poor surface chemistry, which significantly limit its application for removing pollutants from wastewater (Shaaban et al., 2013). Therefore, numerous studies have attempted to surface-dope biochar to enhance its efficiency and versatility in removing a wide range of metals and organic pollutants (T. Wang et al., 2024).

10.3.3 Co-precipitation

The co-precipitation method is among the wet-chemical methods used to convert waste into FNs. This method presents excellent performance for the preparation of oxide-based nanomaterials in solution by harnessing nucleation, growth, and agglomeration processes to obtain well-defined oxide nanostructures (Madji et al., 2023). In the co-precipitation method, factors such as temperature, precursor ratio, surface ligand, and pH influence the size, composition, and morphology of the precipitates (Atul et al., 2022). Co-precipitation method constitutes two main routes for the synthesis of FNs from waste: (i) First route involves extracting metals from solid waste (e.g., industrial waste) into liquid phase, which then serves as the precursor of metal oxides in the co-precipitation process; (ii) Second, carbon materials derived from waste are used as a support for metal hydroxides/oxides to prepare waste-derived carbon (WDC)/metal hydroxides/oxides nanocomposites through co-precipitation method (Chen et al., 2022). For example, Zhao and co-workers prepared magnetic reed biochar through co-precipitation for the adsorption of florfenicol (H. Zhao & Lang, 2018).

10.3.4 Sol-gel method

Sol-gel is an effective and versatile method for preparing oxide nanomaterials. Since its discovery in the 1800s, it has been used to prepare materials for various applications, such as catalysis, water purification, fibres, chemical sensors, and engineering (Parashar et al., 2020). The sol-gel method involves transforming suspended colloidal particles into a continuous gel network, which is then calcined at elevated temperatures to obtain oxide nanomaterials. Compared to hydrothermal materials under similar conditions, those prepared by the sol-gel method are highly crystalline, homogeneous, pure, and modifiable (Zanotti et al., 2021). Sol-gel mainly depends on operational parameters such as temperature, solution concentration, pH, and reaction time. In the sol-gel, biomass or e-wastes are converted into liquefied waste, generating a solution containing reactive species, which serves as a precursor (sol) for forming a gel (Nsubuga et al., 2018). For example, Yao and co-workers prepared $LiMn_2O_4$ from lithium-ion battery waste via the sol-gel method (Yao et al., 2021). In a separate study, Zhang and co-workers developed a biochar-doped $MnFe_2O_4$ nanocomposite via sol-gel. In this nanocomposite, biochar acted as a support to allow effective dispersion of $MnFe_2O_4$ to facilitate adsorption Pb (II) and Cd (II) (L. Zhang et al., 2019).

10.3.5 Ball milling

Ball milling is a commonly used top-down approach for reducing particle size and enhancing particle reactivity. The typical ball milling process involves the use of mechanical forces to grind materials into fine pieces and thoroughly mix them (Wei et al., 2023). Ball milling is a versatile technique used to prepare well-defined powders, synthesise complex materials, and disperse nanomaterials. Ball milling offers crucial benefits such as low-temperature operation, reduced processing time, minimal usage of chemicals, and high yields, making it an effective strategy for transforming waste into value-added materials (Reyes-Vallejo et al., 2023). The milling process can be tailored to a specific product outcome by adjusting factors such as rotational speed, milling time, atmospheric conditions, temperature, and ball-to-material ratio.

Ball milling has been widely used to transform solid waste into FNs, with biochar-based nanocomposites serving as notable representatives (Xiao et al., 2021). For example, Hail and co-workers prepared activated biochar from shrimp waste via ball milling for effective CO_2 capture and waste valorisation (Aljomard et al., 2025). A different study milled waste aluminium shavings into fine particles, followed by chemical precipitation to obtain alumina nanoparticles (Oji et al., 2025). Doping nanoparticles into biochar yields nanocomposites with synergistic properties that make them highly effective for the adsorption of a wide variety of pollutants. A recent study demonstrated the efficacy of doping nanoparticles into biochar via pyrolysis-ball milling, in which zinc oxide was successfully incorporated into bamboo-derived biochar for the adsorption of methylene blue (Yu et al., 2021). Thus, transforming solid waste into FNs through ball milling offers an economic and sustainable practice that enables high-yield valorisation of waste.

10.3 REMEDIATION TECHNOLOGIES FOR TRACE METAL POLLUTION IN WATER/WASTEWATER

WDNs have been used to remove trace metals from environmental wastewater using various techniques, including physicochemical methods (adsorption), biological methods (bioremediation and biosorption), and hybrid methods (photocatalysis). Depending on the target pollutants, various techniques coupled with suitable WDNs enable efficient elimination of either trace metals or organic pollutants. Recent and ongoing studies have demonstrated the successful application of WDNs in various treatment techniques to remove trace metals and organic pollutants from wastewater.

10.3.1 Physicochemical methods

10.3.1.1 Adsorption

Adsorption is one of the most widely used physical treatment methods for removing pollutants from wastewater. The performance of the adsorption method mainly depends on the properties of the adsorbents such as surface area, porosity, functional moieties, and particle sizes (Tee et al., 2022). Metals have been reported to have severe adverse effects on aquatic systems, even when present at concentrations in the low ng L^{-1} (Mitra et al., 2022). Consequently, adsorption has emerged as an effective technique, particularly when coupled with WDNs, for removing trace metals from wastewater (R. Sharma et al., 2024). WDNs can interact with trace metals through distinct mechanisms, including electrostatic interactions, ion exchange, metal-π interactions, and surface complexation.

Among WDNs, biomass-derived FNs, such as biochar and its nanocomposites, have demonstrated exceptional performance in removing trace metals. This is because biochar possesses desirable properties, such as oxygen-based functional groups (–COOH, –C=O, and –OH), facile synthesis, and a high surface area. These functional groups are favourable for the adsorption of a wide range of trace metals (Ye et al., 2022). Moureen and colleagues investigated food waste-derived biochar in removing Pb, Cr, and Cd from wastewater. The biochar demonstrated exceptional performance in removing these trace metals, primarily through electrostatic interactions at pH 8 (Moureen et al., 2024). In another study, Madzin and co-workers prepared spent mushroom compost biochar (MCB) to remove trace metals (Cu, Mn, Fe, and Pd). The interactions between MCB and metals were mainly cation exchange, π-complexation, and electrostatic interaction. The study demonstrated a high removal efficiency, exceeding 97.4% for all metals within 60 min (Madzin et al., 2025).

To improve the adsorption performance of biochar-based adsorbents, various approaches have been employed, including salt activation, surface modification and addition of bases, acids, and oxidising agents. Thus far, activation of biochar with NaOH, $ZnCl_2$, K_2CO_3, and KOH increases surface area, porosity, and surface functionality, thereby facilitating the adsorption of trace metals. Acids such as H_2SO_4 can add additional functional moieties to the surface of biochar, thereby enhancing its adsorption efficiency and capacity. Zuhara and McKay used K_2CO_3 as an activating agent for waste-derived activated carbons (WDACs) to remove strontium (Sr) and barium (Ba) from wastewater, with the adsorption mechanism revealing electrostatic attraction, ion exchange, and π–electron interactions (Zuhara & McKay, 2025). Pan and colleagues used biomass-derived biochar activated with H_3PO_4 for effective adsorption of Cr(VI). In the study, H_3PO_4 served as an acidic activator, promoting the formation of pores, which in turn increased the surface area (782.6 m^2/g). In addition, H_3PO_4 added functional moieties (–COOH, –OH, and O–P) onto the surface of activated biochar, which promoted electrostatic attraction between Cr(VI) and activated biochar (Pan et al., 2025). Compared with H_3PO_4, K_2CO_3, and other biochar activators, KOH has demonstrated exceptional performance in biochar activation as Herath and colleagues synthesised KOH-activated biochar, with a high surface area of 1049 m^2/g and adsorption capacity of 127.2 mg g^{-1} for Cr(VI), Cd(II), and Pb(II) (Herath et al., 2021).

In addition to biomass waste, industrial waste-derived nanomaterials (IWDNs) have shown significant potential for adsorbing trace metals. Depending on the source, FNs such as metals and metal nanoparticles, carbon-based materials, silica and silicate nanomaterials, and other functional nanomaterials can be obtained from industrial waste. Hamadeen and colleagues synthesised nanostructured water treatment residual (nWTRs) from industrial waste via ball milling for the removal of Cr(VI) from wastewater. According to the EDX analysis of nWTRs, silicon, aluminium, iron, and calcium were predominant with smaller amounts of manganese, titanium, potassium, and sulfur. The adsorption capacity of nWTRs for Cr(VI) was 40.7 mg g^{-1} after 15 min, which was 15 times higher than that of bulk water treatment residual (nWTRs) (2.8 mg g^{-1}). The dominating interaction mechanisms between nWTRs and Cr(VI) involved pore filling, complexation, electrostatic attraction and reduction process (Hamadeen et al., 2022). In another study, Groszek and co-workers used herbal industry-derived activated carbons (HIDACs) in removing Cd(II) and As(V) from wastewater. The HIDACs nanocomposite was prepared through simultaneous pyrolysis and physical activation using CO_2 and demonstrated reasonable adsorption capacities of 4 mg g^{-1} for As(V) and 8 mg g^{-1} for Cd(II) after 24 h. The low adsorption performance of HIDACs for Cd(II) and As(V) was ascribed to the presence of multiple elements that were not completely removed during pyrolysis and activation, which hindered its performance (Groszek et al., 2025). Therefore, additional synthesis steps and post-modification processes are required to obtain HIDACs with enhanced performance.

E-waste-derived FNs also demonstrated exceptional efficiency in removing metals from wastewater. Metals and metal oxide nanomaterials are the predominant FNs derived from e-waste. Various FNs derived from e-waste have shown remarkable efficiency in removing a spectrum of trace metals. Khan and colleagues reported the use of waste-derived printed circuit board (PCB) non-metallic powder (A-NMP) for the removal of As(V) from wastewater. The A-NMP exhibited extremely low adsorption capacity (0.004 mg g^{-1}) for As(V) due to residual impurities, limited surface functionalities, and inaccessible porosity (M. Khan et al., 2024). In another study, Zou and colleagues used a spent $LiFePO_4$ battery to prepare AlFePO-Li zeolite nanocomposite via hydrothermal, which was used to remove Pb(II) and Cd(II) from wastewater. The performance of the nanocomposite was exceptional as it exhibited adsorption capacities of 723 mg g^{-1} for Pb(II) and 71.4 for Cd(II) (Zou et al., 2021). E-waste-derived metal oxide nanomaterials generally exhibit limited performance for certain trace metals, depending on the physicochemical properties of the e-waste and the treatment processes applied. Nonetheless, they have been widely used and demonstrated remarkable efficiency in energy storage, catalysis, sensing, and the degradation of organic pollutants (Bhattacharya et al., 2021).

In addition to the use of individual nanomaterials for trace metal removal, designing nanocomposites with synergistic effects has demonstrated high performance in trace metal removal. Carbon-doped metal nanocomposites have attracted tremendous attention owing to the complementary merits of their components: carbon provides high surface area, porosity, and high surface functionalities, while metals offer specific surface interactions, high surface activity, magnetism, and tunable surface chemistry. Algethami and colleagues employed a cross-linked shrimp waste-derived chitosan@Al_2O_3 (Cs@Al_2O_3) nanocomposite for the removal of Cr(VI) from wastewater. The Cs@ Al_2O_3 nanocomposite exhibited a high adsorption capacity of 433.1 mg g^{-1} at pH 4 after 120 min (Algethami et al., 2025). Similarly, Elkhatib and co-workers demonstrated the efficacy of combining nWTR with zeolite for the removal of Pb(II) from wastewater. In the study, the zeolite-nWTR nanocomposite achieved an adsorption capacity of 198.7 mg g^{-1}, which was 5.5 and 2.6 times higher than the adsorption capacities of pristine zeolite (36 mg g^{-1}) and nWTR (75 mg g^{-1}), respectively (Elkhatib et al., 2025). Ternary or quaternary nanocomposites can also be formed by combining zeolite/carbon-based materials with metal-organic frameworks (MOFs), sulfides, and metal (hydr)oxides. Such nanocomposites inherit the advantageous features of all pristine materials, thereby enhancing adsorption capacity and overall adsorption performance. For example, Liu and colleagues developed a tea waste-derived porous biochar Fe–Mn bimetallic nanocomposite (FMBC) via pyrolysis, and the nanocomposite was used to remove U(VI) from wastewater (Liu et al., 2021). Billah and co-workers used epichlorohydrin cross-linked shrimp waste-derived chitosan-based calcium phosphate nanocomposite (Cs@Epic/Ca Apatite) for the removal of Pb(II) from aqueous wastewater. The Cs@Epic/Ca Apatite nanocomposite exhibited a high adsorption capacity of 154.9 mg g^{-1}, which demonstrated that integrating multiple pristine waste-derived nanomaterials can significantly improve overall adsorption performance (Billah et al., 2023). Table 10.1 summarises recent representative WDNs and nanocomposites used as effective adsorbents for trace metal removal.

10.3.2 Biological methods and hybrid methods

Biological methods, such as microbial remediation and biosorption, are widely used for trace metal removal from wastewater, while photocatalysis is used as a hybrid method to remove metals by reduction or oxidation. Among these methods, photocatalysis is most widely explored owing to its diverse features, such as minimal secondary pollution, the ability to design green and eco-friendly photocatalysts, and low operating costs (Waleng et al., 2025; J. Ma et al., 2024). Recently, semiconductor-based photocatalysts, including ZnO, TiO_2, and CeO_2, have been widely used to convert toxic metals into more beneficial or more stable forms (Nallaselvam et al., 2023). This is because ZnO, TiO_2, and CeO_2 have superior photochemical properties, including high photostability and non-selectivity and non-toxicity during the photocatalytic transformation process (Khyrun et al., 2025). Thus far, Cr(VI) has been widely used as a model pollutant to explore the photocatalytic performance of semiconductor-based photocatalysts under UV light and visible light (X. Xu et al., 2024; Q. Wang et al., 2025; Z. Xu et al., 2025). While offering advantages, these photocatalysts have demonstrated a few drawbacks such as low quantum efficiency, photocatalyst agglomeration and difficulty in separation. Therefore, doping these photocatalysts with WDNs addresses both the performance and environmental sustainability (Khoirunisa et al., 2025). For example, Nallaselvam and colleagues doped TiO_2 with biochar derived from COVID-19 face mask waste for photoreduction of toxic Cr(VI) to essential Cr(III). The biochar was obtained through co-pyrolysis. The biochar/TiO_2 nanocomposite exhibited high photoreduction performance at low concentration and acidic pH (Nallaselvam et al., 2023). In a different study, Du and colleagues employed TiO_2 doped with biochar derived from wheat straw for photoreduction of Cr(VI) under simulated sunlight illumination (Du et al., 2024). The study also demonstrated that combining nanomaterials into nanocomposites improves overall performance. This was observed when biochar/TiO_2 exhibited higher removal efficiency of 99.9%, which

TABLE 10.1
Application of WDNs and nanocomposites as adsorbents for trace metal removal from environmental wastewater.

[1]WDM	Pollutant	Conditions	Adsorption Capacity/%Removal Efficiency (%RE)	Reference
Food waste-derived biochar	Pd(II), Cr(VI), and Cd(II)	[2]MA (2.5 g/L), pH (8), [3]AT (2 h), Concentration (1000 mg/L)	≥ 90% (all trace metals)	Moureen et al. (2024)
Spent mushroom compost biochar	Cu(II), Mn(II), Fe(II), and Pb(II)	[2]MA (0.2 g), pH (6–10), [3]AT (60 min), Concentration (10 mg/L)	≥97.4% (all trace metals)	Madzin et al. (2025)
Waste-derived activated carbons	Sr(II) and Ba(II)	[2]MA (0.3 g), pH (5.5), [3]AT (9 h), Concentration (100 mg/L)	Sr(II) = 91.0 mg/g and Ba(II) = 99.0 mg/g	Zuhara and McKay (2025)
[4]GNMACs	Cr(VI)	[2]MA (20 mg), pH (2), [3]AT (90 min), Concentration (5 mg/L)	101.0 mg/g	Pan et al. (2025)
Activated Douglas Fir biochar	Cr(VI), Cd(II), and Pb(II)	Pb(II)- [2]MA (25 mg), pH (5), [3]AT (60 min), Concentration (100 mg/L)	140.0 mg/g	Herath et al. (2021)
[5]nWTRs	Cr(VI)	[2]MA (100 mg), pH (5–7), [3]AT (15 min), Concentration (160 mg/L)	40.7 mg/g	Hamadeen et al. (2022)
[6]HIDACs	Cd(II) and As(V)	[2]MA (30 mg), pH (6), [3]AT (24 h), Concentration (200 mg/L)	Cd(II) = 8 mg/g, As(V) = 4 mg/g	Groszek et al. (2025)
[7]A-NMP	As(V)	[2]MA (200 mg), pH (6), [3]AT (30 min), Concentration (1 mg/L)	0.004 mg/g	M. Khan et al. (2024)
AlFePO-Li zeolite derived from lithium battery waste	Cd(II) and Pb(II)	[2]MA (200 mg), pH (5), [3]AT (720 min), Concentration (1 g/L)	Cd(II) = 71.4 mg/g, Pb(II) = 723.8 mg/g	Zou et al. (2021)
IONPs derived from iron oxide waste powder	Cr(VI) and Pb(II)	[2]MA (350 mg), pH (7.5), [3]AT (50 min), Concentration (1 mg/L)	Cr(VI) = 326.8 mg/g, Pb(II) = 417 mg/g	Mwebembezi et al. (2024)

Ag NPs derived from *Murcott Mandarin* peel	Pb(II)	[2]MA (0.15 g), pH (5.5), [3]AT (60 min), Concentration (100 mg/L)	42.7 mg/g	Aboelghait et al. (2024)
[8]PP@ TiO_2	As(III)	[2]MA (25 mg), pH (7), [3]AT (24 h), Concentration (25 mg/L)	76.9 mg/g	Poudel et al. (2020)
SnO_2@CaO derived from chicken eggshell	Cr(VI) and Pb(II)	[2]MA (10 mg), pH 7, [3]AT (5 h), Concentration: Cr(VI) = 5 mg/L, Pb(II) = 160 mg/L	Cr(VI) = 94%, Pb(II) = 88%	Rani et al. (2023)
[9]$CNFs_{WP}$	Co(II)	[2]MA (50 mg), pH 4, [3]AT (25 min), Concentration (1 mg/L)	95.8%	Alatawi et al. (2025)
[10]OWPSA	Cu(II)	[2]MA 20 mg), pH (5), [3]AT (600 min), Concentration (50 mg/L)	47.2 mg/g	Elboughdiri et al. (2024)
[11]MWBC	Cr(VI)	[2]MA (30 mg), pH (2), [3]AT (200 min), Concentration (100 mg/L)	90.7 mg/g	Tejada-Tovar et al. (2023)
[12]SR and BH ACs	Cr(VI) and Zn(II)	[2]MA (320 mg), pH: Cr(VI) (7): Zn(II) (6), [3]AT (130 min), Concentration (0.8 mg/L)	Cr(VI) = 1.6 mg/g, Zn(II) = 1.5 mg/g	Batubara et al. (2023)
Corncob waste-derived AC doped with γ-Fe_3O_4	Pb(II) and Cd(II)	[2]MA (200 mg), pH (5), [3]AT (-), Concentration (100 mg/L)	Pb(II) = 127 mg/g, Cd(II) = 100 mg/g	Arul et al. (2023)
Waste facial masks-derived AC-Fe_3O_4	As(III) and Pb(II)	[2]MA (5 mg), pH (5): As(III) (7): Pb(II) (5), [3]AT (50 min), Concentration (60 mg/L)	As(III) = 186.3 mg/g, Pb(II) = 233.9 mg/g	Dolati et al. (2025)
Orange Peel Cellulose (OPC)	Cr(VI), Cd(II), and Pb(II)	[2]MA (100 mg), pH (4–8), [3]AT (36 h), Concentration (1 mM)	Cr(VI) = 4.9 mg/g Cd(II) = 29 mg/g, Pb(II) = 50.1 mg/g	Rahman et al. (2023)

(Continued)

TABLE 10.1 (*Continued*)
Application of WDNs and nanocomposites as adsorbents for trace metal removal from environmental wastewater.

[1]WDM	Pollutant	Conditions	Adsorption Capacity/%Removal Efficiency (%RE)	Reference
Carbon/alginate/Fe_3O_4	Pb(II)	[2]MA (1.5 g/L), pH (6), [3]AT (80 min), Concentration (10 mM)	37.8 mg/g	Zadeh and Jafari (2024)
Peanut shells-derived activated carbon (PS-AC)	Pb(II)	[2]MA (2.75 g/L), pH (4.5), [3]AT (1 h), Concentration (150 mg/L)	130.9 mg/g	Hashem et al. (2024)
Corn shell-derived activated carbon (CS-AC)	Cd(II) and U(VI)	[2]MA (4 g/L), pH (7), [3]AT (12 h), Concentration (50 mg/L)	Cd(II) = 22.4 mg/g, U(VI) = 28.5 mg/g	Abou-Hadid et al. (2024)

1 WDM—waste-derived material
2 MA—mass of adsorbent
3 AT—adsorption time
4 GNMACs—golden needle mushroom roots activated carbons
5 nWTRs—nanostructured water treatment residual
6 HIDACs—herbal industry-derived activated carbons
7 A-NMP—activated non-metallic powder
8 PP@ TiO2–TiO2 impregnated pomegranate peels
9 CNFsWP—cellulose nanomaterials derived from wastepaper
10 OWPSA—Olive-Waste-Derived Adsorbents
11 MWBC—modified wheat bran-derived biochar
12 SR and BH ACs—sheep remnants and barely husks activated carbons

was 3.5 and 2.9 times higher than that of pure biochar and TiO_2, respectively. From the electrochemical impedance spectroscopy (EIS) results obtained, the biochar exhibited small arc radius than TiO_2 indicating improved electronic conductivity and interface separation efficiency which are consistent with enhanced photoreduction activity (Du et al., 2024). Khoirunisa and co-workers employed TiO_2 doped with Fe/Fe_3O_4 from iron rust waste as a photocatalyst for photoreduction of Cr(VI) to Cr(III). Again, high photoreduction performance of TiO_2–Fe/Fe_3O_4 was observed at acidic pH and low concentrations. This is because, at low pH, H^+ serves as scavenger for electrons preventing recombination of photogenerated charges (electrons and holes). What is more, most photocatalysts are positively charged at low pH, encouraging electrostatic interaction with metallic oxyanions and subsequent photoreduction. Similarly, photoreduction occurs more efficiently at low concentrations because, (i) metals have full access to the photocatalyst active sites, (ii) light penetrates the solution more effectively, allowing photocatalyst activation and (iii) lastly, recombination of photogenerated charge carriers is minimised, promoting better separation and photoreduction. Ma and colleagues prepared biomass carbon dots/TiO_2 modified chitosan hydrogel nanocomposite (B-CDs/TiO_2@CS) from peanut shell powder for photoreduction of Cr(VI) under simulated sunlight (R. Ma et al., 2025). The B-CDs/TiO_2@CS nanocomposite demonstrated remarkable photoreduction efficiency at pH 2 within 30 min. Constantly, most studies on Cr(VI) photoreduction using WDNs have reported that the efficient photoreduction activity is observed in acidic conditions. This is because Cr(VI) exists predominantly as $HCrO_4^-$ and H_2CrO_4 in acidic aqueous solutions (pH < 6.5), while it is present as CrO_4^{2-} in alkaline solutions (pH > 6.5). Therefore, efficient photoreduction is observed at low pH, and since most photocatalysts are negatively charged in alkaline conditions, electrostatic repulsion with CrO_4^{2-} prevails, suppressing overall photocatalytic activity (Du et al., 2024; R. Ma et al., 2025).

Bioremediation is also a promising method for removing metal pollutants. This approach employs microbial fuel cells (MFC) coupled with WDNs cathode catalysts for effective wastewater remediation. Wang and colleagues employed a bioremediation method to remove Cr(VI) from wastewater. In the study, a nano-jarosite biochar cathode catalyst derived from rice husks was incorporated in MFC system to reduce Cr(VI) to Cr(III). An MFC system achieved an exceptional removal efficiency of 91.5% within 6 h at pH 2, further confirming the trend for metal removal in acidic conditions (Y. Wang et al., 2024). Similarly, Zhao and colleagues prepared a magnetic biochar cathode from corn straw, which served as an electron shuttle in an MFC system to reduce Cr(VI) to Cr(III). The Fe NPs/biochar cathode in MFC system exhibited a high power density of 8.285 ± 0.253 mW m^{-2} and a low internal resistance of $5.4 \pm 0.1\ \Omega$, which were consistent with its effective reduction of Cr(VI) (S. Zhao et al., 2024). Similarly, Christy and colleagues prepared biochar from biomass stalks and incorporated it into a soil microbial fuel cell (SMFC) for the reduction of Cr(VI). The performance of SMFC was influenced by catholyte concentration, pH, and bacterial count. Enhanced SMFC performance and Cr(VI) reduction were observed at low catholyte concentrations, high bacterial count, and optimum pH of 7–8 (Christy et al., 2025). Studies reported that adding a catalyst to the MFC improves the overall removal efficiency of the system. For example,

Lin and colleagues added biochar prepared from peanut husk waste to the MFC system for the removal of Cr(VI). With MFC alone, the percentage removal efficiency (%RE) and power density (PD) were 60.9% and 8.8 mW m^{-3}, respectively. When biochar was added to the MFC system, the PD and %RE increased to 55.1 mW m^{-3} and 97.7%, which were 6.3 and 1.6 times the system without biochar (Lin et al., 2023).

Biosorption is also one of the methods used for effective trace metal removal. Biosorption is similar to adsorption, except that it uses a special class of adsorbents of biological origin, such as bacteria, algae, agricultural wastes, and fungi. For example, Al-Ghamdi and colleagues prepared a biosorbent, silver nanoparticles-doped coffee-ground waste (CWAg), for the removal of Cr(VI) from wastewater. The CWAg exhibited a remarkable adsorption capacity of 149.9 mg g^{-1} at pH 2 within 90 min (Al-Ghamdi et al., 2025). Cobo and co-workers employed the fungus *Penicillium* sp. 8L2 for biosorption of Zn(II). The fungus exhibited an adsorption capacity of 52.1 mg g^{-1} within five days at pH 5.6 (Cobo et al., 2025). In another study, Khalaf and colleagues prepared carbon nanodots (CNDs) from walnut peels as a biosorbent for the removal of Cd(II), Zn(II), and Cu(II) from wastewater. The CNDs exhibited exceptional removal efficiency of the metals within 60 min at pH 7 (Khalaf et al., 2025). Biosorption is not the preferred method for removing trace metals because some bioadsorbents, such as algae and fungi, are highly sensitive to pH, exhibit limited reusability, and have relatively slow adsorption kinetics. Therefore, modifying bioadsorbents with chemicals or other functionalities has become a significant research approach to improve their performance. For example, Belcaid and colleagues loaded cellulose-based biochar with iron oxide nanoparticles and manganese dioxide to remove Co(II) and Cr(VI) from wastewater. Iron oxide nanoparticles and manganese dioxide were prepared from cassava peel carbon (Belcaid et al., 2024). Besides the biosorbents presented in Table 10.2 for trace metal removal, other studies in literature have also reported the use of biosorbents for the removal of a wide range of trace metals from wastewater (Cheng et al., 2018; Papari et al., 2021; Ganesan et al., 2025; Han et al., 2021; Shaaban et al., 2013; T. Wang et al., 2024; Madji et al., 2023; Rao et al., 2024). On the other hand, membrane filtration, among other methods, has been used to a limited extent for trace metal removal (H. Zhao & Lang, 2018; Parashar et al., 2020; Zanotti et al., 2021; Gul & Senol-Arslan, 2025). Table 10.2 summarises recent applications of WDNs in photocatalysis, microbial remediation, and biosorption.

10.4 REMEDIATION TECHNIQUES FOR ORGANIC MICRO-POLLUTANTS IN WATER/WASTEWATER

WDNs and their nanocomposites have also been widely used for the removal of organic micro-pollutants from wastewater using physicochemical, biological and hybrid methods. Adsorption and photocatalysis are the most extensively used methods for the remediation of organic micro-pollutants using WDNs. Table 10.3 summarises recent representative WDNs, and their nanocomposites used as effective adsorbents for organic micro-pollutant removal.

TABLE 10.2
Application of WDNs and their nanocomposites for trace metal removal from environmental wastewater.

[1]WDM	Pollutant	Conditions	%Removal/ Adsorption Capacity	Method	Reference
Biochar/TiO_2	Cr(VI)	[2]PD (0.3 g/L), pH (2), [3]PT (180 min), Concentration (10 mg/L)	98.7%	Photocatalysis	Nallaselvam et al. (2023)
TiO_2–Fe/Fe_3O_4 Source of Fe: rusty iron waste	Cr(VI)	[2]PD (0.2 g), pH (3), [3]PT (90 min), Concentration (10 mg/L)	92.8%	Photocatalysis	KHOIRUNISA et al. (2025)
Biochar/TiO_2 Source of biochar: wheat straw	Cr(VI)	[2]PD (0.2 g), pH (3), [3]PT (90 min), Concentration (10 mg/L)	99%	Photocatalysis	Du et al. (2024)
[4]c-SB/TiO_2	Cr(VI)	[2]PD (0.5 g/L), pH (2), [3]PT (180 min), Concentration (10 mg/L)	98.6%	Photocatalysis	Velumani and Jeyadharmarajan (2023)
[5]B-CDs/TiO_2@CS Biomass carbon dots source: peanut shell powder	Cr(VI)	[2]PD (0.2 g), pH (2), [3]PT (30 min), Concentration (50 mg/L)	99.6%	Photocatalysis	R. Ma et al. (2025)
Fe NPs/biochar electrode	Cr(VI)	*PD (8.285 ± 0.253 mW m^{-2}), [6]RS (0.136 mg/L.h) Concentration (20 mg/L)	98.2%	Bioremediation	S. Zhao et al. (2024)
Biochar derived from stalks of biomass	Cr(VI)	Biochar dosage (10 g), Static potential (774 mV)	16.5%	Bioremediation	Christy et al. (2025)
[7]RHJB	Cr(VI)	Biochar dosage (0.32 g), pH (2) *PD (1211 mW m^{-2}), [6]RS (0.302 mg/L.h), Voltage (682.5 mV) [8]R_{ct} (2580 Ω), Concentration (50 mg/L)	91.5%	Bioremediation	Y. Wang et al. (2024)

(Continued)

TABLE 10.2 (*Continued*)
Application of WDNs and their nanocomposites for trace metal removal from environmental wastewater.

[1]WDM	Pollutant	Conditions	%Removal/ Adsorption Capacity	Method	Reference
Biochar-anatase	Cr(VI)	Biochar dosage (0.3 g), pH (2) [9]CD (164 mA m^{-2}), *PD (10.3 W m^{-2}), Concentration (10 mg/L)	97.1%	Bioremediation	Cui et al. (2023)
[10]CWAg	Cr(VI)	MA (1 g/L), pH (2), AT (90 min), Concentration (30 mg/L)	149.9 mg/g	Biosorption	Al-Ghamdi et al. (2025)
[11]F-P sp. 8L2	Zn(II)	MA (0.2 g/L), pH (5.6), AT (5 days), Concentration (50 mg/L)	52.1 mg/g	Biosorption	Cobo et al. (2025)
CNDs Source: walnut peel	Cd(II), Zn(II), and Cu(II)	MA (0.2 g), pH (7), AT (60 min), Concentration (100 mg/L)	≥ 90% for all analytes	Biosorption	Khalaf et al. (2025)

* PD—power density
[1] WDM—waste derived material
[2] PD—photocatalyst dosage
[3] PT—photoreduction time
[4] c-SB/TiO2—chemically activated sludge biochar/titanium oxide
[5] B-CDs/TiO2@CS—biomass carbon dots/TiO2 modified chitosan hydrogel
[6] RS—removal speed
[7] RHJB—Rice husks-derived jarosite biochar
[8] Rct—charge transfer resistance
[9] CD—current density
[10] CWAg—Coffee-ground waste doped silver nanoparticles
[11] F-P sp. 8L2—fungus Penicillium sp. 8L2

TABLE 10.3
Application of WDNs, and their nanocomposites as adsorbents for organic pollutant removal from environmental wastewater.

WDM	Pollutant	Conditions	Adsorption Capacity/% Removal Efficiency (%RE)	Reference
AC/SiO_2	Phenol and 3,4-dimethoxybenzyl alcohol	MA (0.1 g), pH (-), AT (24 h), Concentration (5400 mg/L)	Phenol = 39.5 mg/g 3,4 dimethoxybenzyl alcohol = 66.9 mg/g	León et al. (2025)
CCNs	[1]AD	MA (0.05 g), pH (11), AT (3 h), Concentration (20 mg/L)	78.1 mg/g	Kargule et al. (2025)
ZnO	MO and MB	MA (0.5 g (MB) and 0.6 g (MO)), pH (13 (MB) and 6 (MO)), AT (105 min), Concentration (50 mg/L)	MO = 83.7%, MB = 87.9%	Hasan et al. (2024)
AMD-mag@ PET-MPAC-25%	CIP, LEV, NOR, and ENR	MA (30 mg), pH (7), AT (30 min), Concentration (1 mg/L)	CIP = 285 mg/g, LEV = 218 mg/g, NOR = 302 mg/g, ENR = 337 mg/g	Mvala et al. (2025)
CNOs	MB	MA (2 g/L), pH (10), AT (20 min), Concentration (20 mg/L)	43.1 mg/g	Patel, Tripathi, et al. (2024)
MJRS	MG	MA (0.6 g/L), pH (8), AT (45 min), Concentration (100 mg/L)	440.5 mg/g	S. Khan et al. (2025)
CB-MOF	IBU	MA (0.3 g/L), pH (3), AT (120 min), Concentration (300 mg/L)	196.1 mg/g	A et al. (2025)
Hydrochar	CBZ and NPX	MA (0.1 g), pH (4), AT (15 h), Concentration (1 mg/L)	CBZ = 49.1 mg/g, NPX = 14.4 mg/g	Saqib et al. (2024)
Activated rice husk nanocomposite	[2]TCN	MA (2 g/L), pH (6), AT (120 min), Concentration (150 mg/L)	78.5 mg/g	Okey-Onyesolu et al. (2025)
CuO/biochar	DOX	MA (2 g/L), pH (7), AT (2 h), Concentration (30 mg/L)	430 mg/g	Kar et al. (2023)

(Continued)

TABLE 10.3 (*Continued*)
Application of WDNs, and their nanocomposites as adsorbents for organic pollutant removal from environmental wastewater.

WDM	Pollutant	Conditions	Adsorption Capacity/% Removal Efficiency (%RE)	Reference
Tomato waste-derived biochar-doped Fe_3O_4/MnO_2	[3]ACT	MA (0.3 g/L), pH (2), AT (20 min), Concentration (30 mg/L)	294.1 mg/g	Stegarescu et al. (2025)
Ag_2O-CuO/biochar	[3]ACT and [4]AMX	MA (0.3 mg), pH (10), AT (30 min), Concentration (600 mg/L)	PCT = 55.2 mg/g AMX = 64.9	Allwar et al. (2023)
Banana peel biochar	IBU	MA (20 mg), pH (9), AT (60 min), Concentration (125 mg/L)	4.5 mg/g	Islam et al. (2024)
Polypyrrole-GO-Biochar	CBZ	MA (1.4 g/L), pH (7.8), AT (90 min), Concentration (1 mg/mL)	45.05 mg/g	P et al. (2024)
MgO/ Al_2O_3-BC700 BC—biochar derived from rice husk	[3]ACT and [5]TMP	MA (0.05 g), pH (6–10), AT (8 h), Concentration (15 mg/L)	ACT = 137.4 mg/g TMP = 269 mg/g	Chaubey et al. (2023)
Acid-activated coconut husk biochar	[4]AMX	MA (4.8 mg), pH (7), AT (6 min), Concentration (50 mg/L)	598 mg/g	Adu-Poku et al. (2024)
nano-MgO-modified biochar derived from castor wood	NPX	MA (1.5 g/L), pH (4), AT (120 min), Concentration (10 mg/L)	123.34 mg/g	Bano et al. (2024)

Activated carbon derived-ginkgo leaves	IBU	MA (0.04 g), pH (3), AT (120 min), Concentration (50 mg/L)	160 mg/g	Su et al. (2025)
Bamboo biochar	[6]CIP and NPX	MA (7.5–10 g/L), pH (3–4), AT (90 min), Concentration (10–20 mg/L)	CIP = 50.8 mg/g, NPX = 56.5 mg/g	Pham et al. (2025)
ZnONP-modified biochar derived from breadfruit seed hull	AMX	MA (0.03 g), pH (7), AT (120 min), Concentration (100 mg/L)	90.67 mg/g	Amaku and Mtunzi (2024)
Prosopis juliflora biochar	[7]SMX and [6]CIP	MA (1 g/L), pH (SMX (8), CIP (5)), AT (120 min), Concentration (50 mg/L)	SMX = 91.4 mg/g, CIP = 49.8 mg/g	Ashebir et al. (2024)

1 AD—aniline dye
2 TCN—tetracycline
3 ACT—acetaminophen
4 AMX—amoxicillin
5 TMP—trimethoprim
6 CIP—ciprofloxacin
7 SMX—sulfamethoxazole

10.4.1 Physicochemical methods

10.4.1.1 Adsorption

Adsorption prevails as one of the extensively used methods for removing organic pollutants from wastewater. Numerous studies have shown that using WDNs as adsorbents addresses both economic and environmental sustainability issues. In addition, WDNs have exhibited exceptional removal efficiency and high adsorption capacity for organic pollutants. This is because WDNs prepared from waste exhibit well-defined tunable structures with increased surface area. Like metals, WDNs and their nanocomposites have also shown exceptional performance in removing organic pollutants. For example, León and colleagues prepared activated carbon (AC) from cocoa shell biomass through pyrolysis for the removal of phenol and 3,4-dimethoxybenzyl alcohol from wastewater. In the study, AC was doped with SiO_2 which exhibited adsorption capacities of 39.5 mg g^{-1} for phenol and 66.9 mg g^{-1} for 3,4-dimethoxybenzyl alcohol, attributed to its mesoporosity and high surface area of 800 m^2/g (León et al., 2025). In a different study, Mvala and colleagues prepared magnetic mesoporous AC derived from PET waste plastic and acid mine drainage (AMD) (AMD-mag@PET-MPAC-25%) for the removal of four fluoroquinolones (FQs) from wastewater. The AMD-mag@PET-MPAC-25% displayed exceptional adsorption capacities ranging from 233 to 385 mg g^{-1} for four FQs (levofloxacin (LEV), norfloxacin (NOR), ciprofloxacin (CIP), and enrofloxacin (ENR)) (Mvala et al., 2025). In another study, Kargule and co-workers prepared calcium carbonated nanoparticles (CCNs) from eggshells for the removal of aniline dye from wastewater. CCNs exhibited high removal efficiency and adsorption capacity of 78.1 mg g^{-1} for aniline dye. The interaction mechanism was predominantly governed by electrostatic attraction (Kargule et al., 2025). Patel and colleagues prepared carbon nano-onions (CNOs) from waste frying oil via pyrolysis for the elimination of methylene blue (MB) from wastewater. The CNOs exhibited a removal efficiency of 99.8% and adsorption capacity of 43.1 mg g^{-1} at pH 10. In the study, the point of zero charge (PZC) of CNOs was 1.75, and the RE% increased with an increasing pH indicating that the CNOs interacted effectively with MB through electrostatic attraction (Patel, Tripathi et al., 2024). Khan and co-workers employed co-precipitation method to prepare magnetically separable nanocomposite derived from lignocellulosic *Juglans regia* for sequestration of malachite green (MG) dye from wastewater. A novel magnetic *Juglans regia* shell (MJRS) displayed an exceptional adsorption capacity of 440.5 mg g^{-1} for MG at pH, facilitated by multiple interaction mechanisms including electrostatic attraction, hydrogen bonding, Yoshida hydrogen bonding, and others (S. Khan et al., 2025). Saqib and colleagues employed a hydrothermal method to synthesise hydrochar from food waste, which was used as an adsorbent for the removal of carbamazepine (CBZ) and naproxen (NPX) from wastewater. The hydrochar demonstrated adsorption capacities of 49.10 mg g^{-1} for CBZ and 14.35 mg g^{-1} for NPX at pH 4 (Saqib et al., 2024). Waste-derived nanocomposites have also been used in removing organic pollutants from wastewater. Coupling of WDNs with other nanomaterials addresses performance shortfalls and enhances economic viability. For example, Suresh and colleagues prepared AC from sugarcane bagasse and doped it with bimetallic (Fe and Zn) metal-organic framework (MOF) for the removal of ibuprofen (IBU) from wastewater. A low surface area of 13.04 m^2/g was obtained for CB-MOF likely due to insufficient cleaning of AC during preparation. Nonetheless, activated carbon-doped bimetallic MOF (CB-MOF) displayed a removal efficiency of 93% and monolayer adsorption capacity of 196.1 mg g^{-1} for IBU (A et al., 2025). In another study, Kar and colleagues combined copper oxide nanoparticles with fabric waste-derived biochar to remove doxycycline hydrochloride (DOX) from wastewater. The CuO/biochar nanocomposite has a high removal efficiency of 97% and adsorption capacity of 430 mg g^{-1} for DOX. In the study, the CuO/biochar nanocomposite achieved a %RE of 97%, while a %RE of 58% was obtained with pristine biochar, demonstrating that doping WDNs with other nanomaterials improves overall performance of the nanocomposite (Kar et al., 2023). Hasan and colleagues employed ZnO derived from battery waste for the removal of MB and methyl orange (MO) dyes from wastewater. The ZnO nanomaterial exhibited remarkable removal efficiencies of 83.7% and 87.9% for MO and MB, respectively. Methyl orange is an anionic dye whereas methylene blue is a cationic dye. Therefore, their interaction with metal oxide nanomaterials primarily occurs though electrostatic attraction, as observed in the study (Hasan et al., 2024).

10.4.1.2 Photocatalysis

WDNs and their nanocomposites have also been widely used as photocatalysts for degradative removal of organic pollutants from wastewater. The photocatalytic degradative process primarily relies on the semiconductor type and its related energy band gap. This determines the type of irradiation required to activate the semiconductor and initiate the degradation process. What is more, oxidants like hydrogen peroxide (H_2O_2), peroxymonosulfate (PMS), and others, play a crucial role in semiconductor-mediated degradation, as they can be activated to generate reactive radicals that facilitate the mineralisation of organic pollutants into less toxic substances such as H_2O and CO_2. Table 10.4 summarises recent advances in the use of WDNs and their nanocomposites as photocatalysts for the removal of organic pollutants from wastewater. For example, Boruah and colleagues prepared magnetic mixed metal oxide nanomaterials from industrial waste via ball milling, and the nanocomposite was labelled IOW-9 h. The IOW-9 h nanocomposite was used as a photocatalyst for the degradation of MB, rhodamine B (RhB), and Congo red (CR) under sunlight irradiation in wastewater. The IOW-9 h nanocomposite has exhibited degradation efficiencies of 100%, 93%, and 99% for MB, RhB, and CR, respectively. Incorporating Fe_2O_3 (2.2 eV) into TiO_2 (3.2 eV) lowers the energy band gap of the nanocomposite, promoting activation under sunlight irradiation. Sunlight is the cheapest light source, reducing costs and the need for special photoreactors. In the study, hydroxide radicals (HO•) were the main oxygen reactive species facilitating degradation of the dyes (Boruah et al., 2020).

In another study, Nath and colleagues prepared a green $ZnO–Fe_2O_3$ photocatalyst from waste-expanded polystyrene (EPS), pea-eggplant, and ion scrap waste. Again, incorporating Fe_2O_3 (2.05 eV) into ZnO (3.11) reduced the energy band gap of the $ZnO–Fe_2O_3$ (2.39 eV), which facilitated photocatalyst activation under visible light. In the study, three photocatalysts with different physicochemical properties, namely $ZnO–Fe_2O_3$-EPS ball, $ZnO–Fe_2O_3$-EPS film, and $ZnO–Fe_2O_3$ powder, were prepared

TABLE 10.4
Summary of recent advances in the use of WDNs, and their nanocomposites as photocatalysts for the removal of organic pollutants from wastewater.

WDM	Pollutant	Conditions	%Degradation Efficiency	Reference
IWO-9 h	MB, RhB, and CR	[1]PD (1 g/L), pH (5), [2]IT ((60 min (MB and CR), 150 min (RhB)) Concentration (MB (80 mg/L), RhB (47.9 mg/L), CR (69.7 mg/L)	MB = 100 RhB = 93 CR = 99	Boruah et al. (2020)
ZnO–Fe_2O_3 powder	MB	[1]PD (50 mg), pH (-), [2]IT (240 min) Concentration (20 mg/L)	94.6	Nath et al. (2025)
AgNPs@PPC	CV	[1]PD (25 mg), pH (7), [2]IT (30 min) Concentration (80 mg/L)	98	Kamal et al. (2025)
CQDs/CNTs	RhB and MO	[1]PD (2 mg), pH (7), [2]IT RhB (30 min), MO (40 min), Concentration (10 mg/L)	RhB = 98.4 MO = 99.1	Meraj et al. (2025)
[3]SS-CQDs	MB	[1]PD (20 mg), pH (-), [2]IT (90 min) Concentration (10 mg/L)	94	Thamilselvan et al. (2025)
$CuFe_2O_4$	CIP	[1]PD (40 mg), pH (7), [2]IT (120 min) Concentration (20 mg/L)	97	Alqassem et al. (2025)
ZnO/rGO	Naphthol	[1]PD (0.02 g), pH (7), [2]IT (120 min) Concentration (10 mg/L)	95	Faaizatunnisa et al. (2025)
TiO_2-orange peel-derived biochar composite	ACT	[1]PD (0.8 g/L), pH (4), [2]IT (100 min) Concentration (20 mg/L)	94	Mohtaram et al. (2024)
Orange peel biochar/clay/titania nanocomposites	TCN	[1]PD (50 mg), pH (7), [2]IT (120 min) Concentration (20 mg/L)	Under UV = 92, Under Solar light = 89	Adesina et al. (2024)
Wood flour derived-2D/3D biochar modified g-C_3N_4	TCN	[1]PD (0.4 g/L), pH (9), [2]IT (60 min) Concentration (10 mg/L)	98	F. Feng et al. (2025)

PET-based Zn-MOF on orange peel biochar	ACT	[1]PD (0.1 g/L), pH (6.07), [2]IT (60 min) Concentration (50 mg/L)	99	Vigneshwaran et al. (2024)
Diospyros texana seed biomass-derived TiO_2/AC	[4]OTC and AMX	[1]PD (0.1 g), pH (AMX (3), OTC (7)) [2]IT (5 min) Concentration (50 mg/L))	OTC = 89, AMX = 91	Bilal et al. (n.d.)
Fe_3O_4@SnO_2 modified wheat straw biochar	TCN	[1]PD (0.01 g), pH (7), [2]IT (3 h) Concentration (25 mg/L)	91.8	C. Zhang et al. (2025)

1 PD—Photocatalyst dosage
2 IT—irradiation time
3 SS-CQDs—Shrimp shell carbon quantum dots
4 OTC—Oxytetracycline

and used for MB degradation in wastewater. These photocatalysts displayed degradation efficiencies of 87.5%, 65.7%, and 94.6%, respectively. Superoxide ($\bullet O_2^-$) and hydroxyl (HO•) radicals were the main reactive species facilitating MB degradation (Nath et al., 2025). Kamal and co-workers combined AgNPs from medical X-ray sheets and activated carbon from potato peels to form a photocatalyst for degradation of crystal violet (CV) in wastewater. The AgNPs potato peels carbon (AgNPs@PPC) exhibited 98% degradation efficiency within 30 min under UV light, compared to AC with 63% alone, demonstrating improved performance by coupling AC with semiconductors. Activated carbon is amorphous in nature and does not have a defined energy band gap due to localised π-electrons. Therefore, AC serves as a support material in photocatalysis, offering exceptional features such as high surface area, and enhanced electron transfer when combined with semiconductors like TiO_2. In the study, HO• radicals were main active species for CV degradation (Kamal et al., 2025).

Numerous studies report the application of WDNs as photocatalysts for degradation of dyes. However, WDNs have also demonstrated efficacy in degrading other classes of organic pollutants, including pharmaceuticals. For example, Alqassem and co-workers prepared *bio*-$CuFe_2O_4$/DS nanoparticles via sol-gel method using plant waste. The *bio*-$CuFe_2O_4$/DS achieved 97% ciprofloxacin (CIP) degradation efficiency facilitated by Fenton/peroxymonosulfate system under UV-Vis illumination. In the study, trapping experiments revealed that $SO_4^{\bullet -}$, HO•, $\bullet O_2-$ were the active players for CIP degradation. The photocatalytic activity of WDNs in degrading organic pollutants was also evaluated in the study where photolysis, bio-$CuFe_2O_4$/DS, and PMS + light were used for CIP degradation. The results obtained indicate that bio-$CuFe_2O_4$/DS alone achieved a degradation efficiency of 84%, which increased to 97% in the presence of PMS. This demonstrated that WDNs combined with oxidants play a crucial role in the degradation of organic pollutants (Alqassem et al., 2025). Faaizatunnisa and colleagues prepared ZnO/rGO nanocomposite from palm kernel shell waste and used it as a photocatalyst for degradation of naphthol in wastewater. The ZnO/rGO photocatalyst achieved a degradation efficiency of 95% within 120 min, outperforming ZnO alone with degradation efficiency of 92% (Faaizatunnisa et al., 2025).

Studies have shown that coupling WDNs with other semiconductors improves overall photocatalytic performance of resulting nanocomposites. The synergistic features of individual components promote the formation of heterostructures with enhanced charge separation, charge transfer, and enhanced degradation efficiency. Adesina and colleagues prepared a composite constituting orange peel biochar, clay, and titania for degradation of tetracycline (TCN). The composite exhibited degradation efficiencies of 92% and 89% under UV and solar light exposure, respectively (Adesina et al., 2024). In a different study, Feng and colleagues demonstrated that coupling g-C_3N_4 with biochar (2D/3D BC-CN60) effectively mitigated rapid charge recombination and poor light absorption in g-C_3N_4, thereby enhancing photocatalytic activity for TCN degradation. This was observed in the study when 2D/3D BC-CN60 exhibited high degradation efficiency, which is 5.27 times higher than that of pristine g-C_3N_4. The $^1O_2$, $\bullet SO_4^-$, and •OH were the main reactive species facilitating TCN degradation (F. Feng et al., 2025). Meraj and colleagues prepared CQDs derived from okra waste via hydrothermal method and conjugated it with carbon nanotubes (CNTs) to form CQDs/CNTs nanocomposite. The CQDs/CNTs nanocomposite was used as a visible light-driven photocatalyst for RhB and MO degradation in wastewater. Pristine CQDs achieved degradation efficiencies of 48% for RhB, and 50% for MO, whereas CNTs exhibited degradation efficiencies of 39% for RhB, and 21% for MO. On the contrary, the CQDs/CNTs nanocomposite achieved degradation efficiencies of 99.1% for MO and 98.4% for RhB, demonstrating that combining WDNs with other semiconductors significantly improves photoelectrochemical properties and overall photocatalytic performance (Meraj et al., 2025). A considerable number of studies have reported the use of WDNs and their nanocomposites as photocatalysts for the degradation of organic pollutants in wastewater.

10.4.1.3 Biological methods, membrane filtration, and electrochemical methods

Biological methods, membranes, and electrochemical methods have been widely used for the removal of organic pollutants from wastewater, as they rely on low-cost, environmentally sustainable materials. Extensive research has explored the use of biological, membrane technologies, and electrochemical methods integrated with WDNs to remove organic pollutants from wastewater (A. Sharma et al., 2025; Akinyemi et al., 2023; Castro-Fernández et al., 2025). For example, Idris and colleagues prepared three biomass-derived substrates (corncob flour (CCF), rotten rice flour (RRF), and potato peel waste (PPW)) for the bioremediation of benzene in a single-chamber MFC. In the study, RRF-fed MFCs achieved the highest bioremediation efficiency of 84%, followed by CCF-fed MFCs (80%), and PPW-fed MFCs (78%). This is because RRF-fed MFCs exhibited high-power density (4.76 mW m^{-2}) compared to CCF-fed MFCs (3.22 mW m^{-2}) and PPW-fed MFCs (2.49 mW m^{-2}) (Idris et al., 2025). In another study, He and colleagues prepared N-rich 3D microporous polypyrrole derived from biomass (PPy-DSC/CF) and used it as the MFC anode for the removal of RhB from wastewater. The PPy-DSC/CF-MFC system achieved a removal efficiency of 96.4% within 72 h, with a power density of 5184 mW m^{-3}. The (PPy-DSC/CF)-MFC system was estimated to cost $2.59 per unit, demonstrating its economic viability and sustainability (X. He et al., 2025). Table 10.5 summarises recent advances in the use of WDNs and their nanocomposites in bioremediation, membrane-based, and electrochemical remediation of organic pollutants in wastewater.

Combining WDNs with other materials has proven an effective strategy for improving MFC performance and enhancing power density. For example, Priyadarshini and co-workers combined biochar derived from waste-banana peels with the MIL-53(Fe) metal-organic framework (MIL-53(Fe)/BC) for the removal of salicylic acid in MFC. The MIL-53(Fe)/BC-MFC achieved a high-power density of 142.2 ± 1.5 mW/m^2, which is 2.34 times higher than that of pristine MFC-MIL-53(Fe) (60.6 ± 1.8 mW/m^2), while also attaining a removal efficiency of 93.8% (Priyadarshini et al., 2024). Idris and colleagues prepared palm kernel shell biowaste-derived graphene oxide (PKS-GO) and reduced graphene oxide (PKS-rGO) for the remediation of formaldehyde in MFC. The PKS-rGO-MFC system achieved a power density of 24.47 mW/m^2, which was more than that of the PKS-GO-MFC system (13.77 mW/m^2).

TABLE 10.5
Summary of recent advances in the use of WDNs and their nanocomposites in bioremediation, membrane-based, and electrochemical treatment of organic pollutants in wastewater.

WDM	Pollutant	Conditions	%Removal/Adsorption Capacity	Method	Reference
RRF-fed MFCs	Benzene	Dosage (60 g), [1]RT (35 days), *PD (4.76 mW/m^2), Voltage (90 mV) Concentration (5 mg/L)	84%	Bioremediation	Idris et al. (2025)
PPy-DSC/CF	RhB	[1]RT (70 h), *PD (5184 mW/m^3), Voltage (0.4 V) Concentration (40 mg/L)	96.4%	Bioremediation	X. He et al. (2025)
MIL-53(Fe)/BC	Salicylic acid	[1]RT (3 days), *PD (142.2 ± 1.5 mW/m^2), Concentration (50 mg/L)	93.8%	Bioremediation	Priyadarshini et al. (2024)
PKS-rGO	Formaldehyde	[1]RT (70 days), *PD (24.47 mW/m^2), [2]CD (84.98 mA/m^2) Concentration (50 mg/L)	87%	Bioremediation	Omenesa Idris et al. (2023)
PKS-rGO/ZnO	Naphthalene	[1]RT (70 days), *PD (45.49mW/m^2), [2]CD (114.0 mA/m^2) Concentration (5 mg/L)	94%	Bioremediation	Omenesa Idris et al. (2024)
GQDs-PET membrane	Phenol	[1]RT (4 h), [3]HWF (24 L/m^2.hr.bar), pH (9.5) Concentration (20 mg/L)	94%	Membrane	Raza et al. (2025)
PVDF/SEBS/A-$CBp_{3wt\%}$	MB and RhB	[1]RT (120 min), Concentration (10 mg/L)	MB = 79.1% RhB = 94.6%	Membrane	Ebah et al. (2025)
ON11@PSU	MG, RhB and MB	[1]RT (60 min), pH (6) Concentration (2.5 mg/L)	MG = 98% MB = 67% RhB = 92%	Membrane	A. Sharma et al. (2025)
LDPE/ZnO	ACT	Dosage (0.1 g), [1]RT (30 min), pH (7) Concentration (0.2 mg/L)	83%	Membrane	Vijay et al. (2025)
[4]H-PFCM	Ofloxacin, tetracycline and sulfadiazine	Dosage (25 g), [1]RT (40 min), pH (7) Concentration (5 mg/L)	Ofloxacin = 26.6 mg/g Tetracycline = 23 mg/g Sulfadiazine = 23.6 mg/g	Membrane	Hajyani et al. (2025)

Waste-derived cathodes and IrO_2–Ta_2O_5|Ti (70:30) anodes	AMX	[5]DT (30 min), [6]EC (0.1096 and 0.0895 kWh m^{-3} $order^{-1}$ for PA and SR, respectively), [2]CD (2.54 mA/m^2)	100%	Electrochemical oxidation	Castro-Fernández et al. (2025)
Battery waste-derived carbon electrode	MB	[5]DT (15 min), pH (1), Concentration (50 mg/L)	100%	Electrochemical oxidation	Hakim et al. (2024)
E-Co/SBC@NF-PMS	SMX	[5]DT (60 min), [6]EC (0.41 kWh/m^3), pH (7), [2]CD (5 mA/m^2)	96.8%	Electrochemical oxidation	Wan et al. (2025)
[7]A-BC/Fe	Dimethyl phthalate	[5]DT (4 h), *PD (124.03 ± 5.64 mW/m^2), pH (7), Concentration (50 mg/L)	94.8%	Electrochemical oxidation	Raj et al. (2024)

* PD—power density
1 RT—removal time
2 CD—current density
3 HWF—high-water reflux
4 H-PFCM—hole-punched flat ceramic membrane based on clay and hydroxyapatite in a fixed bed module
5 DT—degradation time
6 EC—energy consumption
7 A-BC/Fe—waste-iron electrocoagulated algal biomass

Moreover, the PKS-rGO-MFC system exhibited a current density of 84.98 mA/m^2 compared to 63.74 mA/m^2 for the PKS-GO-MFC system, demonstrating greater energy output, stronger microbial activity, and improved electron transfer that facilitated formaldehyde degradation (87%) (Omenesa Idris et al., 2023). In a separate study, Idris and colleagues improved the performance of PKS-rGO by incorporating ZnO particles, enabling efficient naphthalene degradation in MFCs. In the study, the PKS-rGO/ZnO nanocomposite achieved a degradation efficiency of 94.3% with an enhanced power density of 45.49 mW/m^2, which was 2 times higher than that of PKS-rGO. Moreover, the current density of PKS-rGO/ZnO reached 114.0 mA/m^2, exceeding that of PKS-rGO alone (Omenesa Idris et al., 2024). This demonstrates that WDNs are effective for removing organic pollutants in MFC systems while also promoting economic and environmental sustainability.

Membranes are also widely employed as an effective method for removing organic pollutants from wastewater. Membranes act as selective barriers that allow certain pollutants to pass through, while retaining non-target pollutants. The performance of membranes relies on key factors such as selectivity, permeability, chemical and thermal stability, and mechanical strength (Akinyemi et al., 2023). Recent studies have reported several challenges and limitations associated with membranes, including high operational costs, membrane fouling and scaling, and elevated energy demand (Mallah et al., 2024). To overcome these issues, researchers have made significant advances by coupling membranes with WDNs, thereby improving their performance and addressing these limitations. For instance, Raza and colleagues synthesised graphene quantum dots from waste carton paper and incorporated them into PET membranes derived from waste plastic bottles for the removal of phenol from wastewater. The GQDs-PET membrane exhibited a high-water reflux of 24 $L/(m^2.h.bar)$ compared to the PET membrane alone (9.9 $L/(m^2.h.bar)$), indicating enhanced water permeability and good overall membrane performance. Moreover, the GQDs-PET membrane achieved a 94% removal efficiency under alkaline conditions. This demonstrates that incorporating WDNs into membranes improves the overall membrane performance (Raza et al., 2025).

In another study, Ebah and co-workers fabricated a polyvinylidene fluoride (PVDF) membrane and incorporated styrene-ethylene-butylene-styrene (SEBS) and 2D carbon nanoparticles derived from tyre rubber (GTR) for efficient removal of MB and RhB in wastewater. The membrane achieved a removal efficiency of 79.1% for MB and 94.6% for RhB. In the reusability studies, RhB was used as a model analyte, and the results showed a decrease in removal efficiency from 94.6% to 80.6% after 10 cycles, indicating reduced membrane fouling and improved overall membrane performance (Ebah et al., 2025). Sharma and colleagues prepared carbon nanoparticles derived from onion peel and incorporated them into polysulfone membranes (ON11@PSU) for the removal of malachite green (MG), RhB, and MB dyes. The ON11@PSU achieved removal efficiencies of 67%, 92%, and 98% for MB, RhB, and MG, respectively (A. Sharma et al., 2025). Vijay and colleagues fabricated a low-density polyethene membrane from plastic waste and hybridised it with ZnO nanoparticles (LDPE/ZnO) for the removal of paracetamol in wastewater. The LDPE/ZnO membrane achieved a removal efficiency of 83% at pH 7, indicating good membrane performance (Vijay et al., 2025). This indicates that membranes coupled with WDNs can also be reliably used to remove pollutants from wastewater.

Electrochemical oxidation has also been used to remove organic pollutants from wastewater. It is considered an advanced oxidation process, in which an electric current is applied to the electrodes to destroy organic pollutants either through direct electron transfer at the anode or via indirect oxidation by reactive species such as superoxide radicals, hydroxyl radicals, and other strong oxidants (S. Zhang et al., 2024). In direct oxidation, the mechanism involves the adsorption of pollutants onto the anode surface, where they donate electrons to the electrode, initiating the oxidation process. In indirect oxidation, the electrode generates strong reactive species in situ, which subsequently attack and oxidise organic pollutants present in the solution (Chokejaroenrat et al., 2025). Typical electrodes include anodes that produce HO•, and cathodes that generate H_2O_2, promoting reduction reactions. However, the practical application of electrochemical oxidation has been limited by drawbacks such as high-energy consumption and the high cost of electrode materials, such as boron-doped diamond. To address these challenges and support a circular economy, researchers have focused on using electrodes derived from waste or enhancing electrode performance by doping them with WDNs. For instance, Hakim and colleagues prepared a carbon electrode from battery waste for the electrochemical oxidation of MB in wastewater. The electrode achieved complete MB degradation within 15 min, facilitated by HO•, demonstrating high efficiency and reliability as a treatment method (Hakim et al., 2024). In another study, Wan and co-workers designed a novel cobalt oxide-modified sludge biochar (SBC) composite electrode (Co/SBC@NF) supported on nickel foam for the electrochemical degradation of sulfamethoxazole (SMX) in wastewater. The established electro-enhanced Co/SBC@nickel foam-peroxymonosulfate system (E-Co/SBC@NF-PMS) achieved 96.8% SMX removal at pH 7. The removal rate of SMX was (k_{obs} = 0.0870 min^{-1}) for the E-Co/SBC@NF-PMS system, which was 4.7 and 9.3 times higher than Co/SBC@NF-PMS (k_{obs} = 0.0186 min^{-1}) and E-Co/SBC@NF (k_{obs} = 0.0094 min^{-1}), indicating the effect of SBC for electrochemical activation of PMS. The HO•, singlet oxygen ($^1O_2$), and $SO_4^{\bullet-}$ were the main reactive species for SMX degradation. For energy consumption evaluation, the E-Co/SBC@NF-PMS system demonstrated an energy consumption of 0.41 kWh·m^{-3}, representing a >93% reduction compared to 5.88 kWh·m^{-3} for the E-SBC@NF system and 3.71 kWh·m^{-3} for the E-SBC@NF-PMS system (Wan et al., 2025).

Castro-Fernández and colleagues coupled waste-derived cathodes with IrO_2–Ta_2O_5|Ti (70:30) anodes for synergistic degradation of amoxicillin (AMX). The activated carbon doped on the cathode surface is derived from *Spergularia rubra* (SR) and *Phragmites australis* (PA). The study showed improved H_2O_2 production on carbon-modified cathodes via the two-electron oxygen reduction reaction, which promoted the production of reactive HO• that ultimately mineralised AMX. The synergistic system achieved complete degradation of AMX in 30 min, with reduced energy consumption of 0.0895 and 0.1096 kWh·m^{-3}·$order^{-1}$ for SR and PA, respectively, at a current density of 2.54 mA cm^{-2}. In comparison, traditional platinum cathodes required 4.9844 kWh·m^{-3}·$order^{-1}$. Thus, the waste-derived cathodes coupled with IrO_2–Ta_2O_5|Ti (70:30) anodes offer lower energy consumption while maintaining high AMX removal efficiency (Castro-Fernández et al., 2025). Raj and colleagues prepared waste-iron electrocoagulated algal biomass (A-BC/Fe) as an efficient bio-electro-Fenton for removing dimethyl phthalate (DMP) in wastewater. The A-BC/Fe displayed a removal efficiency of 94.8 within 4 h (Raj et al., 2024).

10.5 CHALLENGES AND PROSPECTS OF REMEDIATION TECHNIQUES

Rapid industrialisation and urbanisation continue to introduce toxic organic and inorganic pollutants into the environment. This poses a serious risk to aquatic and terrestrial organisms, thereby threatening biodiversity. Various remediation techniques, including adsorption, photocatalysis, bioremediation, membrane filtration, and electrochemical oxidation, have been used to remediate organic pollutants and trace metals from the environment. These technologies have been successfully coupled with WDNs to effectively eliminate pollutants from wastewater (Rashid et al., 2021; Al-Sayed et al., 2023). Despite significant advances in these technologies, there are still some limitations and challenges that require urgent attention. For example, the development of an adsorption system requires deep insights into the adsorbent's selectivity, recovery and reusability, economic viability, stability, adsorption behaviour, environmental regulations and challenges. Other techniques, such as photocatalysis, require materials with specific heterostructures and favourable photoelectrochemical properties, including a low-energy band gap that enables utilisation of solar light, which is more abundant than UV light. In addition, adsorption and photocatalysis lack selectivity; i.e., they are effective for some pollutants but less effective for others, depending on the materials used (Crini & Lichtfouse, 2019). Thus far, per- and polyfluoroalkyl (PFAS) compounds, antibiotics, and other endocrine disruptors require specialised techniques for removal due to their resistance and persistence in wastewater treatment processes (Aryal et al., 2020). Some techniques produce secondary pollution by generating toxic by-products or by using materials that are prone to leaching due to instability. The complexity of wastewater matrices often makes it difficult for some techniques to efficiently remove pollutants; therefore, multiple treatment stages are required to eliminate interfering species. This not only complicates the process but also increases operational costs. Techniques such as AOPs, membrane filtration, bioremediation, and biosorption are energy-intensive and have hindered application in resource-limited countries. The membrane filtration method is prone to membrane fouling, which reduces membrane efficiency and increases maintenance costs. Bioremediation and biosorption require specialised techniques and trained personnel to grow and isolate target microorganisms, which are then incorporated into treatment processes to eliminate pollutants. A major drawback of many treatment techniques is that some are promising at the laboratory-scale but often fail at the industrial scale. Therefore, addressing these challenges is crucial to improving both economic viability and environmental sustainability.

Exploiting the synergistic effects of hyphenating multiple wastewater treatment systems addresses the shortcomings of individual techniques, thereby enabling the removal of a broad range of pollutants with high efficiency (Bello & Raman, 2019; D. Wang et al., 2023). Incorporating cost-effective, renewable, eco-friendly, and scalable WDNs into wastewater treatment systems reduces costs and aligns with circular economy principles. In wastewater treatment, achieving consistent adsorption or photocatalytic performance is a tremendous challenge. Nonetheless, the use of WDNs as either adsorbents or photocatalysts has demonstrated remarkable resilience and consistency in removing pollutants (Belhaj et al., 2025; Kamal et al., 2025). Moreover, WDNs have functional moieties that can be tailored to optimise their performance in any treatment system of interest. Therefore, WDNs are a promising candidate for wastewater remediation, offering both economic and environmental sustainability. Integrating WDNs with tailored nanomaterials or microorganisms offer several features such as: (i) reducing the energy band gap in photocatalysts to promote activation by solar light; (ii) increasing surface area to promote adsorption of a wide range of pollutants; (iii) encouraging the adsorption of target pollutants with enhanced selectivity; (iv) boosting microorganisms with WDNs improves stability, adaptability, and action towards target pollutants. Statistical methods should also be explored to predict the behaviour of the wastewater treatment system, considering the properties of the pollutants relative to the proposed WDNs.

10.6 CONCLUSION

The application of WDNs for the removal of trace metals and organic pollutants from wastewater offers a promising, cost-effective, and sustainable approach to water treatment. Converting waste into functional materials such as catalysts, electrode adsorbents, or sensors addresses two environmental problems simultaneously: (i) the remediation of contaminated water matrices and (ii) the reduction of solid waste accumulation. Studies conducted show that WDNs possess fascinating properties, including high surface area, abundant and tunable surface moieties, and favourable physicochemical features that enable efficient interactions with trace metals and micro-organic pollutants. Furthermore, their performance can be further improved through modification strategies such as surface functionalisation, chemical activation, or the formation of composite materials, thereby enhancing catalytic activity, adsorption capacity, biological activity, and electrochemical reactivity.

The use of WDNs contributes to their circular economy by transforming industrial by-products, biomass, and electronic waste into value-added products. This approach attenuates reliance on expensive commercial materials, lowers treatment expenses, and minimises environmental impact. Moreover, WDNs demonstrate good reusability and stability, which further supports their potential for application at larger, practical scales. Nevertheless, several challenges remain in the application of WDNs, such as difficulty in scaling up production, ensuring consistent material quality, and assessing long-term environmental safety. Therefore, further research is required to integrate WDN-based systems into pilot and full-scale wastewater treatment systems, as well as to investigate hybrid systems that couple WDNs with electrochemical or biological processes to achieve synergistic pollutant removal.

Overall, WDNs stand out as eco-friendly, versatile, and cost-effective materials that bridge waste management with water treatment. Their continued development and optimisation hold significant potential for advancing sustainable wastewater treatment technologies, contributing to environmental protection and safeguarding human health.

REFERENCES

Aboelghait, K. M., Abdallah, W. E., Abdelfattah, I., & El-Shamy, A. M. (2024). Green synthesis of silver nanoparticles by waste of Murcott Mandarin peel as a sustainable approach for efficient heavy metal removal from metal industrial wastewater. *Separation and Purification Technology, 347*, 127609.

Abou-Hadid, A. F., El-Behairy, U. A., Elmalih, M. M., Amdeha, E., Naggar, A. M. A. El, Taha, M. H., & Hussein, A. E. M. (2024). Conversion of corn shell as biomass solid waste into carbon species for efficient decontamination of wastewater via heavy metals adsorption. *Biomass Conversion and Biorefinery, 14*(14), 16435–16449.

Achilias, D. S., Antonakou, E. V., Koutsokosta, E., & Lappas, A. A. (2009). Chemical recycling of polymers from waste electric and electronic equipment. *Journal of Applied Polymer Science, 114*(1), 212–221.

Adesina, M. O., Alfred, M. O., Seitz, H., Brennenstuhl, K., Rawel, H. M., Wessig, P., Kim, J., Wedel, A., Koopman, W., & Günter, C. (2024). Orange peel biochar/clay/titania composites: Low cost, high performance, and easy-to-reuse photocatalysts for the degradation of tetracycline in water. *Environmental Science: Water Research & Technology, 10*(6), 1432–1450.

Adu-Poku, D., Saah, S. A., Sakyi, P. O., Bandoh, C. K., Agyei-Tuffour, B., Azanu, D., Oteng-Peprah, M., Hawawu, I., Azibere, S., & Affram, K. A. (2024). Acid-activated biochar for efficient elimination of amoxicillin from wastewater. *Journal of Chemistry, 2024*(1), 3648098.

Akinyemi, B. T., Ogundele, O. D., & Afolabi, A. B. (2023). Advancements in sustainable membrane technologies for enhanced remediation and wastewater treatment: A comprehensive review. *Acadlore Transactions on Geosciences, 2*(4), 196–207.

Alatawi, I. S. S., Almughathawi, R., Madkhali, M. M. M., Alshammari, N. M., Alaysuy, O., Mogharbel, A. T., Hosni, M., & El-Metwaly, N. M. (2025). Sustainable waste-derived cellulose-based nanosensor for cobalt ion detection, removal, and recovery from industrial effluents and battery wastes. *Journal of Water Process Engineering, 70*, 106974. https://doi.org/10.1016/j.jwpe.2025.106974

Algethami, J. S., Billah, R. E. K., Alhamami, M. A. M., Alamrani, N. A., Bahsis, L., Gomase, V., Bambal, A., & Jugade, R. (2025). Confiscation of Cr(VI) by cross-linked shrimp waste-derived chitosan @Al2O3 nanocomposite: Experimental and theoretical approach. *Journal of Molecular Structure, 1321*, 140182. https://doi.org/10.1016/j.molstruc.2024.140182

Al-Ghamdi, A. A., Ibrahim, S. M., Lu, X., Xin, J., El-Sayed, I. E.-T., Galhoum, A. A., & Wageh, S. (2025). Biosorbent silver nanoparticles decorated coffee-ground waste composite for cleaning water and antimicrobial applications. *Environmental Science and Pollution Research, 32*(5), 2280–2297.

Aljomard, H., Awayssa, O., & Ghenai, C. (2025). Activated carbon derived from shrimp waste enhanced by ball milling: A green solution for CO 2 capture and waste valorization. *Nanoscale, 17*(24), 14696–14708.

Allwar, A., Herawati, M., Wardana, F. S., Khoirunnisa, A., & Anugrah, Z. M. (2023). Composite of Ag2O-CuO/biochar as an adsorbent for removal of amoxicillin and paracetamol from aqueous solution. *International Journal of Environmental Science and Technology, 20*(12), 13411–13422.

Alqassem, B., Torad, N. L., Palmisano, G., Banat, F., & Abu Haija, M. (2025). Green synthesis of CuFe2O4 nanoparticles using plant waste for photocatalytic degradation of ciprofloxacin. *Separation and Purification Technology, 368*, 132903. https://doi.org/10.1016/j.seppur.2025.132903

Al-Sayed, A., Hassan, G. K., Al-Shemy, M. T., & El-Gohary, F. A. (2023). Effect of organic loading rates on the performance of membrane bioreactor for wastewater treatment behaviours, fouling, and economic cost. *Scientific Reports*, *13*(1), 15601.

Amaku, J. F., & Mtunzi, F. M. (2024). Adsorption of amoxicillin onto breadfruit seed hull biochar modified using ZnONPs: Disinfection and decontamination. *Biomass Conversion and Biorefinery*, 1–17.

Arul, A., Kavitha, S., Anand Babu Christus, A., Surya, V. J., Ravikumar, A., & Sivalingam, Y. (2023). Enhanced removal of Pb (II) and Cd (II) ions from aqueous systems using coated magnetic nanoparticles in activated carbon derived from corncob waste. *Surfaces and Interfaces*, *40*, 103095. https://doi.org/10.1016/j.surfin.2023.103095

Aryal, N., Wood, J., Rijal, I., Deng, D., Jha, M. K., & Ofori-Boadu, A. (2020). Fate of environmental pollutants: A review. *Water Environment Research*, *92*(10), 1587–1594.

Ashebir, H., Nure, J. F., Worku, A., & Msagati, T. A. M. (2024). Prosopis juliflora biochar for adsorption of sulfamethoxazole and ciprofloxacin from pharmaceutical wastewater. *Desalination and Water Treatment*, *320*, 100691. https://doi.org/10.1016/j.dwt.2024.100691

Aslam, J., Aslam, R., & Hussain, C. M. (2024). *Waste-derived nanoparticles: Synthesis, applications and sustainability*. Elsevier.

Aththanayaka, S., Thiripuranathar, G., & Ekanayake, S. (2025). Sustainable approach for fabrication of pineapple agro-waste mediated cellulose nanocrystals embedded with Ag/Ag2O/ZnO nanocomposites for efficient removal of waterborne pathogens in wastewater. *International Journal of Biological Macromolecules*, 143272.

Atul, A. K., Srivastava, S. K., Gupta, A. K., & Srivastava, N. (2022). Synthesis and characterization of NiO nanoparticles by chemical co-precipitation method: An easy and cost-effective approach. *Brazilian Journal of Physics*, *52*(1), 2.

Bano, A., Aziz, M. K., Ameen, F., Singh, K., Prasad, B., Nandan, Dave, H., Ravi, R., Manjhi, J., & Kumari, M. (2024). Adsorptive removal of naproxen onto nano magnesium oxide-modified castor wood biochar: Treatment of pharmaceutical wastewater via sequential Fenton's-adsorption process. *IUBMB Life*, *76*(12), 1106–1124.

Batubara, A. S., Adress Hasan, H. M., Abel Moniem, M. A., Masoud, M. S., Mostafa, A.-E. R., Gamal, M., & Elsayed, M. A. (2023). Usage of natural wastes from animal and plant origins as adsorbents for the removal of some toxic industrial dyes and heavy metals in aqueous media. *Journal of Water Process Engineering*, *55*, 104204. https://doi.org/10.1016/j.jwpe.2023.104204

Bayrakdar, M., Palas, B., & Ersöz, G. (2025). Utilization of electroplating industry waste-derived layered double hydroxides for the removal of environmental contaminants: Ni and Zn recovery and adsorption of pesticides. *International Journal of Environmental Research*, *19*(5), 1–16.

Belcaid, A., Beakou, B. H., Bouhsina, S., & Anouar, A. (2024). Biosorption of cobalt and chromium from wastewater using manganese dioxide and iron oxide nanoparticles loaded on cellulose-based biochar: Modeling and optimization with machine learning (artificial neural network). *International Journal of Biological Macromolecules*, *282*, 136855.

Belhaj, J., Khiari, R., García-Caballero, V., Romero, A. A., & García, A. (2025). Agricultural waste-derived cellulose/ZnO composites: Dual photocatalytic and adsorptive action for textile dye removal. *Polymers*, *17*(13), 1737.

Bello, M. M., & Raman, A. A. A. (2019). Synergy of adsorption and advanced oxidation processes in recalcitrant wastewater treatment. *Environmental Chemistry Letters*, *17*, 1125–1142.

Bhattacharya, G., Fishlock, S. J., McLaughlin, J. A., & Roy, S. S. (2021). Metal-oxide nanomaterials recycled from E-waste and metal industries: A concise review of applications in energy storage, catalysis, and sensing. *International Journal of Energy Research*, *45*(5), 8091–8102.

Bilal, M., Behroozi, A. H., Jahan, R., Ali, J., & Vatanpour, V. (n.d.). Sustainable TiO2/ Diospyros texana seed-derived activated carbon composites for rapid photodegradation of oxytetracycline and amoxicillin under visible light. *Journal of Chemical Technology & Biotechnology*, *100*(11), 2371–2382.

Billah, R. E. K., Şimşek, S., Majdoubi, H., Kaya, S., Agunaou, M., Soufiane, A., & Katin, K. P. (2023). Removal of lead (II) from aqueous solution using epichlorohydrin cross-linked shrimp waste-derived chitosan based@ calcium phosphates biocomposite: Experimental study and computational approach. *Journal of Molecular Liquids*, *389*, 122872.

Boruah, P. K., Yadav, A., & Das, M. R. (2020). Magnetic mixed metal oxide nanomaterials derived from industrial waste and its photocatalytic applications in environmental remediation. *Journal of Environmental Chemical Engineering*, *8*(5), 104297. https://doi.org/10.1016/j.jece.2020.104297

Castro-Fernández, J. A., Ramírez, Á., Muñoz-Morales, M., Bustos, E., & Llanos, J. (2025). Synergetic electrochemical degradation of amoxicillin (AMX) with sustainable transition metal oxides-anode and waste-derived carbon-cathode system from a circular economy approach. *Electrochimica Acta*, *525*, 146181. https://doi.org/10.1016/j.electacta.2025.146181

Chaubey, A. K., Patel, M., Pittman, C. U., & Mohan, D. (2023). Acetaminophen and trimethoprim batch and fixed-bed sorption on MgO/Al2O3-modified rice husk biochar. *Colloids and Surfaces A: Physicochemical and Engineering Aspects*, *677*, 132263. https://doi.org/10.1016/j.colsurfa.2023.132263

Chen, Z., Wei, W., Chen, H., & Ni, B.-J. (2022). Recent advances in waste-derived functional materials for wastewater remediation. *Eco-Environment & Health*, *1*(2), 86.

Cheng, Y., Luo, F., Jiang, Y., Li, F., & Wei, C. (2018). The effect of calcination temperature on the structure and activity of TiO2/SiO2 composite catalysts derived from titanium sulfate and fly ash acid sludge. *Colloids and Surfaces A: Physicochemical and Engineering Aspects*, *554*, 81–85.

Chokejaroenrat, C., Poomipuen, K., Techauay, K., Weaoseng, P., Jermnak, U., Chokthaweepanich, H., Suriyachai, N., Lin, Y., Imsilp, K., & Sakulthaew, C. (2025). Fered-Fenton enhanced boron-doped diamond electrochemical oxidation for wastewater treatment: Antibiotic degradation, kinetics, and environmental effects. *Journal of Water Process Engineering*, *72*, 107543.

Christy, K. D., Sengottuvelan, N., Sathiyamootthy, J., Edison, T. N. J. I., & Senthilkumar, A. (2025). Power benefitted bioremediation of hexavalent chromium ions in biochar blended soil microbial fuel cell. *Biomass Conversion and Biorefinery*, *15*(4), 5739–5752.

Cobo, A. J. M., Lozano, F. E., Vilar, M. M., Valenzuela, C. M., & Ramos, E. R. (2025). Biotechnological applications of the ubiquitous fungus Penicillium sp. 8L2: Biosorption of Zn (II) and synthesis of ZnO nanoparticles as biocidal agents. *Sustainability*, *17*(6), 2379.

Crini, G., & Lichtfouse, E. (2019). Advantages and disadvantages of techniques used for wastewater treatment. *Environmental Chemistry Letters*, *17*, 145–155.

Cui, X., Miao, Q., Shi, X., Zheng, P., & Li, H. (2023). Accelerating electricity generation and Cr (VI) removal using anatase–biochar-modified cathode microbial fuel cells. *Sustainability*, *15*(16), 12276.

da Silva Araújo, F. P., de Souza Cupertino, G., de Cássia Superbi de Sousa, R., de Castro Santana, R., & Pereira, A. F. (2024). Use of biochar produced from brewer's spent grains as an adsorbent. *Biomass Conversion and Biorefinery*, 1–16.

Dehkordi, M. M., Nodeh, Z. P., Dehkordi, K. S., Khorjestan, R. R., & Ghaffarzadeh, M. (2024). Soil, air, and water pollution from mining and industrial activities: Sources of pollution, environmental impacts, and prevention and control methods. *Results in Engineering*, *23*, 102729.

Devi, M., Rawat, S., & Sharma, S. (2021). A comprehensive review of the pyrolysis process: From carbon nanomaterial synthesis to waste treatment. *Oxford Open Materials Science*, *1*(1), itab014.

Dolati, E., Tavakoli, S., Khazaie, A., & Hashemzadeh, F. (2025). Efficient removal of arsenic and lead from water using a low-cost adsorbent derived from used face masks. *Journal of Molecular Liquids*, *426*, 127243. https://doi.org/10.1016/j.molliq.2025.127243

Du, Y., Ye, X., Hui, Z., Jiao, D., Xie, Y., Chen, S., & Ding, J. (2024). Synergistic effect of adsorption-photocatalytic reduction of Cr (vi) in wastewater with biochar/TiO 2 composite under simulated sunlight illumination. *Physical Chemistry Chemical Physics*, *26*(22), 15891–15901.

Dunnigan, L., Morton, B. J., Ashman, P. J., Zhang, X., & Kwong, C. W. (2018). Emission characteristics of a pyrolysis-combustion system for the co-production of biochar and bioenergy from agricultural wastes. *Waste Management*, *77*, 59–66.

Ebah, K., Bensalah, H., Nekhlaoui, S., Raji, M., Bouhfid, R., & Qaiss, A. el kacem. (2025). Smart and flexible PVDF/SEBS membranes reinforced with pyrolytic carbon black nanoparticles from waste tires for enhanced wastewater remediation. *Journal of Water Process Engineering*, *72*, 107469. https://doi.org/10.1016/j.jwpe.2025.107469

Ejraei, A., Aroon, M. A., & Saravani, A. Z. (2019). Wastewater treatment using a hybrid system combining adsorption, photocatalytic degradation and membrane filtration processes. *Journal of Water Process Engineering*, *28*, 45–53.

Elboughdiri, N., Ferkous, H., Rouibah, K., Boublia, A., Delimi, A., Yadav, K. K., Erto, A., Ghernaout, D., Salih, A. A. M., & Benaissa, M. (2024). Comprehensive investigation of Cu2+ adsorption from wastewater using olive-waste-derived adsorbents: Experimental and molecular insights. *International Journal of Molecular Sciences*, *25*(2), 1028.

Elkhatib, E. A., Moharem, M. L., Saad, A. F., & Abdelhamed, S. (2025). Effective elimination of lead from polluted wastewater utilizing a novel nanocomposite derived from byproducts of drinking water industry. *BMC Chemistry*, *19*(1), 188.

Faaizatunnisa, N., Ediati, R., Yusof, E. N. M. D., Fadlan, A., Karelius, K., Kulsum, U., & Ariesta, M. N. (2025). Optimized photocatalytic degradation of 2-naphthol using ZnO/rGO nanocomposites synthesized from palm kernel shell waste. *Journal of Water Process Engineering*, *70*, 106977.

Feng, A., Zhu, X., Chen, Y., Liu, P., Han, F., Zu, Y., Li, X., & Bi, P. (2024). Functional biomass-derived materials for the development of sustainable batteries. *ChemElectroChem*, *11*(13), e202400086.

Feng, F., Zhang, X., Xu, X., Huang, Q., Yin, H., Li, R., Wu, G., & Xing, W. (2025). Construction of 2D/3D biochar modified g-C3N4 for efficient removal of tetracycline hydrochloride via photocatalytic PMS activation. *Optical Materials*, *167*, 117246. https://doi.org/10.1016/j.optmat.2025.117246

Ganesan, A., Rezazgui, O., Langlois, S., Boussabbeh, C., & Barnabé, S. (2025). Pyrolytic conversion of construction, renovation, and demolition (CRD) wood wastes in Québec to biochar: Production, characterization, and identifying relevant stability indices for carbon sequestration. *Science of the Total Environment*, *965*, 178650.

Gao, P., Song, S., Wang, M., Yao, M., & Xue, J. (2025). Oxygen atmosphere enhances ball milling remediation of petroleum-contaminated soil and reuse as adsorptive/catalytic materials for wastewater treatment. *Journal of Environmental Sciences*, *147*, 652–664.

Groszek, M., Wójcik, G., Wiśniewska, M., & Nowicki, P. (2025). Application of environmentally friendly activated carbons derived from herbal industry waste for water purification: A study on the removal of selected organic and inorganic pollutants. *Journal of Water Process Engineering*, *75*, 107952.

Gul, A., & Senol-Arslan, D. (2025). Eco-friendly fabrication of cellulose-derived polyvinylidene fluoride membranes from wastepaper for efficient Fe (II) removal from mine wastewater. *Polymer International*, *74*(12), 1109–1119.

Guo, T., Chen, X., & Yin, L. (2024). Recent advancements in modified SnO 2–Sb electrodes for electrochemical treatment of wastewater. *Journal of Materials Chemistry A*, *12*(8), 4397–4420.

Hajyani, Z., Mousavi, Z., Soleimanbeigi, M., Wong, Y. J., & Beni, A. A. (2025). Antibiotic adsorption from real pharmaceutical factory wastewater using a hole-punched flat ceramic membrane based on clay and hydroxyapatite in a fixed bed module. *Results in Engineering*, 105364.

Hakim, S., Manurung, T. W., Hermayantingsih, D., & Ariefin, M. (2024). Electrochemical oxidation of methylene blue using carbon electrode from battery waste. *Helium: Journal of Science and Applied Chemistry*, *4*(2), 33–38.

Hamadeen, H. M., Elkhatib, E. A., & Moharem, M. L. (2022). Optimization and mechanisms of rapid adsorptive removal of chromium (VI) from wastewater using industrial waste derived nanoparticles. *Scientific Reports*, *12*(1), 14174.

Han, R., Fang, Y., Sun, P., Xie, K., Zhai, Z., Liu, H., & Liu, H. (2021). N-doped biochar as a new metal-free activator of peroxymonosulfate for singlet oxygen-dominated catalytic degradation of acid orange 7. *Nanomaterials*, *11*(9), 2288.

Hasan, M. A., Hossain, R., & Sahajwalla, V. (2024). Utilization of battery waste derived ZnO in the removal of dye from aqueous solution: A waste to wealth approach. *Journal of Environmental Management*, *356*, 120461.

Hashem, H. M., El-Maghrabey, M., & El-Shaheny, R. (2024). Inclusive study of peanut shells derived activated carbon as an adsorbent for removal of lead and methylene blue from water. *Scientific Reports*, *14*(1), 13515.

He, H., Zhang, R., Zhang, P., Wang, P., Chen, N., Qian, B., Zhang, L., Yu, J., & Dai, B. (2023). Functional carbon from nature: Biomass-derived carbon materials and the recent progress of their applications. *Advanced Science*, *10*(16), 2205557.

He, X., Ma, S., Ma, P., Cao, R., Tian, X., Lu, Y., Li, J., Liang, Y., Wang, Z., & Lu, X. (2025). Waste biomass durian shell carbon to derive N-rich porous microbial fuel cell anode for simultaneous dye degradation and electricity generation. *Bioresource Technology*, 132984.

Herath, A., Layne, C. A., Perez, F., Hassan, E. I. B., Pittman Jr, C. U., & Mlsna, T. E. (2021). KOH-activated high surface area Douglas Fir biochar for adsorbing aqueous Cr (VI), Pb (II) and Cd (II). *Chemosphere*, *269*, 128409.

Huang, M., Xiang, Y., Ni, J., Zhang, H., Wei, R., & Chen, W. (2024). The carbon stability and energy characteristics of tea waste-derived biochar: Effects of pyrolytic temperature and co-pyrolysis with nanoscale zero-valent iron. *Biomass and Bioenergy*, *191*, 107458.

Idris, M. O., Ibrahim, M. N. M., Yaqoob, A. A., Suleiman, A. I., Zakari, D. A., Azmi, S. N. H., Min, K., & Almutairi, B. O. (2025). Investigating the influence of biomass-derived organic substrates on power generation with benzene bioremediation in microbial fuel cell. *Biomass Conversion and Biorefinery*, 1–16.

Islam, M. R., Wang, Q., Sharmin, S., & Enyoh, C. E. (2024). Adsorption of ibuprofen from water using banana peel biochar: Experimental investigation and machine learning algorithms. *Water*, *16*(23), 3469.

Jain, M., Kumar, D., Chaudhary, J., Kumar, S., Sharma, S., & Verma, A. S. (2023). Review on E-waste management and its impact on the environment and society. *Waste Management Bulletin*, *1*(3), 34–44.

Judith, J. V., & Vasudevan, N. (2022). Synthesis of nanomaterial from industrial waste and its application in environmental pollutant remediation. *Environmental Engineering Research*, *27*(2).

Kaithwas, A., Prasad, M., Kulshreshtha, A., & Verma, S. (2012). Industrial wastes derived solid adsorbents for CO2 capture: A mini review. *Chemical Engineering Research and Design*, *90*(10), 1632–1641.

Kamal, A., Hassane, A. M. A., An, C., Deng, Q., Hu, N., Abolibda, T. Z., Altaleb, H. A., Gomha, S. M., Selim, M. M., & Shenashen, M. A. (2025). Developing a cost-effective and eco-friendly adsorbent/photocatalyst using biomass and urban waste for crystal violet removal and antimicrobial applications. *Biomass Conversion and Biorefinery*, *15*(7), 10089–10107.

Kar, S., Dey, S., Chowdhury, K. B., Ghosh, S. K., Mukhopadhyay, J., Kumar, S., Ghosh, S., & Majumdar, S. (2023). Phyto-assisted synthesis of CuO/industrial waste derived biochar composite for adsorptive removal of doxycycline hydrochloride and recycling of spent biochar as green energy storage device. *Environmental Research*, *236*, 116824. https://doi.org/10.1016/j.envres.2023.116824

Kargule, B. B., Al-Asadi, M., Al-Anssari, S., Aljibori, H. S. S., Hamzah, H. T., Tastambek, K. T., Abdullah, T. A., & Abdullah, O. I. (2025). Sustainable removal of dyes from wastewater using eggshell-derived calcium carbonate nanoparticles: Adsorption isotherms, kinetics, and thermodynamic analysis supporting Sustainable Development Goals (SDGs). *ASEAN Journal of Science and Engineering*, *5*(2), 369–394.

Katika, R. M., Boddu, S., & Mandapati, R. N. (2025). Hydrothermal Synthesis and Photocatalytic Efficiency of Turmeric Leaves Biochar/TiO2 Composite for Photooxidation of Congo Red Dye. *Periodica Polytechnica Chemical Engineering*.

Khalaf, S. A., Shartooh, S. M., & Shihan, M. A. (2025). Assessing the loading capacity of walnut peels as a nanobiomass for the biosorption of certain heavy metals from wastewater. *Journal of Ecological Engineering*, *26*(2).

Khan, M., Khan, I. A., Farooqi, A., & Malik, R. N. (2024). Removal of arsenic and fluoride ions from aqueous solutions using electronic waste-derived adsorbent. *Materials Chemistry and Physics*, *327*, 129889.

Khan, S., Ahamad, Z., & Nasar, A. (2025). Lignocellulosic Juglans regia waste-derived magnetically separable nanocomposite for enhanced malachite green dye removal. *International Journal of Biological Macromolecules*, *318*, 144989. https://doi.org/10.1016/j.ijbiomac.2025.144989

Khandaker, T., Islam, T., Nandi, A., Anik, M. A. A. M., Hossain, M. S., Hasan, M. K., & Hossain, M. S. (2025). Biomass-derived carbon materials for sustainable energy applications: A comprehensive review. *Sustainable Energy & Fuels*, *9*(3), 693–723.

Khoirunisa, K., Lestari, N. D., Wahyuni, E. T., & Natsir, T. A. (2025). Enhanced photocatalytic reduction of Cr (VI) under visible light a magnetically separable TiO2-Fe/Fe3O4 photocatalyst prepared from iron rusty waste. *Journal of Metals, Materials and Minerals*, *35*(1), e2162–e2162.

Khyrun, S. M. F., Christy, A. J., & Sagadevan, S. (2025). Enhanced photocatalytic performance of CeO2/ZnO/TiO2 nanocomposite: Role of reduced bandgap and core-shell heterojunction morphology. *Optical Materials*, *163*, 116990.

Kshirsagar, D., Thanekar, P., Balapure, K., & Bhandari, V. M. (2023). Biomass-derived adsorbents and nanocomposites for wastewater treatment. *Materials Today: Proceedings*, *90*, 18–29.

Kumar, R., Kaushal, S., Kumar, P., Kumar, V., Thakur, N., & Kumar, K. (2025). Industrial waste-derived carbon nanomaterials. In *Waste derived carbon nanomaterials* (Vol. 1, pp. 179–201). ACS Publications.

León, A. Y., Contreras-Arenas, J. D., Garnica-Fuentes, C. F., Jiménez-Caballero, M. A., Pinto-Hernández, D. F., Ariza-León, E., Peña-Ballesteros, D. Y., & Molina-Velasco, D. R. (2025). Removal of organic compounds in wastewater using cocoa shell-based activated carbon–SiO2 nanoparticles. *Environment, Development and Sustainability*, *27*(3), 6811–6830.

Lestari, R., Kamiya, Y., Wahyuningsih, T. D., & Kartini, I. (2025). Recycling of plastic bag waste into carbon quantum dots using optimized pyrolysis-hydrothermal methods for selective Fe (III) sensing. *Carbon Research*, *4*(1), 51.

Li, J., Xu, L., Sun, P., Zhai, P., Chen, X., Zhang, H., Zhang, Z., & Zhu, W. (2017). Novel application of red mud: Facile hydrothermal-thermal conversion synthesis of hierarchical porous AlOOH and Al2O3 microspheres as adsorbents for dye removal. *Chemical Engineering Journal*, *321*, 622–634.

Liang, L., Xi, F., Tan, W., Meng, X., Hu, B., & Wang, X. (2021). Review of organic and inorganic pollutants removal by biochar and biochar-based composites. *Biochar*, *3*(3), 255–281.

Lin, C.-W., Jhan, Y.-C., Zhu, T.-J., & Liu, S.-H. (2023). Enhancement of chromium (VI) removal and power generation by adding biochar to a single-medium sediment-based microbial fuel cell. *Journal of Water Process Engineering*, *53*, 103612. https://doi.org/10.1016/j.jwpe.2023.103612

Liu, J., Ge, Y., Wang, G., Liu, Y., & Xu, X. (2021). Highly efficient removal of U (VI) in aqueous solutions by tea waste-derived biochar-supported iron-manganese oxide composite. *Journal of Radioanalytical and Nuclear Chemistry*, *330*(3), 871–882.

Ma, J., Fan, L., Wang, X., Li, L., Zhang, Y., Zhao, G., Chai, B., & Gao, J. (2024). Defect engineering on metal-organic frameworks for enhanced photocatalytic reduction of Cr (VI). *Surfaces and Interfaces*, *54*, 105174.

Ma, R., Zhu, J., Xu, Z., Shi, Q., Zhang, L., & Fan, J. (2025). Novel biomass carbon dots/chitosan hydrogel-modified TiO2: An effective reduction of Cr(VI) in various wastewater under sunlight. *Separation and Purification Technology*, *357*, 130082. https://doi.org/10.1016/j.seppur.2024.130082

Madji, S., Belmedani, M., Mekatel, E., & Trari, M. (2023). Co-precipitation synthesis and characterization of the sillenite Bi12CoO20: Application to photoactivity for water treatment on the heterojunction with ZnO. *Journal of Materials Science: Materials in Electronics*, *34*(31), 2093.

Madzin, Z., Zahidi, I., Talei, A., Raghunandan, M. E., Hermawan, A. A., & Karam, D. S. (2025). Optimising spent mushroom compost biochar for heavy metal removal: Mechanisms and kinetics in mine water treatment. *Journal of Water Process Engineering*, *69*, 106829.

Mallah, N. B., Shah, A. A., Pirzada, A. M., Ali, I., Khan, M. I., Jatoi, A. S., Ullman, J. L., & Mahar, R. B. (2024). Advanced control strategies of membrane fouling in wastewater treatment: A review. *Processes*, *12*(12), 2681.

Mecha, C. A., Monchari, C., Chollom, M. N., & Bakare, B. F. (2025). Applications of solid waste-derived carbon nanomaterials in water treatment. In *Waste derived carbon nanomaterials: ACS publications series,* (Vol. 2, pp. 85–101).

Meraj, U., Laiq, E., Bushra, R., Qurtulen, Ahmad, M., Gupta, S., & Mujahid, M. (2025). Hydrothermal synthesis of okra waste-derived CQDs-CNTs nanocomposites: Potent catalysts for pollutant degradation of RhB and MO dyes and its interaction with lysozyme protein. *Biomass Conversion and Biorefinery*, 1–21.

Mitra, S., Chakraborty, A. J., Tareq, A. M., Emran, T. Bin, Nainu, F., Khusro, A., Idris, A. M., Khandaker, M. U., Osman, H., & Alhumaydhi, F. A. (2022). Impact of heavy metals on the environment and human health: Novel therapeutic insights to counter the toxicity. *Journal of King Saud University-Science*, *34*(3), 101865.

Mo, J., Yang, Q., Zhang, N., Zhang, W., Zheng, Y., & Zhang, Z. (2018). A review on agro-industrial waste (AIW) derived adsorbents for water and wastewater treatment. *Journal of Environmental Management*, *227*, 395–405.

Mohtaram, M. S., Mohtaram, S., Sabbaghi, S., You, X., Wu, W., Jia, L., Muzammil, K., Alraee, N. A., Islam, S., & Aryanfar, Y. (2024). Photocatalytic degradation of acetaminophen using a novel TiO2-orange peel-derived biochar composite: Synthesize, characterization and optimization of key factors. *Journal of Water Process Engineering*, *58*, 104884. https://doi.org/10.1016/j.jwpe.2024.104884

Moureen, A., Waqas, M., Khan, N., Jabeen, F., Magazzino, C., Jamila, N., & Beyazli, D. (2024). Untapped potential of food waste derived biochar for the removal of heavy metals from wastewater. *Chemosphere*, *356*, 141932.

Mvala, B. A., Munonde, T. S., Bambo, M. F., Matabola, K. P., & Nomngongo, P. N. (2025). Reusable waste-derived magnetic mesoporous activated carbon for efficient preconcentration and adsorptive removal of fluoroquinolones. *Chemical Engineering Journal Advances*, *23*, 100771. https://doi.org/10.1016/j.ceja.2025.100771

Mwebembezi, T., Wakatuntu, J., Jjagwe, J., Kanyesigye, C., Kulabako, R. N., & Olupot, P. W. (2024). Synthesis, characterization and application of steel waste-based iron oxide nanoparticles for removal of heavy metals from industrial wastewaters. *Heliyon*, *10*(6).

Nallaselvam, T., Rajamohan, S., Kalaiarasu, B., & Hoang, A. T. (2023). High efficient COVID-19 waste co-pyrolysis char/TiO2 nanocomposite for photocatalytic reduction of Cr (VI) under visible light. *Environmental Science and Pollution Research*, *30*(43), 97178–97194.

Nath, A., Mahato, D. K., Singh, L. R., Kuiri, P. K., & Mahato, M. (2025). Waste-derived green ZnO–Fe2O3 photocatalyst for degradation of MB dye pollutant in water. *Inorganic and Nano-Metal Chemistry*, *55*(2), 178–191.

Ndagijimana, P., Cui, B., Zhang, X., Nkinahamira, F., Rong, H., Guo, D., Rugabirwa, B., Hakizimana, J. C., Ndokoye, P., & Nizeyimana, J. C. (2025). Advances in carbon-based materials for solar-driven steam generation, desalination and water treatment. *Desalination*, *593*, 118192.

Nsubuga, H., Basheer, C., Haider, M. B., & Bakdash, R. (2018). Sol-gel based biogenic silica composite as green nanosorbent for chemometric optimization of micro-solid-phase extraction of beta blockers. *Journal of Chromatography A*, *1554*, 16–27.

Oji, B., Borisade, S. G., & Owoeye, S. S. (2025). Effects of acid concentration on the synthesis of alumina nanoparticles from waste aluminum shavings using chemical precipitation technique. *Sustainable Chemistry One World*, 100079.

Okey-Onyesolu, C. F., Nwokedi, I. C., Edeoga, C. A., & John, S. U. (2025). Rice husk biomass-derived nanoparticles: A green strategy for tetracycline removal. *UNIZIK Journal of Engineering and Applied Sciences*, *5*(1), 2180–2203.

Omenesa Idris, M., Asshifa Md Noh, N., Nasir Mohamad Ibrahim, M., & Ali Yaqoob, A. (2023). Sustainable microbial fuel cell functionalized with a bio-waste: A feasible route to formaldehyde bioremediation along with bioelectricity generation. *Chemical Engineering Journal*, *455*, 140781. https://doi.org/10.1016/j.cej.2022.140781

Omenesa Idris, M., Nasir Mohamad Ibrahim, M., Asshifa Md Noh, N., Ali Yaqoob, A., Najihah Mohamad Daud, N., Hazwan Hussin, M., & AlObaid, A. A. (2024). Assessment of biomass material as valuable electrode for high energy performance in microbial fuel cell with biodegradation of organic pollutant. *Fuel*, *371*, 132059. https://doi.org/10.1016/j.fuel.2024.132059

P., A., S., V., G, S., & M., R. (2024). Sustainable development and analysis of a novel bio-derived (biochar) nanocomposite for the remediation of carbamazepine from aqueous solution. *Chemosphere*, *347*, 140696. https://doi.org/10.1016/j.chemosphere.2023.140696

Pan, X., Zhou, C., Wang, H., Ruan, F., Wang, H., & Feng, Q. (2025). Biomass waste-derived activated carbon for effective Cr(VI) removal from textile wastewater. *Inorganic Chemistry Communications*, *181*, 115196. https://doi.org/10.1016/j.inoche.2025.115196

Panneerselvam, C., Suresh, U., Alatawi, F. A., Angayarkanni, J., & Murugan, K. (2022). Cigarette butt waste derived activated carbon incorporated silver nanoparticle (AC-Ag): Effective nanocomposite for anti-bacterial and anti-larval activity in wastewater remediation. *Materials Letters*, *313*, 131809.

Papari, S., Bamdad, H., & Berruti, F. (2021). Pyrolytic conversion of plastic waste to value-added products and fuels: A review. *Materials*, *14*(10), 2586.

Parashar, M., Shukla, V. K., & Singh, R. (2020). Metal oxides nanoparticles via sol-gel method: A review on synthesis, characterization and applications. *Journal of Materials Science: Materials in Electronics*, *31*(5), 3729–3749.

Patel, D., Kulwant, M., Shirin, S., Varshney, R., Pandey, G., & Yadav, A. K. (2024). Fluoride removal from aqueous solution using iron-based materials: Preparation, characterization, and applications. In *Advanced treatment technologies for fluoride removal in water: Water purification* (pp. 71–92). Springer.

Patel, D., Tripathi, K. M., & Sonwani, R. K. (2024). Waste-derived carbon nano-onions for the removal of organic dye from wastewater and phytotoxicity studies. *ACS Omega*, *9*(28), 30834–30845.

Pham, T. D., Nguyen, D. T., Nguyen, H. L., Nguyen, M. Q., Tran, T. M., Nguyen, M. V., Nguyen, T. L., Ngo, T. M. V., Namakamura, K., & Tsubota, T. (2025). Adsorption characteristics of ciprofloxacin and naproxen from aqueous solution using bamboo biochar. *Biomass Conversion and Biorefinery*, *15*(2), 3071–3082.

Pineda-Vásquez, T., Rendón-Castrillón, L., Ramírez-Carmona, M., & Ocampo-López, C. (2023). From e-waste to high-value materials: Sustainable synthesis of metal, metal oxide, and MOF nanoparticles from waste printed circuit boards. *Nanomaterials*, *14*(1), 69.

Poudel, B. R., Aryal, R. L., Bhattarai, S., Koirala, A. R., Gautam, S. K., Ghimire, K. N., Pant, B., Park, M., Paudyal, H., & Pokhrel, M. R. (2020). Agro-waste derived biomass impregnated with TiO2 as a potential adsorbent for removal of as (III) from water. *Catalysts*, *10*(10), 1125.

Praveen, S., Jegan, J., Bhagavathi Pushpa, T., Gokulan, R., & Bulgariu, L. (2022). Biochar for removal of dyes in contaminated water: An overview. *Biochar*, *4*(1), 10.

Priya, A. K., Muruganandam, M., & Suresh, S. (2024). Bio-derived carbon-based materials for sustainable environmental remediation and wastewater treatment. *Chemosphere*, *362*, 142731.

Priyadarshini, M., Ahmad, A., Shinde, A., Das, I., & Madhao Ghangrekar, M. (2024). Anodic degradation of salicylic acid and simultaneous bio-electricity recovery in microbial fuel cell using waste-banana-peels derived biochar-supported MIL-53(Fe)-metal-organic framework as cathode catalyst. *Journal of Electroanalytical Chemistry*, *967*, 118451. https://doi.org/10.1016/j.jelechem.2024.118451

Rahman, A., Yoshida, K., Islam, M. M., & Kobayashi, G. (2023). Investigation of efficient adsorption of toxic heavy metals (chromium, lead, cadmium) from aquatic environment using orange peel cellulose as adsorbent. *Sustainability*, *15*(5), 4470.

Raj, R., Dhanda, A., Kumar, S., Das, S., & Ghangrekar, M. M. (2024). Repurposing waste-iron electrocoagulated algal biomass as effective heterogenous (bio) electro-fenton catalyst for phthalate removal from wastewater. *Scientific Reports*, *14*(1), 23932.

Rani, M., Keshu, Usha, & Shanker, U. (2023). Chicken egg shell waste derived calcium oxide based nanohybrid for rapid removal of heavy metal ions from water: Green synthesis, kinetics and reusability. *ChemistrySelect*, *8*(16), e202203540.

Rao, V. D., Raghutu, R., Rao, K. S., Kumar, R., Padma, D. V., & Sastry, S. (2024). Efficient biosorption of cadmium ions from wastewater using iron oxide–maize shell activated bio-carbon nanocomposites. *Biomass Conversion and Biorefinery*, 1–12.

Rashid, R., Shafiq, I., Akhter, P., Iqbal, M. J., & Hussain, M. (2021). A state-of-the-art review on wastewater treatment techniques: The effectiveness of adsorption method. *Environmental Science and Pollution Research, 28*(8), 9050–9066.

Raza, M. K., Ali, J. K., Shetty, D., & Alhseinat, E. (2025). Waste to wastewater treatment: Graphene quantum dots embedded poly(ethylene terephthalate) membranes for efficient phenol removal. *Environmental Technology & Innovation, 38*, 104163. https://doi.org/10.1016/j.eti.2025.104163

Reyes-Vallejo, O., Sánchez-Albores, R., Maldonado-Alvarez, A., Ashok, A., Duran-Alvarez, J. C., & Subramaniam, V. (2023). Calcium–Magnesium oxide by the ball-milling method using eggshell as calcium source: Its study for photodegradation of methylene blue. *Journal of Materials Science: Materials in Electronics, 34*(8), 770.

Robinson, B. H. (2009). E-waste: An assessment of global production and environmental impacts. *Science of the Total Environment, 408*(2), 183–191.

Saqib, N. U., Naqvi, M., Li, B., Sarmah, A. K., & Munir, M. T. (2024). From scraps to purification: Innovative use of food waste-derived hydrochar in eradicating pharmaceutical pollutants. *Environmental Science and Pollution Research*, 1–16.

Seif, R., Salem, F. Z., & Allam, N. K. (2023). E-waste recycled materials as efficient catalysts for renewable energy technologies and better environmental sustainability. *Environment, Development and Sustainability*, 1.

Shaaban, A., Se, S.-M., Mitan, N. M. M., & Dimin, M. F. (2013). Characterization of biochar derived from rubber wood sawdust through slow pyrolysis on surface porosities and functional groups. *Procedia Engineering, 68*, 365–371. https://doi.org/10.1016/j.proeng.2013.12.193

Sharma, A., Datta, S., Sanjana, R. K., Pooja, B. M., Bose, S., & Hegde, G. (2025). Onion peel derived carbon nanoparticles incorporated polysulfone membranes: Enhanced dye removal from water. *RSC Advances, 15*(10), 7786–7798.

Sharma, R., Agrawal, P. R., Chankit, Chanchal, Ittishree, Kashyap, V., Sharma, A. K., & Alagesan, V. (2024). Industrial waste-derived materials for adsorption of heavy metals from polluted water. *Remediation of Heavy Metals: Sustainable Technologies and Recent Advances*, 169–197.

Singh, E., Kumar, A., Mishra, R., You, S., Singh, L., Kumar, S., & Kumar, R. (2021). Pyrolysis of waste biomass and plastics for production of biochar and its use for removal of heavy metals from aqueous solution. *Bioresource Technology, 320*, 124278.

Singh, J., Bhattu, M., Brar, S. K., & Jadeja, R. (2025). Emerging trends in waste-derived biochar-based sustainable platform for the removal of organic and inorganic water pollutants. *Waste and Biomass Valorization*, 1–34.

Soundari, L., & Prasanna, K. (2025). Optimum usage of biochar derived from agricultural biomass in removing organic pollutant present in pharmaceutical wastewater. *Sustainable Chemistry for the Environment*, 100259.

Stegarescu, A., Lung, I., Cârdan, A., Boçşa, M., Turza, A., Lazar, M. D., Dan, M., Tripon, S., Kacso, I., & Pintea, S. (2025). Composite materials based on biochar obtained from tomato wastes and Fe3O4/MnO2 used for paracetamol adsorption. *Materials, 18*(16), 3914.

Su, Q., Ren, F., Zhang, Y., Wang, Y., Cao, S., Li, Z., Shen, Y., Li, H., Su, Y., & Wang, Y. (2025). Adsorption of ibuprofen using biomass carbon derived from one-step pyrolysis of ginkgo leaves. *Industrial Crops and Products, 226*, 120739.

Suresh, A., Sundararaman, S., Benzzley, C., & Jefrin, P. (2025). Effective ibuprofen removal utilizing agricultural waste derived activated carbon doped bimetallic MOF using an adsorptive technique emphasizing environment sustainability. *Next Materials, 9*, 101035. https://doi.org/10.1016/j.nxmate.2025.101035

Tanveer, A., Pradhan, S., Tong, Y., Pasha, M., Shetty, A. R., Al Emadi, A., Al-Ansari, T., & McKay, G. (2025). Valorization of orange peel via slow pyrolysis: Effect of particle size on biochar characteristics for wastewater remediation. *Results in Engineering*, 106612.

Tee, G. T., Gok, X. Y., & Yong, W. F. (2022). Adsorption of pollutants in wastewater via biosorbents, nanoparticles and magnetic biosorbents: A review. *Environmental Research*, *212*, 113248.

Tejada-Tovar, C., Villabona-Ortiz, A., & Ortega-Toro, R. (2023). Removal of metals and dyes in water using low-cost agro-industrial waste materials. *Applied Sciences*, *13*(14), 8481.

Thamilselvan, V., Balu, S., Ganapathy, D., Atchudan, R., Arya, S., Hazra, S., & Sundramoorthy, A. K. (2025). Utilization of biomass waste derived carbon quantum dots intercalated ZnO for effective photocatalytic degradation of methylene blue. *Results in Surfaces and Interfaces*, *19*, 100520. https://doi.org/10.1016/j.rsurfi.2025.100520

Ullah, S., Shah, S. S. A., Altaf, M., Hossain, I., El Sayed, M. E., Kallel, M., El-Bahy, Z. M., ur Rehman, A., Najam, T., & Nazir, M. A. (2024). Activated carbon derived from biomass for wastewater treatment: Synthesis, application and future challenges. *Journal of Analytical and Applied Pyrolysis*, *179*, 106480.

Velumani, M., & Jeyadharmarajan, J. (2023). Conversion of novel tannery sludge-derived biochar/TiO2 nanocomposite for efficient removal of Cr (VI) under UV light: Photocatalytic performance and mechanism insight. *Environmental Science and Pollution Research*, *30*(10), 28173–28191.

Vigneshwaran, S., Kim, D.-G., & Ko, S.-O. (2024). Porous biochar supported PET plastic waste MOF heterostructure as a novel, efficient and recyclable catalyst for acetaminophen degradation. *Biochar*, *6*(1), 94.

Vijay, M., Kour, M., Jadoun, S., Jangid, N. K., Dwivedi, J., & Sharma, S. (2025). Plastic waste-derived modified low-density polyethylene membrane for solar-assisted degradation of antipyretic drug paracetamol from water resources. *ACS ES&T Water*, *5*(5), 2684–2697.

Waleng, N. J., Munonde, T. S., Mpupa, A., Moremedi, T., Zhang, Y., & Nomngongo, P. N. (2025). Photoelectrochemical properties of magnetic amine-based MIL-101 (Cr) hybrid material and its application in the degradation of acebutolol in water. *RSC Advances*, *15*(23), 17986–17999.

Wan, H., Jin, R., Tian, Z., Qiu, X., Zhai, S., Niu, J., Yang, S., & Zhao, Y. (2025). Electro-activation of peroxymonosulfate by novel Co3O4/sludge biochar cathode for sulfamethoxazole removal: Cobalt-mediated synergistic effect and mechanism. *Environmental Research*, *279*, 121742. https://doi.org/10.1016/j.envres.2025.121742

Wang, D., Li, T., Yan, C., Zhou, Y., & Zhou, L. (2023). A novel bio-flocculation combined with electrodialysis process: Efficient removal of pollutants and sustainable resource recovery from swine wastewater. *Separation and Purification Technology*, *304*, 122330.

Wang, Q., Zhou, Q., Wang, P., Wang, E., Chen, S., Shi, Y., Deng, H., Liu, A., Du, H., & Li, Z. (2025). S-scheme POM/MIL-101 (Fe) heterojunction for photocatalytic decontamination of Cr (VI). *Separation and Purification Technology*, *361*, 131498.

Wang, T., Kumar, A., Sharma, G., Wang, S., Jia, J., Zheng, J., & Shi, H. (2024). Synergistic mechanisms of novel Z-Scheme N, S co-doped biochar-based Ag3PO4 composites for efficient removal of norfloxacin. *NPJ Clean Water*, *7*(1), 97.

Wang, Y., Wang, Q., Zhao, X., Zhang, C., Zhou, Y., Xie, W., & Ren, G. (2024). Carbon skeleton dispersed nano-jarosite for efficient Cr (VI) degradation: A bioinspired MFC cathode catalyst. *Journal of Environmental Chemical Engineering*, *12*(2), 112003. https://doi.org/10.1016/j.jece.2024.112003

Wei, L. K., Abd Rahim, S. Z., Al Bakri Abdullah, M. M., Yin, A. T. M., Ghazali, M. F., Omar, M. F., Nemeş, O., Sandu, A. V., Vizureanu, P., & Abdellah, A. E. (2023). Producing metal powder from machining chips using ball milling process: A review. *Materials*, *16*(13), 4635.

Xiao, Y., Lyu, H., Yang, C., Zhao, B., Wang, L., & Tang, J. (2021). Graphitic carbon nitride/biochar composite synthesized by a facile ball-milling method for the adsorption and photocatalytic degradation of enrofloxacin. *Journal of Environmental Sciences*, *103*, 93–107.

Xu, X., Sui, Y., Chen, W., Li, X., Huang, W., Chai, L., Li, Y., & Zhong, H. (2024). Photocatalytic Cr (VI) reduction by metal-free photocatalysts under visible-light irradiation. *Journal of Environmental Chemical Engineering*, *12*(6), 114306.

Xu, Z., Su, X., Yang, P., Zhong, J., Li, M., Burda, C., & Dou, L. (2025). Efficient photocatalytic CO2 and Cr (VI) reduction on carbon quantum dots/carbon nitride heterojunctions. *Fuel*, *381*, 133285.

Yao, L., Xi, Y., Han, H., Li, W., Wang, C., & Feng, Y. (2021). LiMn2O4 prepared from waste lithium ion batteries through sol-gel process. *Journal of Alloys and Compounds*, *868*, 159222.

Ye, Q., Li, Q., & Li, X. (2022). Removal of heavy metals from wastewater using biochars: Adsorption and mechanisms. *Environmental Pollutants and Bioavailability*, *34*(1), 385–394.

Yu, F., Tian, F., Zou, H., Ye, Z., Peng, C., Huang, J., Zheng, Y., Zhang, Y., Yang, Y., & Wei, X. (2021). ZnO/biochar nanocomposites via solvent free ball milling for enhanced adsorption and photocatalytic degradation of methylene blue. *Journal of Hazardous Materials*, *415*, 125511.

Yunus, A. I., & Fu, G. Y. (2024). Selecting the effective biochar for removal of pollutants from highway stormwater runoff. *Journal of Environmental Engineering*, *150*(7), 4024027.

Zadeh, K. K., & Jafari, D. (2024). Activated carbon/alginate/Fe3O4 magnetic nanocomposite as a superior functional material for removal of lead from aqueous media. *Biomass Conversion and Biorefinery*, *14*(16), 19025–19043.

Zanotti, K., Igal, K., Colombo Migliorero, M. B., Gomes Zuin, V., & Vázquez, P. G. (2021). Synthesis of silica-based materials using bio-residues through the sol-gel technique. *Sustainable Chemistry*, *2*(4), 670–685.

Zhang, C., Zhao, S., Huang, Q., Liu, J., & Zhang, Q. (2025). Adsorptive-photocatalytic removal of tetracycline from wastewater by Fe3O4-and SnO2-containing biochar nanocomposites. *Biochar*, *7*(1), 1–19.

Zhang, L., Guo, J., Huang, X., Wang, W., Sun, P., Li, Y., & Han, J. (2019). Functionalized biochar-supported magnetic MnFe 2 O 4 nanocomposite for the removal of Pb (ii) and Cd (ii). *RSC Advances*, *9*(1), 365–376.

Zhang, S., Leng, W., Zhang, S., Lu, H., Xu, X., Zang, Z., & Wu, D. (2024). In-situ active sites analysis of bifunctional metal-organic frameworks for coupled adsorption and electrochemical oxidation of PPCPs. *Chemical Engineering Journal*, *486*, 150322.

Zhang, Y., Li, Y., Cheng, M., Yao, A., & Jing, Z. (2025). A novel approach to return sludge to nature: Hydrothermal conversion of full sludge components into humic acid and functional minerals. *Journal of Cleaner Production*, 145562.

Zhao, H., & Lang, Y. (2018). Adsorption behaviors and mechanisms of florfenicol by magnetic functionalized biochar and reed biochar. *Journal of the Taiwan Institute of Chemical Engineers*, *88*, 152–160.

Zhao, S., Liu, S., Sumpradit, T., Zhou, J., & Qu, J. (2024). Magnetic nanoparticles doped biochar cathode in a two-chamber microbial fuel cell for the adsorption-reduction of hexavalent chromium. *International Journal of Hydrogen Energy*, *63*, 163–172. https://doi.org/10.1016/j.ijhydene.2023.08.105

Zhao, Y., Luo, M., Yang, R., Yao, S., Zhang, Q., Yu, W., Feng, Z., Yan, Y., & Xu, Y. (2025). Hydrothermal synthesis of biomass waste derived graphene quantum dots with high ion detection ability. *Diamond and Related Materials*, *155*, 112363.

Zou, W., Feng, X., Wei, W., Zhou, Y., Wang, R., Zheng, R., Li, J., Luo, S., Mi, H., & Chen, H. (2021). Converting spent LiFePO4 battery into zeolitic phosphate for highly efficient heavy metal adsorption. *Inorganic Chemistry*, *60*(13), 9496–9503.

Zuhara, S., & McKay, G. (2025). Single and binary pollutant adsorption of strontium and barium on waste-derived activated carbons: Modelling, regeneration and mechanistic insights. *Environmental Technology & Innovation*, 104220.

11 Application of waste-derived nanomaterials in the food, pharmaceutical, and cosmetic industries

11.1 INTRODUCTION

As an emerging field dealing with nanoscale materials, nanotechnology, due to its various applications, plays an important role in human daily lives (Abd Elkodous et al., 2021; Malode et al., 2024). By definition, structures of matter with dimensions 1×10^{-9} of a metre are considered nanomaterials (Chadha et al., 2022; Sharma & Hussain, 2020). The miniaturisation of matter has opened new avenues of research into the preparation, characterisation, and application of nanomaterials.

Compared to their bulk counterparts, nanomaterials have remarkable physical properties. This is due to their high aspect ratios, which give them an enhanced surface-to-volume ratio, rendering improved reactivity and efficiency (Judith & Vasudevan, 2022). Over the years, researchers have shown competence in developing nanosized materials for almost every compound. Noteworthy contributions from nanotechnology span from displays, solar cells, and nanosized sensor assemblies, among others (Jeet & Kaur, 2023).

Organic and inorganic nanomaterials have been developed. These include carbon-based materials such as carbon nanotubes (CNTs), biochar, graphene, carbon nanofibres (CNFs), carbon dots, and inorganic nanomaterials. There are several ways these can be prepared, including physical, chemical, and biological synthesis routes. This chapter is focused mainly on nanoparticles synthesised via the biological route (Abid et al., 2022; Chopra et al., 2022; Omar et al., 2024). Nanomaterials prepared using biological methods are relatively non-toxic, clean, safe, and eco-friendly compared to other methods. Waste-derived nanomaterials can be considered as nanomaterials made from biological methods (Chopra et al., 2022; Omar et al., 2024). Thus, the aim of this chapter was to review the application of waste-derived nanomaterials in industries such as the food, pharmaceuticals, and cosmetics. The appropriate studies were compared according to the biomass source, the nanomaterial prepared, and the application of the as-prepared nanomaterials.

11.2 APPLICATION OF WASTE-DERIVED NANOMATERIALS IN THE FOOD INDUSTRY

Food spoilage is a major challenge worldwide when conventional preservation techniques fail. There is a need for innovative methods for food preservation (Dubey

DOI: 10.1201/9781003745846-11

et al., 2025; Sarker et al., 2024). Hence, come waste-derived nanomaterials. These offer two distinct advantages: reducing waste disposal and producing value-added products. In the food industry, waste-derived nanomaterials have been used for their antimicrobial properties for packaging (Aksu & Güzdemir, 2025)

The application of nanomaterials in the food industry has gained popularity in recent years. This application in the food industry spans from packaging, preservation, and fortification. For packaging, the use of nanomaterials enables the protection of food products from potential hazards, ensuring the safety of food (Omar et al., 2024). Food packaging is often divided into primary and secondary packaging, where primary packaging is in direct contact with the product, and secondary packaging is focused on storage (Baraketi & Khwaldia, 2024). Primary packaging has been a subject of research, as it is linked to preserving the packaged product (Chadha et al., 2022). This is also termed as active packaging. Examples of studies showing the applicability of waste-derived nanoparticles in the food industry are summarised in Table 11.1.

Waste-derived nanomaterials, such as quantum dots, have the potential to replace traditional detection techniques, such as chromatography and spectroscopy. They have been used in food safety testing in the detection of metal ions, food additives, microbes, pesticides, and flavour. These are often used as fluorescent probes. Aksu and Güzdemir covered in depth these applications as well as the synthetic techniques, precursors, and properties of CQDs from food waste (Aksu & Güzdemir, 2025).

TABLE 11.1
Summary of selected studies based on the application of waste-derived nanomaterials in the food industry.

Source	Nanoparticle	Application	Reference
Pomegranate and Kinnow fruit peel	Ag nanoparticles	Cellulose wrapper for packaged bread	Gopalakrishnan et al. (2023)
Rice husks	Lignin nanoparticles	Polylactic acid (PLA)-based food packaging	Daassi et al. (2023)
Orange peel	Nanocellulose	Polyvinyl alcohol (PVA)-based film	Espinosa et al. (2022)
Mango peel	Cellulose nanocrystal	PVA-based films for fruit packaging	Xing et al. (2021)
Grapefruit seed	ZnO nanoparticles	Cellulose nanofibre	Roy et al. (2021)
Prawn shell	Chitin nanocrystals	Pickering emulsion	Barkhordari and Fathi (2018)
Crab shell	Chitin nanoparticles	Stabilisation oil-in-water Pickering emulsion	Liu et al. (2021)
Prawn shell	Nanocrystals and nanofibres	Pickering emulsion	Larbi et al. (2018)
Shrimp shell	α-chitin	Saltiness enhancer	Somsak et al. (2021)
Squid pens	Chitin nanofibres	Saltiness enhancer	Hsueh et al. (2017)

Gopalakrishnan et al. showed the valorisation of pomegranate and Kinnow for the green synthesis of Ag nanoparticles. These nanoparticles were then used to prepare a cellulose-based bread packaging. The prepared wrapper, compared with unmodified packaging, showed superior characteristics, including lower oxygen and water vapour permeability coefficients and lower water adsorption. Upon storage experiments, the modified cellulose wrapper showed delayed growth of microbes, yeast and mould over a 7-day period, confirming its potential for increasing shelf life (Gopalakrishnan et al., 2023).

Other studies where valorisation of rice husks, orange peels, mango peels, and grapefruit seeds were used by different researchers (Daassi et al., 2023; Espinosa et al., 2022; Roy et al., 2021; Xing et al., 2021) prepared lignin nanoparticles for PLA-based food packaging. These innovative materials address product safety issues by improving material performance. The study by Daassi et al. (Daassi et al., 2023) showed reduced optical transmittance and increased antioxidant properties of the modified PLA (10 to 12 times higher than unmodified PLA). Similarly, Espinosa et al. (Espinosa et al., 2022) observed the same using orange peel nanocellulose. The valorisation of mango peel into cellulose nanocrystals increased the thermostability of PVA films, increased tensile strength, and rendered good antifungal activity. The review by Baraketi and Khwaldia (2024) thoroughly discusses the different preparation methods of various waste-derived nanomaterials for food packaging.

Chitin, a natural polymer found in crustacean shells, is widely available in shell waste. As reported by previous researchers, improper disposal of crustacean shell waste can have adverse environmental impacts (Sampath et al., 2022; Yan & Chen, 2015). However, recently these have been transformed into valuable products through valorisation. Chitin-based nanomaterials are some of these valorisation products (Younes & Rinaudo, 2015). Chitin can be extracted from crustacean shells using chemical and biological methods in three steps: demineralisation, deproteinisation, and decolourisation (Dayakar et al., 2022; Sampath et al., 2022).

In the food industry, chitin-based nanomaterials have been used as emulsifiers (Pickering emulsions), food additives, and salt enhancers, among other applications (Table 11.1). Pickering emulsions reported by Bai et al. (2019), Barkhordari and Fathi (2018), and Larbi et al. (2018) are used to reduce surface tension and thus stabilise oil–water products, using chitin nanocrystals, nanoparticles, and nanofibres from prawn, crab, and shrimp shells, respectively. These were shown to increase emulsion storage stability by up to four weeks.

Additionally, chitin-based nanomaterials have been used as food additives to extend shelf life. Similarly, in the saltiness enhancement of salt-cured tilapia fillets, chitin nanofibres have been shown to be applicable (Hsueh et al., 2017; Somsak et al., 2021).

11.3 APPLICATION OF WASTE-DERIVED NANOMATERIALS IN THE PHARMACEUTICAL INDUSTRY

In the pharmaceutical industry, nanomaterials are used for multiple applications such as drug delivery, imaging, and therapeutics (Sen et al., 2025). This is due to their unique properties to bind, carry, and absorb molecules, DNA, RNA, and proteins (Bala et al., 2023). Table 11.2 summarises some of the applications of different

TABLE 11.2
Application of waste-derived nanomaterials in the pharmaceutical industry.

Source	Nanoparticle	Preparation Method	Application	Reference
Banana pseudostem fibres	Microcrystalline cellulose	Acid hydrolysis	Delivery of isoniazid	Diarsa and Gupte (2021)
Cotton waste	Cellulose nanoparticles	Acid hydrolysis	Delivery of Doxorubicin	Gülsu and Yüksektepe (2021)
Pine needle	Oxidised cellulose nanoparticles	Alkali treatment	Delivery of Doxorubicin	Kumari et al. (2018)
Rice bran	PBA-chitosan	Water extraction	Delivery of curcumin	Peng et al. (2017)
Egg shells	Nano-CaO	Acid hydrolysis	Pectin-g-P(AN) delivery	Priyadarsini et al. (2021)
Squilla shells	Chitosan nanoparticles	Acid hydrolysis	Delivery of diclofenac	Balde et al. (2020)
Sugarcane bagasse	Carbon nanotubes	Base hydrolysis	Delivery of sulfadiazine	Tohamy (2024)
Marigold plant	N-GQDs	High-temperature pyrolysis	Detection of Fe^{3+} and bioimaging	Zhang et al. (2019)
Rice waste	Carbon Quantum Dots (CQDs)	Acid treatment	in vitro and in vivo bioimaging	Anthony et al. (2020)
Willow catkin	Nitrogen/S-CQDs	Acid treatment	Fluorescent detection of Fe	Cheng et al. (2019)
Lignin biowaste	GQDs	Oxidised cleavage and aromatic refusion	Nanoprobes for multicolour bioimaging	Ding et al. (2018)
White goosefoot plant	Reduced graphene oxide	-	Antimicrobial and anticancer	Umar et al. (2020)
Tridax procumbens	Reduced graphene oxide	-	Antioxidant and antibiotic	Thiyagarajulu et al. (2020)
Kitchen soot	Carbon nanoparticles	Combustion	Antibacterial and antidiabetic	Varghese et al. (2013)
Cissus arnotiana plant	Copper nanoparticles	-	Antibacterial and antioxidant	Rajeshkumar et al. (2019)
Tabernaemontana divaricate	Zinc oxide nanoparticles	-	Antimicrobial	Raja et al. (2018)
Chicken feather	Tragacanth gum	Reduction-hydrolysis	Wound healing and antimicrobial	Mansouri Shirazi et al. (2021)

waste-derived nanomaterials in the pharmaceutical industry, including drug delivery systems (Kumar et al., 2025), imaging, and therapeutics (Malode et al., 2024).

There are several side effects that are related to the use of drugs. These are mainly due to the untargeted nature of drugs (Kumar et al., 2025). As such, research has led to the development of drug delivery systems. Researchers have developed innovative methods for carrying drugs using waste-derived nanomaterials. This approach offers a way of solving the problem of drug delivery while also using up waste and producing a value-added product (Bala et al., 2023). Table 11.2 shows studies that aimed to address the challenge of drug delivery using waste-derived nanomaterials.

Diarsa and Gupte (2021) prepared microcrystalline cellulose from banana pseudostem using an acid hydrolysis method. The prepared microcrystalline cellulose was used to deliver isoniazid (Diarsa & Gupte, 2021). Similarly, Gülsu and Yüksektepe 2021 prepared a similar vehicle for the transport of doxorubin using cotton waste. Drug-delivery experiments showed that 97% of the drug was up to 34 days. A similar study using pine needles as a starting material was done by Kumari et al. earlier; however, the vector prepared was stimulus-responsive (Gülsu & Yüksektepe, 2021).

Other studies on the delivery of drugs using waste-derived nanoparticles were conducted by several researchers. These were made from various starting materials such as rice bran by Peng et al. (2017) for the delivery of curcumin, egg shell chitosan nanoparticles for pectin-g-P(AN) delivery (Priyadarsini et al., 2021), squilla shells for diclofenac delivery, as well as carbon nanotubes from sugarcane bagasse for sulfadiazine delivery (Balde et al., 2020).

A similar study by Chung et al. (2020) used sugarcane bagasse as a precursor to prepare carbon dots paired with hydroxyapatite for acetaminophen delivery. This pairing was shown to increase the surface area from 41 to 78 m^2/g. Additionally, the modified nanocomposite showed high loading capacity and faster acetaminophen release. The release of acetaminophen was observed to be governed by Higuchi kinetics and diffusion mechanisms (Chung et al., 2020).

Drug delivery plays an important role in modern pharmaceutical research (Fuhrmann, 2023). Several formulations have been developed to demonstrate the success of drug delivery (van der Meel et al., 2019). The use of waste-derived nanomaterials in the preparation of drug delivery vehicles is an exciting prospect in the field (Singh & Ray, 2023). Cellulose-based materials have been the leading materials in drug delivery. Kamel and colleagues cover in depth cellulose of different origins as drug delivery vehicles for various drugs. The review covers sources, isolation, pretreatment, and preparation techniques from agricultural waste. There is still much research required on the application of these materials in drug delivery applications (Kamel et al., 2020).

Though thoroughly advanced, traditional microscopy often lacks the accuracy to observe biological components due to their small size (Abd Elkodous et al., 2021). Nanomaterials as fluorescent dyes provide the opportunity to track and evaluate the functionality of cells (Joo et al., 2008; van Ham et al., 2010). This is due to the use of nanomaterials. Nanomaterials have the necessary properties, such as the ability to target cells, fluorescence, and small size, which makes them attractive for bioimaging (Abd Elkodous et al., 2021).

Consequently, NP fluorescence imaging is currently used in several pharmaceutical applications, such as gene detection, enzyme activity evaluation, cell tracking, protein analysis, real-time monitoring of therapeutic effects, and early detection of diseases (Abd Elkodous et al., 2021; Dubreil et al., 2017). Several studies have shown the use of waste valorised nanomaterials for use in bioimaging (Abd Elkodous et al., 2021). Some noteworthy studies include those by Zhang et al., in which graphene quantum dots (GQDs) were synthesised from the marigold plant and doped with nitrogen. These were subsequently shown to be applicable for the detection of Fe^{3+} in water environments and the field of bioimaging. The prepared quantum dots showed good stability without any significant change in their fluorescence intensity under changing pH conditions and long ultraviolet irradiation (Zhang et al., 2019). This study not only shows the potential of waste-derived nanoparticles for bioimaging but also the versatility of the prepared quantum dots.

Studies by Anthony and colleagues, Cheng and co-workers, and Ding et al. also explored the use of different biomass precursors, namely rice waste, willow catkin, and lignin biowaste, respectively, and prepared different types of quantum dots for bioimaging applications. Anthony et al. recycled CQDs from rice wastes and imaged A549 cells in vitro and in vivo (injected inside rat models) (Anthony et al., 2020). Cheng et al. prepared nitrogen- and sulfur-co-doped carbon dots (N/S-CDs) with a high quantum yield of 13.3%, synthesised via a one-step combustion treatment of cellulose-based biowaste from willow catkins. After characterisation with high-resolution TEM, FTIR spectroscopy, XPS spectroscopy, and Raman spectroscopy, the prepared material demonstrated superior photostability, low cytotoxicity, pH stability, and could be used for selective and sensitive intracellular detection of Fe^{3+} ions (Cheng et al., 2019).

In 2018, Ding et al. synthesised GQDs from lignin biowaste using a two-step method based on oxidised cleavage and aromatic refusion. The prepared nanomaterials were then introduced to RAW 264.7 cells. Due to their bright fluorescence, up-conversion properties, long-term photostability, water solubility, and biocompatibility, these high-quality GQDs showed potential for application as nanoprobes for multicolour bioimaging (Ding et al., 2018).

The use of waste-derived nanoparticles in the pharmaceutical industry spans from diagnosis to therapeutics. As such, several studies have been carried out in this regard. Earlier studies by Varghese and colleagues used kitchen soot to prepare carbon nanoparticles. After surface and chemical characterisation, the particles were tested for their antibacterial activity against various pathogenic bacterial strains, including Gram-negative *Proteus refrigere* and *Pseudomonas aeruginosa*, and Gram-positive *Staphylococcus aureus* and *Streptococcus haemolyticus*, using the disk diffusion method. This study supports the ancient use of kitchen soot as an "old woman's remedy."

Later studies by Rajeshkumar and colleagues and by Raja and colleagues also used different precursors to prepare copper and zinc oxide nanoparticles, respectively. Rajeshkumar and colleagues used the *Cissus arnotiana* medicinal plant to reduce and cap copper nanoparticles (CuNPs) and later evaluated their antibacterial activity against Gram-negative and Gram-positive bacteria. The biosynthesised CuNPs showed better antibacterial activity against the Gram-negative bacteria,

Escherichia coli, with an inhibition zone of 22.20 ± 0.16 mm at 75 μg/ml. The antioxidant property observed was comparatively equal to that of the standard antioxidant agent ascorbic acid at a maximum concentration of 40 μg/ml (Rajeshkumar et al., 2019), while Raja and co-workers investigated the green synthesis of zinc oxide nanoparticles (ZnO NPs) using aqueous *Tabernaemontana divaricata* leaf extract. The prepared ZnO NPs were compared to a standard pharmaceutical formulation, showing superior antibacterial activity against three bacterial strains, *Escherichia coli* and *Staphylococcus aureus* and lesser activity against *Salmonella paratyphi* (Raja et al., 2018).

More recently, studies have been conducted by Umar et al. (2020) and Thiyagarajulu et al. (2020), who prepared reduced graphene oxide (RGO) from White goosefoot and *Tridax procumbens*, respectively. Umar et al. demonstrated the typical morphology of RGOX as rippled, silky, and transparent stacked layers, with antibacterial activity demonstrated by analysing minimal inhibitory concentration values, agar diffusion assays, and fluorescence detection techniques. Enhanced activity was shown against Gram-positive and Gram-negative bacteria. The RGOX was also shown to be a potential anticancer agent against human breast cancer cell lines (Umar et al., 2020). Thiyagarajulu and colleagues produced RGOX from the leaf extract of *Tridax procumbens* and tested its in-vitro biological activities. Similarly, the prepared nanomaterial was tested against Gram-positive pathogens rather than Gram-negative bacteria, highlighting its potential to act as a green reducing agent for the successful biosynthesis of rGO nanosheets in vitro, biological activities (Thiyagarajulu et al., 2020).

Keratin protein from chicken feathers was used for its antimicrobial wound healing properties by Mansouri Shirazi et al. (2021). The keratin protein was extracted using a reduction-hydrolysis method, and different nanogel composite ratios were produced via chemical cross-linking. Cinnamon was loaded into the nanogels, increasing the observed size from 80 nm for the free nanogel to 85 and 105 nm for 5% and 10% extract-loaded nanogels, respectively. The antibacterial test showed strong activity against both Gram-positive and Gram-negative bacteria. Finally, MTT assay approved the biocompatibility of the produced nanogels for potential use in medical textile applications

11.4 APPLICATION OF WASTE-DERIVED NANOMATERIALS IN THE COSMETIC INDUSTRY

Cosmetics can be defined as formulations that are used to maintain the health of the epidermis of the skin (Debnath & Aneja, 2025). These formulations are marketed and sold as solids, liquids, and semisolids with the primary objective of cleansing, moisturisation, hydration, and protection from external damage (Salvioni et al., 2021). The use of nanomaterials in cosmetics has resulted in the use of the term "nanocosmetics," referring to the use of cosmetic products made by the inclusion of appropriate nanoparticles (Mordorski et al., 2016; Shende et al., 2020). Nanocosmetics have advantages such as enhanced penetration and increased bioavailability of active ingredients due to the presence of nanomaterials (Morganti, 2020; Salvioni et al., 2021). Debnath and Aneja provided a detailed review of the preparation of cosmetics using pomegranate peel agro-wastes, specifically for skincare formulations (Debnath & Aneja, 2025). Some of the studies in the review highlighted the ability of pomegranate-mediated nanoparticle synthesis not only to enhance therapeutic effects but also to suggest lower toxicities of the prepared materials. The most widely researched application of waste valorised nanomaterials in cosmetics is the production of sunscreens, as seen in the review and summarised in Table 11.3.

The use of waste-derived nanomaterials for cosmetic applications has resulted in several studies conducted, whereby sunscreens are prepared as described by the review of Nirmal et al. (2024) using pomegranate material. Similarly, a study by Kokabi and Ebrahimi (2020) showed the green synthesis of zinc oxide nanoparticles and evaluated the sunscreen potential of the nanoparticles after applying them in water-in-oil emulsions. The nanoparticles showed relatively good antioxidant activity. However, there are studies that use different precursors, such as *Phoenix dactylifera* L. seeds, to prepare cyclodextrin complexes with ultraviolet-absorbing properties. Though showing great potential, the authors further suggested that clinical validation is encouraged to fully translate these botanically derived agents into effective and ethically sound sun care innovations (Siroukane et al., 2025). *Prionace glauca* skin waste was used as a source for collagen/cellulose nanofibres to counter UV-induced skin damage. The material demonstrated enhanced thermal stability and interfacial hydrophobicity, remaining stable for up to 14 days. Upon application, the emulsion showed great UV-blocking performance as well as antioxidant activity (Wang et al., 2025).

TABLE 11.3
Application of waste-derived nanomaterials in cosmetic products.

Source	Nanoparticle	Application	Reference
Pomegranate peels	Zinc oxide	Sunscreen formulation	Kokabi and Ebrahimi (2020)
Phoenix dactylifera L. seeds	Cyclodextrin complexes	Dermocosmetic photoprotection	Siroukane et al. (2025)
Fishery waste	Collagen/cellulose nanofibre	UV-resistant formulations	Wang et al. (2025)
Crustaceans and plant biomass	Chitin nanofibril (CN) and nanolignin (LG)	Beauty masks	Morganti (2020)
Bamboo	Carbon nanoparticles	Beauty masks	Naig et al. (2025)
Goldfish	Chitosan	Anti-dandruff shampoo	Firmansyah et al. (2022)
Crab shell powder	Chitosan and surface deacylated chitin nanofibrils	Hair care	Azuma et al. (2019)
	Alginate/chitosan/fucoidan	Hair care	Hao et al. (2020)
Comb-pen (*Atrina pectinata*) shell	Chitosan	Hand and body cream	Fauzi et al. (2021)
Shrimp shell	Fluoridated chitosan	Oral biofilm	Rahayu et al. (2022)
Chitosan	Mud crab shells	Acne treatment	Luthfiyana et al. (2022)

Chitin and carbon nanoparticles have been used by Naig et al. (2025) and Wang et al. (2025) to prepare biodegradable beauty masks targeting oily and acne-prone skin. The prepared masks were shown to be promising alternatives to skincare products, made with biodegradable materials, offering an eco-friendly solution for oily skin.

Similar to the food and pharmaceutical applications, the use of chitosan-based nanomaterials finds a wealth of applications in the cosmetic industry, as seen in Table 11.3. This may be driven by chitosan's status as a biopolymer. This reduces the likelihood of negative effects on the products. In several studies, as shown in Table 11.3, chitosan/chitin from different sources has been employed for various applications.

The study by Firmansyah et al. (2022) presented a method for preparing anti-dandruff shampoo from goldfish scales, a source of chitosan, via ionic gelation. The study demonstrated that the formulated shampoo (3 g of chitosan) exhibited strong inhibition of *Candida albicans* (27.30 ± 0.98 mm) and acceptable physical properties for commercial products. Another study by Azuma et al. (2019) evaluated the use of chitosan and surface-deacetylated chitin nanofibrils (SDACNFs) on hair growth. The chitosan was prepared from crab shell powder. The study showed that chitosan could increase cell growth in human follicular dermal papilla cells in vitro. Overall, the study showed that chitosan could promote hair growth and demonstrated its potential as a therapeutic agent for the treatment of hair loss in patients. Additional studies by Fauzi et al. (2021), Hao et al. (2020), and Rahayu et al. (2022) demonstrated the use of chitosan in skin repair. The chitosan used was from crustacean shells. A study by Luthfiyana et al. (2022) also showed the application of chitosan in the preparation of a hand and body cream from small crab shells.

11.5 CONCLUSION

The use of waste-derived nanoparticles offers several advantages, making them attractive candidates for the food, pharmaceutical, and cosmetic industries. The studies presented, as well as other reviews by a number of researchers, highlighted their uses in food packaging, food preservation, and as enhancers. Additionally, in the pharmaceutical industry, they have been employed as drug delivery vehicles, used for bioimaging and for their therapeutic potential. Lastly, in the cosmetics industry, different research groups have explored applications such as sunscreen formulations, facial masks, and hand and body creams with antioxidant potential. Across these applications, nanoparticle safety remains a major concern. This is evident by the concentrated use of chitin/chitosan-based materials when compared to other materials. This could be driven by the fact that chitin/chitosan are the more eco-friendly and biodegradable options.

REFERENCES

Abd Elkodous, M., El-Husseiny, H. M., El-Sayyad, G. S., Hashem, A. H., Doghish, A. S., Elfadil, D., Radwan, Y., El-Zeiny, H. M., Bedair, H., & Ikhdair, O. A. (2021). Recent advances in waste-recycled nanomaterials for biomedical applications: Waste-to-wealth. *Nanotechnology Reviews*, *10*(1), 1662–1739.

Abid, N., Khan, A. M., Shujait, S., Chaudhary, K., Ikram, M., Imran, M., Haider, J., Khan, M., Khan, Q., & Maqbool, M. (2022). Synthesis of nanomaterials using various top-down and bottom-up approaches, influencing factors, advantages, and disadvantages: A review. *Advances in Colloid and Interface Science*, *300*, 102597. https://doi.org/10.1016/j.cis.2021.102597

Aksu, M., & Güzdemir, Ö. (2025). Food waste-derived carbon quantum dots and their applications in food technology: A critical review. *Food and Bioprocess Technology*, 1–26.

Anthony, A. M., Murugan, R., Subramanian, R., Selvarangan, G. K., Pandurangan, P., Dhanasekaran, A., & Sohrab, A. (2020). Ultra-radiant photoluminescence of glutathione rigidified reduced carbon quantum dots (r-CQDs) derived from ice-biryani for in vitro and in vivo bioimaging applications. *Colloids and Surfaces A: Physicochemical and Engineering Aspects*, *586*, 124266.

Azuma, K., Koizumi, R., Izawa, H., Morimoto, M., Saimoto, H., Osaki, T., Ito, N., Yamashita, M., Tsuka, T., & Imagawa, T. (2019). Hair growth-promoting activities of chitosan and surface-deacetylated chitin nanofibers. *International Journal of Biological Macromolecules*, *126*, 11–17.

Bai, L., Huan, S., Xiang, W., Liu, L., Yang, Y., Nugroho, R. W. N., Fan, Y., & Rojas, O. J. (2019). Self-assembled networks of short and long chitin nanoparticles for oil/water interfacial superstabilization. *ACS Sustainable Chemistry & Engineering*, *7*(7), 6497–6511.

Bala, S., Garg, D., Sridhar, K., Inbaraj, B. S., Singh, R., Kamma, S., Tripathi, M., & Sharma, M. (2023). Transformation of agro-waste into value-added bioproducts and bioactive compounds: Micro/nano formulations and application in the agri-food-pharma sector. *Bioengineering*, *10*(2), 152.

Balde, A., Hasan, A., Joshi, I., & Nazeer, R. A. (2020). Preparation and optimization of chitosan nanoparticles from discarded squilla (Carinosquilla multicarinata) shells for the delivery of anti-inflammatory drug: Diclofenac. *Journal of the Air & Waste Management Association*, *70*(12), 1227–1235.

Baraketi, S., & Khwaldia, K. (2024). Nanoparticles from agri-food by-products: Green technology synthesis and application in food packaging. *Current Opinion in Green and Sustainable Chemistry*, *49*, 100953.

Barkhordari, M. R., & Fathi, M. (2018). Production and characterization of chitin nanocrystals from prawn shell and their application for stabilization of Pickering emulsions. *Food Hydrocolloids*, *82*, 338–345.

Chadha, U., Bhardwaj, P., Selvaraj, S. K., Arasu, K., Praveena, S., Pavan, A., Khanna, M., Singh, P., Singh, S., & Chakravorty, A. (2022). Current trends and future perspectives of nanomaterials in food packaging application. *Journal of Nanomaterials*, *2022*(1), 2745416.

Cheng, C., Xing, M., & Wu, Q. (2019). A universal facile synthesis of nitrogen and sulfur co-doped carbon dots from cellulose-based biowaste for fluorescent detection of Fe3+ ions and intracellular bioimaging. *Materials Science and Engineering: C*, *99*, 611–619.

Chopra, H., Bibi, S., Singh, I., Hasan, M. M., Khan, M. S., Yousafi, Q., Baig, A. A., Rahman, M. M., Islam, F., & Emran, T. Bin. (2022). Green metallic nanoparticles: Biosynthesis to applications. *Frontiers in Bioengineering and Biotechnology*, *10*, 874742.

Chung, H. K., Wongso, V., Sambudi, N. S., & Isnaeni. (2020). Biowaste-derived carbon dots/hydroxyapatite nanocomposite as drug delivery vehicle for acetaminophen. *Journal of Sol-Gel Science and Technology*, *93*(1), 214–223.

Daassi, R., Durand, K., Rodrigue, D., & Stevanovic, T. (2023). Optimization of the electrospray process to produce lignin nanoparticles for PLA-based food packaging. *Polymers*, *15*(13), 2973.

Dayakar, B., Xavier, K. A. M., Ngasotter, S., Layana, P., Balange, A. K., Priyadarshini, B., & Nayak, B. B. (2022). Characterization of spray-dried carotenoprotein powder from Pacific white shrimp (Litopenaeus vannamei) shells and head waste extracted using papain: Antioxidant, spectroscopic, and microstructural properties. *LWT, 159*, 113188.

Debnath, M., & Aneja, D. (2025). Waste valorization utilizing green nanotechnology: A sustainable approach for pomegranate peel agro wastes in skincare formulations. *Bioresources and Bioprocessing, 12*(1), 152.

Diarsa, M., & Gupte, A. (2021). Preparation, characterization and its potential applications in Isoniazid drug delivery of porous microcrystalline cellulose from banana pseudostem fibers. *3 Biotech, 11*(7), 334.

Ding, Z., Li, F., Wen, J., Wang, X., & Sun, R. (2018). Gram-scale synthesis of single-crystalline graphene quantum dots derived from lignin biomass. *Green Chemistry, 20*(6), 1383–1390.

Dubey, A., Singh, P., & Singh, S. (2025). Waste-derived nanomaterials as smart coatings to extend shelf life of perishables. *Nature Science e-Magazine, 2*, 3–6.

Dubreil, L., Leroux, I., Ledevin, M., Schleder, C., Lagalice, L., Lovo, C., Fleurisson, R., Passemard, S., Kilin, V., & Gerber-Lemaire, S. (2017). Multi-harmonic imaging in the second near-infrared window of nanoparticle-labeled stem cells as a monitoring tool in tissue depth. *ACS Nano, 11*(7), 6672–6681.

Espinosa, E., Rincón, E., Morcillo-Martín, R., Rabasco-Vílchez, L., & Rodríguez, A. (2022). Orange peel waste biorefinery in multi-component cascade approach: Polyphenolic compounds and nanocellulose for food packaging. *Industrial Crops and Products, 187*, 115413.

Fauzi, F., Sulmartiwi, L., & Saputra, E. (2021). The potential of chitosan from comb-pen (Atrina pectinata) shell waste on the characteristics of hand body cream. *IOP Conference Series: Earth and Environmental Science, 679*(1), 012028.

Firmansyah, E., Hardiyanti, R., Samosir, A. K., Sitepu, M. P. B., & Siagian, N. B. (2022). Preparation and characterization of chitosan from goldfish scales for anti-dandruff shampoo. *Rasayan Journal of Chemistry, 15*(2).

Fuhrmann, G. (2023). Drug delivery as a sustainable avenue to future therapies. *Journal of Controlled Release, 354*, 746–754.

Gopalakrishnan, K., Chandel, M., Gupta, V., Kaur, Kuljinder, Patel, A., Kaur, Kamaljit, Kishore, A., Prabhakar, P. K., Singh, A., & Prasad, J. S. (2023). Valorisation of fruit peel bioactive into green synthesized silver nanoparticles to modify cellulose wrapper for shelf-life extension of packaged bread. *Food Research International, 164*, 112321.

Gülsu, A., & Yüksektepe, E. (2021). Preparation of spherical cellulose nanoparticles from recycled waste cotton for anticancer drug delivery. *ChemistrySelect, 6*(22), 5419–5425.

Hao, Y., Zhao, W., Zhang, L., Zeng, X., Sun, Z., Zhang, D., Shen, P., Li, Z., Han, Y., & Li, P. (2020). Bio-multifunctional alginate/chitosan/fucoidan sponges with enhanced angiogenesis and hair follicle regeneration for promoting full-thickness wound healing. *Materials & Design, 193*, 108863.

Hsueh, C.-Y., Tsai, M.-L., & Liu, T. (2017). Enhancing saltiness perception using chitin nanofibers when curing tilapia fillets. *LWT, 86*, 93–98.

Jeet, K., & Kaur, R. (2023). A novel approach for recycling agricultural wastes in the synthesis of nanomaterials and their composites. *Materials Research Express, 10*(10), 105006.

Joo, C., Balci, H., Ishitsuka, Y., Buranachai, C., & Ha, T. (2008). Advances in single-molecule fluorescence methods for molecular biology. *Annual Review of Biochemistry, 77*(1), 51–76.

Judith, J. V., & Vasudevan, N. (2022). Synthesis of nanomaterial from industrial waste and its application in environmental pollutant remediation. *Environmental Engineering Research, 27*(2).

Kamel, R., El-Wakil, N. A., Dufresne, A., & Elkasabgy, N. A. (2020). Nanocellulose: From an agricultural waste to a valuable pharmaceutical ingredient. *International Journal of Biological Macromolecules*, *163*, 1579–1590.

Kokabi, M., & Ebrahimi, S. N. (2020). Polyphenol enriched extract of pomegranate peel: A novel precursor for the biosynthesis of zinc oxide nanoparticles and application in sunscreens. *Pharmaceutical Sciences*, *27*(1), 102–110.

Kumar, R. R., Krishnan, K., Burle, G. S., Samathoti, P., Kumar, J. P., Bagade, O. M., Reddy, K. T. K., & Jahnavi, P. (2025). Biochemical engineering of green nanomaterials for targeted drug delivery and therapeutic applications. *Journal of Chemical Reviews*, *7*(2), 191–215.

Kumari, S., Ram, B., Kumar, D., Ranote, S., & Chauhan, G. S. (2018). Nanoparticles of oxidized-cellulose synthesized by green method. *Materials Science for Energy Technologies*, *1*(1), 22–28.

Larbi, F., García, A., Del Valle, L. J., Hamou, A., Puiggalí, J., Belgacem, N., & Bras, J. (2018). Comparison of nanocrystals and nanofibers produced from shrimp shell α-chitin: From energy production to material cytotoxicity and Pickering emulsion properties. *Carbohydrate Polymers*, *196*, 385–397.

Liu, Z., Hu, M., Zhang, S., Jiang, L., Xie, F., & Li, Y. (2021). Oil-in-water Pickering emulsion stabilization with oppositely charged polysaccharide particles: Chitin nanocrystals/fucoidan complexes. *Journal of the Science of Food and Agriculture*, *101*(7), 3003–3012.

Luthfiyana, N., Bija, S., Anwar, E., Laksmitawati, D. R., & Rosalinda, G. L. (2022). Characteristics and activity of chitosan from mud crab shells on acne bacteria: Staphylococcus aureus, S. epidermidis and Propionibacterium acnes. *Biodiversitas Journal of Biological Diversity*, *23*(12).

Malode, S. J., Pandiaraj, S., Alodhayb, A., & Shetti, N. P. (2024). Carbon nanomaterials for biomedical applications: Progress and outlook. *ACS Applied Bio Materials*, *7*(2), 752–777.

Mansouri Shirazi, N., Eslahi, N., & Gholipour-Kanani, A. (2021). Production and characterization of keratin/tragacanth gum nanohydrogels for drug delivery in medical textiles. *Frontiers in Materials*, *8*, 720385.

Mordorski, B., Landriscina, A., & Friedman, A. (2016). An overview of nanomaterials in dermatology. *Nanoscience in Dermatology*, 31–46.

Morganti, P. (2020). Nanocosmetics: An introduction. In *Nanocosmetics* (pp. 3–16). Elsevier.

Naig, P. J., Kuo, Z.-Y., Wu, Y.-S., & Hung, K. Y. (2025). Eco-engineered nanofibers for skincare: Triglyceride adsorption efficiency of bamboo-derived carbon nanoparticles in biodegradable facial masks. *Surfaces and Interfaces*, 107506.

Nirmal, N., Demir, D., Ceylan, S., Ahmad, S., Goksen, G., Koirala, P., & Bono, G. (2024). Polysaccharides from shell waste of shellfish and their applications in the cosmeceutical industry: A review. *International Journal of Biological Macromolecules*, *265*, 131119.

Omar, R. A., Talreja, N., Chuhan, D., & Ashfaq, M. (2024). Waste-derived carbon nanostructures (WD-CNs): An innovative step toward waste to treasury. *Environmental Research*, *246*, 118096. https://doi.org/10.1016/j.envres.2023.118096

Peng, H., Gan, Z., Xiong, H., Luo, M., Yu, N., Wen, T., Wang, R., & Li, Y. (2017). Self-assembly of protein nanoparticles from rice bran waste and their use as delivery system for curcumin. *ACS Sustainable Chemistry & Engineering*, *5*(8), 6605–6614.

Priyadarsini, M., Biswal, T., & Dash, S. (2021). Biodegradable superabsorbent with potential biomedical application as drug delivery system of "pectin-gP (AN-co-AM)/chicken eggshell" bio-composite. *Polymer Bulletin*, *78*(11), 6337–6349.

Rahayu, D. P., Draheim, R., Lalatsa, A., & Roldo, M. (2022). Harnessing the antibacterial properties of fluoridated chitosan polymers against oral biofilms. *Pharmaceutics*, *14*(3), 488.

Raja, A., Ashokkumar, S., Marthandam, R. P., Jayachandiran, J., Khatiwada, C. P., Kaviyarasu, K., Raman, R. G., & Swaminathan, M. J. J. (2018). Eco-friendly preparation of zinc oxide nanoparticles using Tabernaemontana divaricata and its photocatalytic and antimicrobial activity. *Journal of Photochemistry and Photobiology B: Biology*, *181*, 53–58.

Rajeshkumar, S., Menon, S., Kumar, S. V., Tambuwala, M. M., Bakshi, H. A., Mehta, M., Satija, S., Gupta, G., Chellappan, D. K., & Thangavelu, L. (2019). Antibacterial and antioxidant potential of biosynthesized copper nanoparticles mediated through Cissus arnotiana plant extract. *Journal of Photochemistry and Photobiology B: Biology*, *197*, 111531.

Roy, S., Kim, H. C., Panicker, P. S., Rhim, J.-W., & Kim, J. (2021). Cellulose nanofiber-based nanocomposite films reinforced with zinc oxide nanorods and grapefruit seed extract. *Nanomaterials*, *11*(4), 877.

Salvioni, L., Morelli, L., Ochoa, E., Labra, M., Fiandra, L., Palugan, L., Prosperi, D., & Colombo, M. (2021). The emerging role of nanotechnology in skincare. *Advances in Colloid and Interface Science*, *293*, 102437.

Sampath, L., Ngasotter, S., Layana, P., Balange, A. K., Nayak, B. B., & Xavier, K. A. M. (2022). Effect of chemical treatment duration on physicochemical, rheological, and functional properties of colloidal chitin. *Food Hydrocolloids for Health*, *2*, 100091.

Sarker, A., Ahmmed, R., Ahsan, S. M., Rana, J., Ghosh, M. K., & Nandi, R. (2024). A comprehensive review of food waste valorization for the sustainable management of global food waste. *Sustainable Food Technology*, *2*(1), 48–69.

Sen, S., Sinha, R., Giri, A. H., Kumari, P., Kumbhar, A. P., & Ranjan, O. P. (2025). Nanomaterials and nanocomposites from wastes for drug delivery and advanced therapy: A review. *BioNanoScience*, *15*(1), 169.

Sharma, D., & Hussain, C. M. (2020). Smart nanomaterials in pharmaceutical analysis. *Arabian Journal of Chemistry*, *13*(1), 3319–3343.

Shende, P., Patel, D., & Takke, A. (2020). Nanomaterial-based cosmeceuticals. In *Handbook of functionalized nanomaterials for industrial applications* (pp. 775–791). Elsevier.

Singh, D., & Ray, S. (2023). A short appraisal of nanodiamonds in drug delivery and targeting: Recent advancements. *Frontiers in Nanotechnology*, *5*, 1259648.

Siroukane, N., Kheniche, A., & Souiki, L. (2025). Valorization of date seed waste for sustainable dermocosmetic sunscreens: Phytochemical insights and formulation advances. *Cosmetics*, *12*(5), 225.

Somsak, P., Sriwattana, S., & Prinyawiwatkul, W. (2021). Ultrasonic-assisted chitin nanoparticle and its application as saltiness enhancer. *International Journal of Food Science and Technology*, *56*(2), 608–617.

Thiyagarajulu, N., Arumugam, S., Narayanan, A. L., Mathivanan, T., & Renuka, R. R. (2020). Green synthesis of reduced graphene nanosheets using leaf extract of tridax procumbens and its potential in vitro biological activities. *Biointerface Research in Applied Chemistry*, *11*(3), 9975–9984.

Tohamy, H.-A. S. (2024). Recycling agricultural wastes for controlled drug delivery systems with DFT calculations. *The International Innovations Journal of Applied Science*, (2).

Umar, M. F., Ahmad, F., Saeed, H., Usmani, S. A., Owais, M., & Rafatullah, M. (2020). Bio-mediated synthesis of reduced graphene oxide nanoparticles from chenopodium album: Their antimicrobial and anticancer activities. *Nanomaterials*, *10*(6), 1096.

van der Meel, R., Sulheim, E., Shi, Y., Kiessling, F., Mulder, W. J. M., & Lammers, T. (2019). Smart cancer nanomedicine. *Nature Nanotechnology*, *14*(11), 1007–1017.

van Ham, T. J., Esposito, A., Kumita, J. R., Hsu, S.-T. D., Schierle, G. S. K., Kaminski, C. F., Dobson, C. M., Nollen, E. A. A., & Bertoncini, C. W. (2010). Towards multiparametric fluorescent imaging of amyloid formation: Studies of a YFP model of α-synuclein aggregation. *Journal of Molecular Biology*, *395*(3), 627–642.

Varghese, S., Kuriakose, S., & Jose, S. (2013). Antimicrobial activity of carbon nanoparticles isolated from natural sources against pathogenic gram-negative and gram-positive bacteria, *Journal of Nanoscience, 2013*(1), 457865.

Wang, C., Deng, H., Cui, L., Liu, Q., Liu, Z., Weng, W., & Zhang, Y. (2025). Application of sustainable marine collagen/cellulose nanofiber Pickering emulsions in skincare: From fishery waste to UV-resistant formulations. *Colloids and Surfaces A: Physicochemical and Engineering Aspects*, 138181.

Xing, Y., Liao, X., Liu, X., Li, W., Huang, R., Tang, J., Xu, Q., Li, X., & Yu, J. (2021). Characterization and antimicrobial activity of silver nanoparticles synthesized with the peel extract of mango. *Materials*, *14*(19), 5878.

Yan, N., & Chen, X. (2015). Sustainability: Don't waste seafood waste. *Nature*, *524*(7564), 155–157.

Younes, I., & Rinaudo, M. (2015). Chitin and chitosan preparation from marine sources. Structure, properties and applications. *Marine Drugs*, *13*(3), 1133–1174.

Zhang, Y.-P., Ma, J.-M., Yang, Y.-S., Ru, J.-X., Liu, X.-Y., Ma, Y., & Guo, H.-C. (2019). Synthesis of nitrogen-doped graphene quantum dots (N-GQDs) from marigold for detection of Fe3+ ion and bioimaging. *Spectrochimica Acta Part A: Molecular and Biomolecular Spectroscopy*, *217*, 60–67.

12 Toxic effects of nanomaterials in the ecosystem

12.1 INTRODUCTION

The rapid expansion in the production and application of nanomaterials (NMs) has raised significant environmental concerns. In recent years, NMs have been widely utilised in various applications, including catalysis, wastewater treatment, sensing, drug delivery and storage, energy conversion, and environmental remediation, largely due to their high surface area, tunable physicochemical properties, and cost-effective synthesis routes (Abdelbasir et al., 2020). Nanomaterials are engineered at the nanoscale and can be produced through synthetic or natural processes, using top-down and bottom-up strategies, as well as physical and biological methods. These synthesis routes yield NMs with distinctive physicochemical characteristics, making them highly suitable for a broad range of applications in environmental, catalytic, and biomedical fields (Aslam et al., 2024).

Nanomaterials have been introduced into the aquatic systems through various pathways, including leaching from contaminated soil or water, uncontrolled disposal of spent NMs or synthesis residues, and application-related discharge. Once in the environment, NMs can accumulate and interact with minerals, organic matter, and aquatic biota. Their bioavailability, mobility, and persistence depend on surface charge, size, functional moieties, and environmental conditions such as ionic strength, temperature, and pH (Toha et al., 2025). The presence of NMs in the aquatic systems has been associated with adverse effects, which can manifest at multiple ecological levels. For example, (i) NMs can disrupt metabolic activity and microbial activity by damaging cell membranes and altering enzyme activity, (ii) they can cause DNA damage, reduce growth and reproduction in fish, algae, and other invertebrates, (ii) they can cause genotoxic effects, inflammation, and cytotoxicity in both humans and animals through occupational contact or food chain transfer (Alizadeh et al., 2025). The toxicity of NMs depends mainly on their type, their origin from contaminated waste streams, and their engineered properties. The limited research on monitoring and regulating NMs poses significant threats to animal and human health. Furthermore, the unregulated use and disposal of NMs may lead to secondary pollution, counteracting the sustainability objectives these materials are meant to promote (Kumah et al., 2023).

Research on the toxicity of NMs in aquatic systems remains scarce. Therefore, this chapter aims to provide an overview of the potential toxicity of NMs in the aquatic systems, focusing on their sources and pathways of entry, distribution

DOI: 10.1201/9781003745846-12

mechanisms, fate and ecotoxicological effects, as well as possible mitigation and management strategies.

12.2 SOURCES AND ENTRY OF NANOMATERIALS INTO THE ENVIRONMENT

Nanomaterials (NMs) have gained widespread application worldwide owing to their fascinating features and significant contribution across various fields, including catalysis, sensors, gas storage and transfer, and wastewater treatment, among others (Zhao et al., 2021). Nanomaterials can be synthesised using two popular approaches: top-down and bottom-up. In the bottom-up approach, large nanostructures are formed by the assembly of small atoms. Conversely, the top-down approach involves reducing large-scale materials to form nanoscale materials. The resulting nanomaterials can be classified based on their physicochemical properties, such as composition, shape, size, and surface features (D. Gupta et al., 2023). Thus far, NMs that have been produced include, but are not limited to, (i) carbon-based materials like graphene, fullerenes, nanodiamonds, activated biochar, and carbon nanotubes; (ii) metal oxides such as titanium oxide (TiO_2), cerium oxide (CeO_2), and zinc oxide (ZnO_2); (iii) quantum points containing reactive nuclei formed by combining two elements: cadmium telluride (CdTe), cadmium selenide (CdSe), zinc selenide (ZnSe), or indium phosphate (InP), which are protected from oxidation by ZnS monolayer or SiO_2; and (iv) multifunctional polymers with controllable features (Al-Harbi & Abd-Elrahman, 2025). Consequently, these NMs are released into the environment through manufacturing, processing, or application.

Sources of NMs' entry into the environment are categorised into anthropogenic and natural processes (Gottschalk & Nowack, 2011). Natural processes have been the leading source of NMs' entry into the environment for millennia. They include dust, sandstorms, forest fires, clustering in gases, salt evaporation, aerosol formation, and volcanic eruptions. Biological substances such as viruses, biofilms, metabolic products, and colloids have also been identified as sources of NMs in the environment (Abbas et al., 2024).

Anthropogenic activities are also recognised as significant sources of NMs in aquatic systems. These human-driven activities contribute NMs to the environment through various pathways, including anthropogenic emissions (pollutants released from industrial processes, vehicles, and agricultural practices) and anthropogenic climate change (global warming resulting from activities such as fossil fuel combustion) (Ding et al., 2022). Rapid industrialisation and the advancement of nanotechnology have led to an increase in the quantity of nanomaterials (NMs) present in the environment. Nanotechnology has expanded across numerous fields due to the remarkable properties and performance of NMs (El-Kalliny et al., 2023). However, this intentional and widespread use has consequently increased the release of NMs into various natural habitats. Agriculture is one of the processes that contributes NMs into the aquatic systems through their application as pesticides, nanofertilisers, seed treatment preparations, preparation of hydroponic solutions, and materials for agro-films (S. Tripathi et al., 2023). Industries and manufacturing companies also contribute NMs to the aquatic systems through direct and indirect pathways. In

can promote deeper absorption. Ingestion is regarded as a less exposed pathway, as many NMs are not recommended for ingestion. However, exposure via hand-to-mouth contact is possible. Once ingested, nanomaterials can pass through the gastrointestinal tract and enter the bloodstream, enabling them to reach vital organs such as the kidneys, liver, and brain. Although the impacts of nanomaterials on organ metabolism are not yet fully understood, their accumulation is believed to disrupt normal physiological functions due to their distinctive physicochemical characteristics (Kunhikrishnan et al., 2015). Recent studies have reported that adverse effects associated with nanomaterials include impacts on the brain, circulatory system, and lungs, as well as genotoxic and potentially carcinogenic effects (Alizadeh et al., 2025).

Inhalation is regarded as a primary route of NMs entering the human body, which can occur primarily in environments where nanoparticle dust is prevalent. NMs can enter the respiratory system and sit on the alveoli, where their small size allows them to evade macrophage-mediated clearance mechanisms that typically remove harmful particles from the lungs. The toxicity of NMs is usually greater than that of bulk particles owing to their small and high surface area, which promotes high chemical reactivity (Wadhawan et al., 2025). Recent studies have shown that inhaling carbon nanotubes (CNTs) can disrupt lung function, while other studies have evaluated brain and liver toxicity related to fullerene exposure. Animals and humans are not designed to fight against NMs, therefore, their vulnerability to damage from such NMs is high (Wadhawan et al., 2025). Several studies have reported the toxicity of NMs on marine life and microorganisms (Abishagu et al., 2024; Shinde et al., 2023). The antibacterial effects of nanomaterials depend on several factors, including the bacterial species, exposure conditions, and the specific type of nanomaterial. Their antibacterial mechanisms include disruption of the cell wall and cytoplasmic membrane, as well as interference with nucleic acid and protein synthesis. For instance, the microbial toxicity of silver nanoparticles is strongly influenced by particle size and shape, with smaller particles exhibiting higher toxicity than larger ones. Triangular-shaped nanoparticles have been reported to be more toxic than spherical forms due to the higher density of reactive atoms along their edges (El-Kalliny et al., 2023). Previous studies have shown that the main antibacterial mechanism of nanomaterials involves their attachment to and penetration of the bacterial cell membrane. Other research has indicated that the antimicrobial activity of nanomaterials can suppress microbial degradation processes, leading to the release of contaminants at elevated concentrations into receiving environments (Kunhikrishnan et al., 2015). Consequently, the discharge of nanomaterials from wastewater treatment plants may pose risks to aquatic ecosystems due to their antimicrobial properties.

12.5 MITIGATION AND MANAGEMENT STRATEGIES

Engineered nanomaterials pose a potential environmental health challenge. Effective risk assessment and management rely on a thorough understanding of both environmental exposure and toxicological impacts. While many studies have investigated the toxicity of nanomaterials, relatively few have addressed their environmental exposure. Wastewater treatment plants (WWTPs) represent a key pathway for the release of nanomaterials into aquatic systems, discharging pollutants into soils, surface waters, and the atmosphere via treated effluents and biosolids (Hendricks et al., 2023). Therefore, gaining a comprehensive understanding of the fate and transformation of nanomaterials during wastewater treatment is crucial, alongside the development of innovative removal strategies. Research aimed at identifying effective methods for nanomaterial removal in WWTPs has only recently begun to emerge.

In traditional activated sludge systems, sorption onto activated sludge is one of the primary mechanisms for eliminating contaminants, including NMs. Kiser et al. (2010) conducted batch sorption isotherm experiments using activated sludge as the sorbent and eight various NMs as sorbates. Microscopic studies revealed clear evidence of the biosorption of fluorescent silica nanomaterials, with sorption increasing as NMs exposure to biomass rose. Furthermore, removal efficiency was strongly dependent on the type of NMs. For instance, exposure to wastewater biomass containing 400 mg L^{-1} total suspended solids resulted in the removal of approximately 97% of Ag nanoparticles, likely due in part to aggregation and sedimentation processes. Contrarily, biosorption accounted for the removal of about 39% of functionalised silver nanoparticles, 88% of aqueous fullerenes, 13% of fullerol nanoparticles, and 23% of nanoscale TiO_2 (Kiser et al., 2010).

Removal of NMs from the environment is strongly influenced by the operational efficiency and design of each process unit, including granular media, sedimentation, and membrane separation. A recent study employed mass-flow modelling to track nanomaterials during wastewater treatment and reported a silver nanoparticle (AgNP) removal efficiency of approximately 99% for particles 100 nm in size, as determined from bench-scale experiments conducted in the presence of biomass (Kunhikrishnan et al., 2015). A recent study reported adsorption of AgNP onto wastewater biosolids in both the effluent and the sludge. In the study, the removal of adsorbed AgNP and suspended biomass was controlled by management (Wang et al., 2016). The WWTP secondary sedimentation unit processes are very beneficial for biomass sedimentation and would ultimately eliminate the adsorbed NMs. This finding explains the high NMs removal efficiencies reported in the batch experiments that formed the basis for mass-flow modelling of NMs. In the same study, the concentrations of Ti in raw sewage were reported to range from 181 to 1233 μg L^{-1}, with a median value of 321 μg L^{-1} across 26 samples. The WWTPs eliminated over 97% of the Ti in the influent, and the effluent Ti concentrations were reduced to below 25 μg L^{-1}. In the study, the authors reported that activated sludge or suspended biological processes and trickling filters played a crucial role in trapping NMs in biomass, which can ultimately be removed by membrane filtration or sedimentation (Westerhoff et al., 2011).

Removal of NMs during wastewater treatment occurs primarily through sorption onto sludge, along with flocculation and coagulation. This was observed in a recent study where Kiser and co-workers reported approximately 85% removal efficiency for TiO_2-nanomaterial (TiO_2-NM) at eight U.S. WWTPs, with TiO_2 being most detected in the settled sludge (Kiser et al., 2010). In another study, Afshinnia and colleagues reported that more than 90% of AgNP remained in the water phase following primary treatment. However, subsequent secondary treatment and aeration in a simulated sequencing batch reactor led to the complete removal of Ag nanoparticles over the course of the 15-day experiment (Afshinnia et al., 2017).

Zahra et al. (2022) demonstrated, through modelling, that the efficiency of nanomaterial removal in wastewater treatment plants strongly determines the release of nanoscale pollutants into the environment. At high removal efficiencies, most nanomaterials are retained from the influent and accumulate in the biosolids. In contrast, when removal efficiency is low, a large fraction of nanomaterials passes through the treatment process and is discharged into the environment via treated effluent. These findings highlight the need to develop effective methods for removing or neutralising nanomaterials in biosolids to enable their safe continued use as fertilisers and/or compost (Zahra et al., 2022).

12.6 CONCLUSION

In summary, NMs pose serious environmental challenges due to their unique features, such as high reactivity, large surface area, and increased release into natural and engineered environments. Their large surface area, variable shapes, and small size enable enhanced bioavailability, mobility, and interaction with biological systems in a way not encountered during natural evolutionary processes. This can greatly affect organisms, including humans, as they lack well-established defence mechanisms against exposure to nanomaterials. Evidence from previous studies indicates that nanomaterials can induce significant adverse effects in plants, humans, invertebrates, microorganisms, and other organisms, with their toxicity strongly influenced by surface chemistry, composition, shape, size, and environmental conditions such as ionic strength, pH, and the presence of natural organic matter. Sedimentation, coating by dissolved organic matter (DOM), and aggregation have been reported to moderately mitigate the toxicity of NMs; however, these processes can complicate exposure assessment and risk prediction.

In WWTPs and other aquatic environments, NMs, particularly those with antibacterial properties such as Ag nanoparticles, can disrupt the microbial community, potentially damaging biological treatment processes and altering ecosystem functions. Furthermore, the stability and mobility of nanomaterials raise concerns about bioaccumulation, trophic transfer, and long-term exposure, even when acute toxicity appears minimal.

Existing risk assessment frameworks and exposure models frequently fall short in accurately representing the complex behaviour and transformation of NMs in the environment. Consequently, more comprehensive assessment methods that account for processes such as dissolution, aggregation, sedimentation, and life-cycle dynamics are urgent. While NMs offer significant technological advantages, their environmental releases warrant precautionary management, strengthened regulatory oversight, and continued long-term ecotoxicological research to ensure their safe and sustainable application.

REFERENCES

Abbas, Q., Liu, G., Ali, M. U., Han, J., El-Naggar, A., & Yousaf, B. (2024). Nanomaterials in environment: Release, compartmental transformation, and nanotoxicity. In *Handbook of nanomaterials* (Vol. 2, pp. 745–789). Elsevier.

Abdelbasir, S. M., McCourt, K. M., Lee, C. M., & Vanegas, D. C. (2020). Waste-derived nanoparticles: Synthesis approaches, environmental applications, and sustainability considerations. *Frontiers in Chemistry*, *8*, 782.

Abishagu, A., Kannan, P., Sivakumar, U., Manikanda Boopathi, N., & Senthilkumar, M. (2024). Metal nanoparticles and their toxicity impacts on microorganisms. *Biologia*, *79*(9), 2843–2862.

Afshinnia, K., Sikder, M., Cai, B., & Baalousha, M. (2017). Effect of nanomaterial and media physicochemical properties on Ag NM aggregation kinetics. *Journal of Colloid and Interface Science*, *487*, 192–200.

Al-Harbi, N., & Abd-Elrahman, N. K. (2025). Physical methods for preparation of nanomaterials, their characterization and applications: A review. *Journal of Umm Al-Qura University for Applied Sciences*, *11*(2), 356–377.

Alizadeh, M., Qarachal, J. F., & Sheidaee, E. (2025). Understanding the ecological impacts of nanoparticles: Risks, monitoring, and mitigation strategies. *Nanotechnology for Environmental Engineering*, *10*(1), 6.

Álvarez-Chimal, R., & Arenas-Alatorre, J. Á. (2023). Green synthesis of nanoparticles: A biological approach. In *Green chemistry for environmental sustainability-prevention-assurance-sustainability (PAS) approach*. IntechOpen.

Aslam, J., Aslam, R., & Hussain, C. M. (2024). *Waste-derived nanoparticles: Synthesis, applications and sustainability*. Elsevier.

Avramescu, M.-L., Casey, K., Levesque, C., Chen, J., Wiseman, C., & Beauchemin, S. (2024). Identification and quantification of trace metal (loid) s in water-extractable road dust nanoparticles using SP-ICP-MS. *Science of the Total Environment*, *924*, 171720.

Bang, S. H., Le, T.-H., Lee, S. K., Kim, P., Kim, J. S., & Min, J. (2011). Toxicity assessment of titanium (IV) oxide nanoparticles using Daphnia magna (Water Flea). *Environmental Health and Toxicology*, *26*, e2011002.

Dey, N., Vinayagam, S., Kamaraj, C., Gnanasekaran, L., Goyal, K., Ali, H., Gupta, G., Hussain, M. S., & Subramaniyan, V. (2025). Ecotoxicological evaluation of nanosized particles with emerging contaminants and their impact assessment in the aquatic environment: A review. *Journal of Nanoparticle Research*, *27*(4), 1–21.

Dibyanshu, K., Chhaya, T., & Raychoudhury, T. (2023). A review on the fate and transport behavior of engineered nanoparticles: Possibility of becoming an emerging contaminant in the groundwater. *International Journal of Environmental Science and Technology*, *20*(4), 4649–4672.

Ding, X., Pu, Y., Tang, M., & Zhang, T. (2022). Environmental and health effects of graphene-family nanomaterials: Potential release pathways, transformation, environmental fate and health risks. *Nano Today*, *42*, 101379.

Djanaguiraman, M., Anbazhagan, V., Dhankher, O. P., & Prasad, P. V. V. (2024). Uptake, translocation, toxicity, and impact of nanoparticles on plant physiological processes. *Plants*, *13*(22), 3137.

El-Kalliny, A. S., Abdel-Wahed, M. S., El-Zahhar, A. A., Hamza, I. A., & Gad-Allah, T. A. (2023). Nanomaterials: A review of emerging contaminants with potential health or environmental impact. *Discover Nano*, *18*(1), 68.

Firestone, M., Kavlock, R., Zenick, H., Kramer, M., & Testing, the U. S. E. P. A. W. G. on the F. of T. (2010). The US Environmental Protection Agency strategic plan for evaluating the toxicity of chemicals. *Journal of Toxicology and Environmental Health, Part B*, *13*(2–4), 139–162.

Gambardella, C., & Pinsino, A. (2022). Nanomaterial ecotoxicology in the terrestrial and aquatic environment: A systematic review. *Toxics*, *10*(7), 393.

Gecer, E. N., & Erenler, R. (2023). Biogenic synthesis of silver nanoparticles using Echium vulgare: Characterisation, quantitative analysis of bioactive compounds, antioxidant activity and catalytic degradation. *Journal of the Indian Chemical Society, 100*(5), 101003.

Gottschalk, F., & Nowack, B. (2011). The release of engineered nanomaterials to the environment. *Journal of Environmental Monitoring, 13*(5), 1145–1155.

Gupta, D., Boora, A., Thakur, A., & Gupta, T. K. (2023). Green and sustainable synthesis of nanomaterials: Recent advancements and limitations. *Environmental Research, 231*, 116316.

Gupta, P., Rai, N., Verma, A., & Gautam, V. (2024). Microscopy based methods for characterization, drug delivery, and understanding the dynamics of nanoparticles. *Medicinal Research Reviews, 44*(1), 138–168.

Hendricks, N., Olatunji, O., Ofori, I., & Gumbi, B. P. (2023). Occurrence, fate, and impact of engineered metal/carbonaceous nanomaterials in the environment, detection, and quantitation methods. *International Journal of Environmental Science and Technology, 20*(11), 12937–12954.

Hischier, R., & Walser, T. (2012). Life cycle assessment of engineered nanomaterials: State of the art and strategies to overcome existing gaps. *Science of the Total Environment, 425*, 271–282.

Huang, X., Auffan, M., Eckelman, M. J., Elimelech, M., Kim, J.-H., Rose, J., Zuo, K., Li, Q., & Alvarez, P. J. J. (2024). Trends, risks and opportunities in environmental nanotechnology. *Nature Reviews Earth & Environment, 5*(8), 572–587.

Jiang, C., Liu, S., Zhang, T., Liu, Q., Alvarez, P. J. J., & Chen, W. (2022). Current methods and prospects for analysis and characterization of nanomaterials in the environment. *Environmental Science & Technology, 56*(12), 7426–7447.

Kiser, M. A., Ryu, H., Jang, H., Hristovski, K., & Westerhoff, P. (2010). Biosorption of nanoparticles to heterotrophic wastewater biomass. *Water Research, 44*(14), 4105–4114.

Kumah, E. A., Fopa, R. D., Harati, S., Boadu, P., Zohoori, F. V., & Pak, T. (2023). Human and environmental impacts of nanoparticles: A scoping review of the current literature. *BMC Public Health, 23*(1), 1059.

Kumar, C. M. V., Karthick, V., Kumar, V. G., Inbakandan, D., Rene, E. R., Suganya, K. S. U., Embrandiri, A., Dhas, T. S., Ravi, M., & Sowmiya, P. (2022). The impact of engineered nanomaterials on the environment: Release mechanism, toxicity, transformation, and remediation. *Environmental Research, 212*, 113202.

Kunhikrishnan, A., Shon, H. K., Bolan, N. S., El Saliby, I., & Vigneswaran, S. (2015). Sources, distribution, environmental fate, and ecological effects of nanomaterials in wastewater streams. *Critical Reviews in Environmental Science and Technology, 45*(4), 277–318.

Liu, Z., Ying, S., Jiang, Y., Takeuchi, H., & Huang, Y. (2023). Environmental carriers for metal nanoparticles: Transport, fate, and eco-risks. *Reviews of Environmental Contamination and Toxicology, 261*(1), 23.

Majerič, P., Jović, Z., Švarc, T., Jelen, Ž., Horvat, A., Koruga, D., & Rudolf, R. (2023). Physicochemical properties of gold nanoparticles for skin care creams. *Materials, 16*(8), 3011.

Mat Lazim, Z., Salmiati, S., Marpongahtun, M., Arman, N. Z., Mohd Haniffah, M. R., Azman, S., Yong, E. L., & Salim, M. R. (2023). Distribution of silver (Ag) and silver nanoparticles (AgNPs) in aquatic environment. *Water, 15*(7), 1349.

Meyer, J. N., Lord, C. A., Yang, X. Y., Turner, E. A., Badireddy, A. R., Marinakos, S. M., Chilkoti, A., Wiesner, M. R., & Auffan, M. (2010). Intracellular uptake and associated toxicity of silver nanoparticles in Caenorhabditis elegans. *Aquatic Toxicology, 100*(2), 140–150.

Petreanu, I., Niculescu, V.-C., Enache, S., Iacob, C., & Teodorescu, M. (2023). Structural characterization of silica and amino-silica nanoparticles by Fourier transform infrared (FTIR) and Raman spectroscopy. *Analytical Letters*, *56*(2), 390–403.

Phogat, P., Jha, R., & Singh, S. (2024). Synthesis of novel ZnO nanoparticles with optimized band gap of 1.4 eV for high-sensitivity photo electrochemical detection. *Materials Today Sustainability*, *27*, 100823.

Pirsaheb, M., Gholami, T., Seifi, H., Dawi, E. A., Said, E. A., Hamoody, A.-H. M., Altimari, U. S., & Salavati-Niasari, M. (2024). Green synthesis of nanomaterials by using plant extracts as reducing and capping agents. *Environmental Science and Pollution Research*, *31*(17), 24768–24787.

Prasad, R. D., Prasad, R. S., Prasad, R. B., Prasad, S. R., Singha, S. B., Singha, D., Prasad, R. J., Sinha, P., Saxena, S., & Vaidya, A. K. (2024). A review on modern characterization techniques for analysis of nanomaterials and biomaterials. *ES Energy & Environment*, *23*, 1087.

Qamar, W., Gulia, S., Athar, M., Ahmad, R., Imam, M. T., Chandra, P., Singh, B. P., Haque, R., Hassan, M. I., & Rahman, S. (2024). An insight into impact of nanomaterials toxicity on human health. *PeerJ*, *12*, e17807.

Quik, J. T. K., Vonk, J. A., Hansen, S. F., Baun, A., & Van De Meent, D. (2011). How to assess exposure of aquatic organisms to manufactured nanoparticles? *Environment International*, *37*(6), 1068–1077.

Radamson, H. H., Hallén, A., Sychugov, I., & Azarov, A. (2023). *Analytical methods and instruments for micro-and nanomaterials* (Vol. 6). Springer.

Roy, I. (2023). Interactions of nanoparticles with plants: Accumulation and effects. *Bioinspired and Green Synthesis of Nanostructures: A Sustainable Approach*, 157–188.

Shanmugam, K., Sakthivel, E., Rajendran, J., & Jeyaraman, P. (2024). Nanoparticles in aquatic ecosystems: Origins, destiny, and ecological consequences. *Environmental Science Archives*, *3*(1), 111–124.

Shinde, S. A., More, P. R., & Ingle, A. P. (2023). Hazardous effects of nanomaterials on aquatic life. In *Nanotechnology in agriculture and agroecosystems* (pp. 423–450). Elsevier.

Silva, L. F. O., Oliveira, M. L. S., Crissien, T. J., Santosh, M., Bolivar, J., Shao, L., Dotto, G. L., Gasparotto, J., & Schindler, M. (2022). A review on the environmental impact of phosphogypsum and potential health impacts through the release of nanoparticles. *Chemosphere*, *286*, 131513.

Singh, S., Prasad, S. M., & Bashri, G. (2023). Fate and toxicity of nanoparticles in aquatic systems. *Acta Geochimica*, *42*(1), 63–76.

Toha, M., Rahman, R.-R., Sikder, S., & Rahman, M. M. (2025). Toxicological and life-cycle perspectives on waste derived carbon nanomaterials. In *Waste derived carbon nanomaterials* (Vol. 1, pp. 279–308). ACS Publications.

Tran, T.-K., Nguyen, M.-K., Lin, C., Hoang, T.-D., Nguyen, T.-C., Lone, A. M., Khedulkar, A. P., Gaballah, M. S., Singh, J., & Chung, W. J. (2024). Review on fate, transport, toxicity and health risk of nanoparticles in natural ecosystems: Emerging challenges in the modern age and solutions toward a sustainable environment. *Science of the Total Environment*, *912*, 169331.

Tripathi, N., & Goshisht, M. K. (2022). Recent advances and mechanistic insights into antibacterial activity, antibiofilm activity, and cytotoxicity of silver nanoparticles. *ACS Applied Biomaterials*, *5*(4), 1391–1463.

Tripathi, S., Mahra, S., Tiwari, K., Rana, S., Tripathi, D. K., Sharma, S., & Sahi, S. (2023). Recent advances and perspectives of nanomaterials in agricultural management and associated environmental risk: A review. *Nanomaterials*, *13*(10), 1604.

Wadhawan, S., Wadhawan, D., Jain, A., & Mehta, S. K. (2025). Toxic implication of nanoparticles: A review of factors, mechanism, exposure and control strategies. *International Journal of Environmental Science and Technology*, *22*(2), 1203–1224.

Wang, P., Menzies, N. W., Dennis, P. G., Guo, J., Forstner, C., Sekine, R., Lombi, E., Kappen, P., Bertsch, P. M., & Kopittke, P. M. (2016). Silver nanoparticles entering soils via the wastewater–sludge–soil pathway pose low risk to plants, but elevated Cl concentrations increase Ag bioavailability. *Environmental Science & Technology*, *50*(15), 8274–8281.

Westerhoff, P., Song, G., Hristovski, K., & Kiser, M. A. (2011). Occurrence and removal of titanium at full scale wastewater treatment plants: Implications for TiO 2 nanomaterials. *Journal of Environmental Monitoring*, *13*(5), 1195–1203.

Wu, Y., Yan, H., Zhu, X., Liu, C., Chu, C., Zhu, X., & Chen, B. (2022). Biochar effectively inhibits the horizontal transfer of antibiotic resistance genes via restraining the energy supply for conjugative plasmid transfer. *Environmental Science & Technology*, *56*(17), 12573–12583.

Yang, Y., Quensen, J., Mathieu, J., Wang, Q., Wang, J., Li, M., Tiedje, J. M., & Alvarez, P. J. J. (2014). Pyrosequencing reveals higher impact of silver nanoparticles than Ag+ on the microbial community structure of activated sludge. *Water Research*, *48*, 317–325.

Yoo, K.-B., Yang, S., Choi, H., & Lee, B.-T. (2024). Ion-exchange resins improve the analysis of metal nanoparticles in wastewater using single-particle inductively coupled plasma–mass spectrometry. *Environmental Science and Pollution Research*, *31*(40), 53090–53099.

Zahra, Z., Habib, Z., Hyun, S., & Sajid, M. (2022). Nanowaste: Another future waste, its sources, release mechanism, and removal strategies in the environment. *Sustainability*, *14*(4), 2041.

Zdziebłowska, S., Czarnecki, M., Ciosek-Skibińska, P., & Ruzik, L. (2024). The microalgae's ability to accumulate selected trace elements studied by ICP-MS/MS and chemometric methods. *Journal of Trace Elements in Medicine and Biology*, *81*, 127351.

Zhang, R., Zheng, X., Fan, W., Wang, X., Zhao, T., Zhao, X., Peijnenburg, W. J. G. M., Vijver, M. G., & Wang, Y. (2025). Fate models of nanoparticles in the environment: A critical review and prospects. *Environmental Science: Nano*, *12*(7), 3394–3412.

Zhao, J., Lin, M., Wang, Z., Cao, X., & Xing, B. (2021). Engineered nanomaterials in the environment: Are they safe? *Critical Reviews in Environmental Science and Technology*, *51*(14), 1443–1478.

Index

Note: Page numbers in *italics* indicate a figure, and page numbers in **bold** indicate a table on the corresponding page.

For Product Safety Concerns and Information please contact our EU representative GPSR@taylorandfrancis.com
Taylor & Francis Verlag GmbH, Kaufingerstraße 24, 80331 München, Germany

www.ingramcontent.com/pod-product-compliance
Ingram Content Group UK Ltd.
Pitfield, Milton Keynes, MK11 3LW, UK
UKHW021507240926
475702UK00035B/550
* 9 7 8 1 0 3 2 6 4 5 4 4 5 *